Lecture Notes in Networks and Systems

Volume 747

The series "Lecture Notes in Networks and Systems" publishes the latest developments in Networks and Systems—quickly, informally and with high quality. Original research reported in proceedings and post-proceedings represents the core of LNNS.

Volumes published in LNNS embrace all aspects and subfields of, as well as new challenges in, Networks and Systems.

The series contains proceedings and edited volumes in systems and networks, spanning the areas of Cyber-Physical Systems, Autonomous Systems, Sensor Networks, Control Systems, Energy Systems, Automotive Systems, Biological Systems, Vehicular Networking and Connected Vehicles, Aerospace Systems, Automation, Manufacturing, Smart Grids, Nonlinear Systems, Power Systems, Robotics, Social Systems, Economic Systems and other. Of particular value to both the contributors and the readership are the short publication timeframe and the world-wide distribution and exposure which enable both a wide and rapid dissemination of research output.

The series covers the theory, applications, and perspectives on the state of the art and future developments relevant to systems and networks, decision making, control, complex processes and related areas, as embedded in the fields of interdisciplinary and applied sciences, engineering, computer science, physics, economics, social, and life sciences, as well as the paradigms and methodologies behind them.

Indexed by SCOPUS, INSPEC, WTI Frankfurt eG, zbMATH, SCImago.

All books published in the series are submitted for consideration in Web of Science.

For proposals from Asia please contact Aninda Bose (aninda.bose@springer.com).

Richard Balogh · David Obdržálek ·
Eftychios Christoforou
Editors

Robotics in Education

Proceedings of the RiE 2023 Conference

Springer

Editors
Richard Balogh
Slovak University of Technology (STU)
Bratislava, Slovakia

Eftychios Christoforou
University of Cyprus
Nicosia, Cyprus

David Obdržálek
Faculty of Mathematics and Physics
Charles University
Prague, Czech Republic

ISSN 2367-3370 ISSN 2367-3389 (electronic)
Lecture Notes in Networks and Systems
ISBN 978-3-031-38453-0 ISBN 978-3-031-38454-7 (eBook)
https://doi.org/10.1007/978-3-031-38454-7

This Springer imprint is published by the registered company Springer Nature Switzerland AG
The registered company address is: Gewerbestrasse 11, 6330 Cham, Switzerland

Paper in this product is recyclable.

Committee

Co-chair persons

Panicos Masouras, Cyprus University of Technology in Limassol, Cyprus
Richard Balogh, Slovak University of Technology in Bratislava, Slovakia
David Obdržálek, Charles University in Prague, Czech Republic

International Programme Committee

Dimitris Alimisis, European Lab for Educational Technology EDUMOTIVA, Greece
Sotiris Avgousti, Cyprus University of Technology
Francisco Bellas, Universidade da Coruna, Spain
Sylvain Bertrand, The French Aerospace Lab-ONERA
Ansgar Bredenfeld, Dr. Bredenfeld UG, Germany
Jenny Carter, University of Huddersfield, UK
Alvaro Castro-Gonzalez, Universidad Carlos III de Madrid, Spain
Dave Catlin, Valiant Technology, UK
Luka Čehovin, University of Ljubljana, Slovenia
Morgane Chevalier, University of Teacher Education in Vaud, Switzerland
Eftychios Christoforou, University of Cyprus
Jean-Daniel Dessimoz, HESSO.HEIG-VD Western Switzerland University of Applied Sciences and Arts
Amy Eguchi, University of California San Diego, USA
Hugo Ferreira, Porto Polytechnic Institute, Portugal
Nuno Ferreira, Coimbra Institute of Engineering, Portugal
João Ferreira, Coimbra Institute of Engineering, Portugal
Paolo Fiorini, University of Verona, Italy
Martin Fislake, University of Koblenz, Germany
Reinhard Gerndt, Ostfalia University of Applied Sciences, Germany

José Gonçalves, ESTiG—Polytechnic Institute of Bragança, Portugal
Grzegorz Granosik, Lodz University of Technology, Poland
Simon Haller-Seeber, University of Innsbruck, Austria
Georg Jäggle, Vienna University of Technology, Austria
Boualem Kazed, University of Blida, Algeria
Tomáš Krajník, Czech Technical University in Prague
Miroslav Kulich, Czech Technical University in Prague
Janika Leoste, Tallinn University of Technology and Tallinn University, Estonia
Andrej Lúčny, Comenius University in Bratislava, Slovakia
Karolina Miková, Comenius University in Bratislava, Slovakia
Marta C. Mora, Universitat Jaume I, Spain
Michele Moro, DEI University of Padova, Italy
Lucio Negrini, University of Applied Sciences and Arts of Southern Switzerland (SUPSI)
Luis Paya, Universidad Miguel Hernandez, Spain
Pavel Petrovič, Comenius University in Bratislava, Slovakia
Alfredo Pina, Public University of Navarra, Spain
David Portugal, University of Coimbra, Portugal
Oscar Reinoso, Universidad Miguel Hernandez, Spain
Mireia Usart, Universitat Rovira i Virgili, Spain
Antonio Valente, UTAD University, Portugal
David Valiente Garcia, Universidad Miguel Hernández, Spain
Anton Yudin, Bauman Moscow State Technical University, Russia

Local Conference Organization

Dr. Sotiris Avgousti, Cyprus University of Technology
Dr. Eftychios Christoforou, University of Cyprus
Dr. Panicos Masouras, Cyprus University of Technology

Preface

It is with great pleasure and honor that we present the publication of the Proceedings of the 14th International Conference on Robotics in Education (RiE 2023). The conference has become an established annual event, which serves as a forum that brings together academics, researchers, educators and industry experts from around the world to present the latest developments in the field of Educational Robotics.

Educational robotics refers to the use of robots as a learning tool for teaching various topics. It has been closely associated with STEM education, which is a systematic and integrated approach to teaching Science, Technology, Engineering and Mathematics. Robotics is a suitable platform for this purpose because it is not only interdisciplinary but also exciting and appealing to students. It provides unique opportunities for hands-on experience and active learning. It inspires creativity and cultivates a variety of skills including critical thinking, problem-solving, communication and presentation together with other social skills, widely known as 21st-century skills. It is for these reasons that robotics courses and projects have been integrated into teaching curricula from kindergarten to university level. Moreover, relevant courses often embrace robotics competitions as an added stimulus to complement school curricula. Robotics competitions have been gaining popularity worldwide.

The field of educational robotics has seen remarkable developments over the past couple of decades with a growing scientific community, relevant literature, as well as the availability of affordable and user-oriented hardware. The broadening scope of educational robotics also reflects developments in the wider field of robotics which now includes anthropomorphic, flying and underwater robots and also embraces advances in artificial intelligence (AI). Such systems provide enhanced opportunities for learning, and flexibility to better serve specific pedagogical needs and inspire further research.

Continuing the tradition of the previous events, RiE 2023 took place in Limassol, Cyprus, from April 19 to 21, 2023. The conference was co-organized by the Cyprus Computer Society, the Cyprus University of Technology, the University of Cyprus and the Slovak University of Technology in Bratislava. It served as a venue for international robotics in the education community to present and discuss theoretical approaches, new trends and innovations, practical experiences and case examples, ongoing research projects and relevant hardware and software tools. In total, 49 papers were submitted of which 35 were accepted for publication (acceptance rate: 0.71) following a systematic review process. The accepted papers were presented at the conference and have been included in the present proceedings volume. The papers reflect the status of the field and current trends. The conference programme covered different thematic areas. The interested reader may find information on robotics at school, the pedagogy of educational robotics, robotics projects, devices and hardware, flying robots, robot programming and the Robot Operating System (ROS), robotic competitions, simulation tools, autonomous robots, machine learning and AI.

We would like to take this opportunity to express our sincere thanks to all the authors who submitted their papers to RiE 2023 and congratulate the authors of the accepted papers. We would like to thank the Conference Co-Chairs, the Local Organizing Committee, the Session Chairs, the International Programme Committee and all the Presenters and Participants for making the conference a success. Finally, we would like to express our gratitude to Springer for the continuing support of the RiE conference through the publication of the proceedings as a special volume in the prestigious Lecture Notes in Networks and Systems series.

Nicosia, Cyprus
May 2023

Eftychios Christoforou
Conference Co-chair

Contents

Workshops, Curricula and Related Aspects

On Autonomous Mobile Robot Exploration Projects in Robotics Course

Jan Faigl, Miloš Prágr, and Jiří Kubík

Abstract Autonomous mobile robot exploration can be considered a representative task where multiple problems need to be addressed and solutions integrated into a software framework that exhibits desired autonomous behavior of the robot. The problem includes online decision-making in selecting new navigational waypoints towards which the robot is autonomously navigated to explore not yet covered part of the environment. A mobile robot's navigation consists of localization, mapping, planning, and execution of the plan by following the path toward the waypoint. For these very reasons, we decided to include mobile robot exploration as one of the tasks in our Artificial Intelligence in Robotics course opened at the Faculty of Electrical Engineering, Czech Technical University in Prague. In this paper, we present our experience running the course, where the students start with relatively small isolated tasks that are then integrated into a full exploration framework. We share the students' feedback on our initial approach for the task that becomes a mandatory part of the course evaluation and grading.

Keywords Robot course design · Autonomous robot exploration

1 Introduction

In 2016, we faced a challenge to update our Computer Science study program with a branch specialized in Artificial Intelligence (AI). Studying AI has a long tradition at the Czech Technical University in Prague (CTU) with an overlap to image processing, computer vision, and machine learning. We have decided to include robotics in

J. Faigl (✉) · M. Prágr · J. Kubík
Department of Computer Science, Faculty of Electrical Engineering,
Czech Technical University in Prague, Prague, Czechia
e-mail: faiglj@fel.cvut.cz

M. Prágr
e-mail: pragrmi1@fel.cvut.cz

J. Kubík
e-mail: kubikji2@fel.cvut.cz

R. Balogh et al. (eds.), *Robotics in Education*, Lecture Notes in Networks
and Systems 747, https://doi.org/10.1007/978-3-031-38454-7_1

the AI curricula to offer students an opportunity to experience hands-on robotics systems, where it is necessary to deal with sources of uncertainty in sensing and acting. We proposed a new course on AI in robotics [2] to provide an overview of robotic paradigms, path and motion planning methods, and environment modeling approaches. The first part of the course targets to combine particular tasks in the autonomous navigation problem, with autonomous exploration selected as one of the central problems of the course.

Mobile robot exploration is a problem to develop a system that operates one or multiple robots in a priory unknown environment intending to model some phenomenon, for example, building a geometrical environment map. From the AI perspective, exploration combines processing of the sensory inputs, environment modeling, reasoning about the future robot actions, planning, and navigation towards the determined waypoints. Besides, multi-robot exploration further extends the opportunity to study approaches for multi-robot coordination [9] and cooperation using centralized, distributed, or decentralized methods [4, 29]. Exploration directly relates to search-and-rescue scenarios, where multi-criteria decision-making can be applied [3]. Besides, considering multi-legged robots [6] opens further challenges in motion planning and locomotion control.

Throughout the history of the course, we have initially started with focused lab exercises that have guided the students through individual building blocks of autonomous exploration. Later, we switched to a block of individual tasks that allowed the students to experience locomotion control, navigation, mapping, and decision-making. Students submit the individual tasks into our automated evaluation system called BRUTE (Bundle for Reservation, Upload, Test, and Evaluation), allowing them to work on the assignments outside the course's scheduled hours. The culmination of the student's effort is the semestral course project, where individual building blocks form an initial exploration framework that needs to be improved by various means. The improvements are mostly expected in decision-making using AI techniques that can be acquainted with the course lectures or selected research papers, allowing the students to understand and apply novel results from the literature. Teachers evaluate the projects, and students are requested to defend the project during the course exam, thus verifying the students' work on the project and their understanding of the topic.

From the beginning, we planned to target the exploration task to our hexapod walking robot depicted in Fig. 1a because students can deploy their solutions on the real robot during the course labs. The robot is built from Robotis Dynamixel AX12A servomotors, using the adaptive locomotion controller [17] capable of negotiating the terrain and supporting the robot's endurance and robustness using the students' code. Besides, we have selected the CoppeliaSim (formerly V-REP) simulation environment [12] because it is a multiplatform, stand-alone simulation environment with multiple interfaces, including C/C++ and Python. A model of the robot is available as shown in Fig. 1b, which supports direct deployment from the simulator to the real robot.

The course is primarily implemented in Python, allowing for rapid prototyping. In 2019, we opted for ROS (Robot Operating System) [25] to support the integration

(a) Real hexapod walking robot

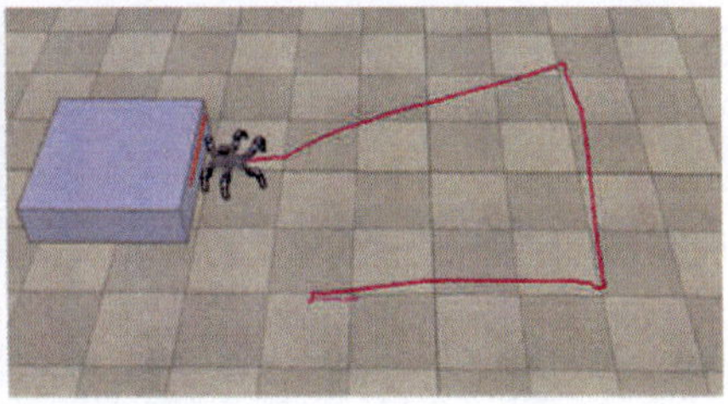

(b) Hexapod robot in the CoppeliaSim

Fig. 1 Hexapod walking robot utilized for the exploration task in the course

of the students' modules. However, in the regular university courses' evaluation, students complained about the initial challenges of setup the ROS environment, despite a dedicated lab for familiarizing with the necessary ROS infrastructure [26] suitable for exploration tasks. We acknowledged the complaints as ROS distracts attention towards the implementation aspects than the principal and algorithmic solutions. The course is in the last year of the AI master studies, where most students do not get in touch with ROS during their studies unless they take dedicated courses on robotics. We found that without ROS, the students progress faster, enjoy the course more, and dedicate more hours to advanced modules enhancing the basic exploration strategy.

The rest of the paper is organized as follows. A brief overview of the mobile robot exploration is summarized in the following section to familiarize the reader with the autonomous exploration framework's principles and basic building blocks. In Sect. 3, a description of the students' tasks and project assignments are presented. Results of the evaluation of the student's achievements are reported in Sect. 4. Finally, the paper is concluded in Sect. 5.

2 Mobile Robot Exploration

We consider exploration a representative task of autonomous behavior with online decision-making, while we restrict it to the problem of creating a map of a priory unknown environment for the course. It consists of fusing sensor measurements into the environment model and making decisions about the next robot's actions. For educational purposes, we limit the task to 2D mapping with an occupancy grid map [24]. Then, we can employ frontier-based exploration [28] to determine the navigational waypoints at the border of known free space and not yet explored part of the environment. The robot explores the environment while being navigated toward frontiers.

The occupancy grid map is a discretized spatial environment representation where each cell is associated with an occupancy probability value. Furthermore, we assume that a robot pose estimate is available. Hence, the sensor measurements are integrated

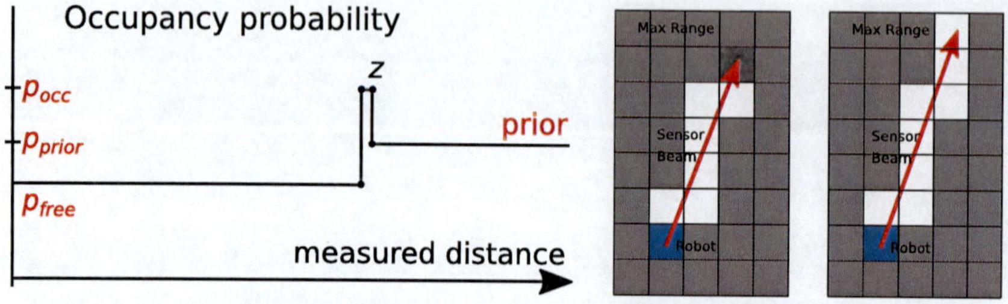

(a) Laser model (b) Ray-casting in a grid map

Fig. 2 Laser sensor model and update of the grid map. The probability values p_{occ} and p_{free} denote the probability of occupied and free, respectively

using the Bayes filter with the sensor model. For simplicity, we utilize LiDAR-based sensors with relatively precise distance measurements to the obstacles; see Fig. 2a. The grid cells of the occupied grid map are updated using laser beam raycasting. The update of the individual grid cells is along a line determined by Bresenham's line algorithm [8] as depicted in Fig. 2b. Thus, the mapping is relatively straightforward and reflects dynamic obstacles as the map is continuously updated. Note that using a 2D grid map, various techniques of path planning can be employed, such as A*, Theta* [13], JPS [19], and D* Lite [21]. Hence, students can deploy various methods and see their impact on online decision-making.

Possible exploration strategies can range from improved frontier-based approaches [23] to information-based methods [1] using mutual information [10]. In multi-robot cases, the problem can be addressed as task allocation [18], or robots can make decisions independently [29]. Furthermore, decisions can be myopic, considering only the immediate reward when selecting the next waypoint, such as the closest waypoint. Alternatively, a suitable option for applying AI techniques is to perform non-myopic decisions considering a longer horizon. It can lead to a solution of the Traveling Salesman Problem (TSP) instance [30] and decisions made by the so-called TSP distance cost [22], which is also applicable in multi-robot exploration [16]. Since obtaining a solution for the TSP can be demanding, heuristics can be considered. Besides, considering individual frontier cells might be unnecessary, and therefore, representatives of the frontiers' free edges can be more suitable [14]. Therefore, determining frontiers representatives provides an opportunity to employ various AI methods such as unsupervised clustering as depicted in Fig. 3.

2.1 Building Blocks of Mobile Robot Exploration Framework

Although autonomous mobile robot exploration can be considered a complex problem, we build on our experience and effort on benchmarking exploration strategies [15] and summarize the problem into the following steps.

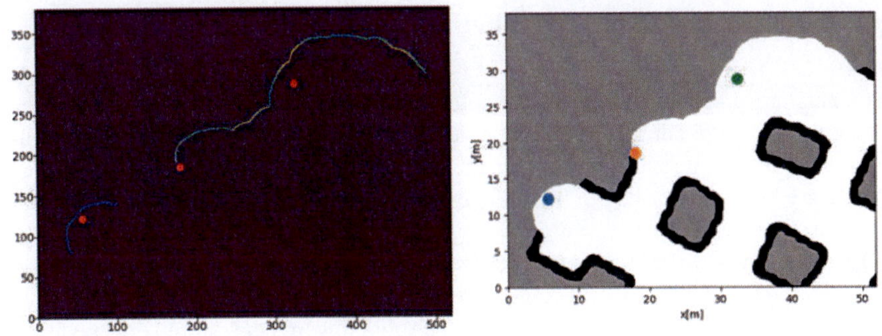

Fig. 3 Representative of frontiers determined as centroids of free-edge clusters

1. Initialization of the occupancy grid map and integrating of the first sensor measurements.
2. Creating a navigation grid map by thresholding the occupancy probability into free space, obstacles, and unknown areas, e.g., using probability threshold values 0.4 and 0.6.
3. Determining the next navigational waypoint, the exploration strategy.
4. Planning a path to the waypoint.
5. Navigating the robot along the planned path and integrating the new sensor measurements into the map.
6. Repeat the procedure from Step 2 when a replanning is triggered.

We can identify three main processes that can run in parallel and that combine building blocks of control, path planning, mapping, and decision-making. The first is mapping, which collects and integrates measurements into the occupancy grid map. It can run relatively fast depending on the robot's velocity. The second is a path following, responsible for determining control commands for the robot to navigate along the planned path. We can follow the hybrid robotics paradigm and employ reactive collision avoidance to ensure safe navigation along the planned path. The path following runs in a loop with a similar frequency as the mapping. Finally, the third process is computing the new plan, which consists of determining the next exploration goal and planning the respective path. The plan computations can run in a loop with a relatively small frequency or be triggered by reaching the previous exploration goal.

We define a sequence of small tasks based on the building blocks, each focusing on a particular subproblem combined in a simple frontier-based exploration framework. The tasks are summarized in the following section.

3 Exploration Task Assignments

Students are prepared to develop an autonomous mobile robot exploration framework through five consecutive tasks guiding them in implementing robot control, mapping, planning, and determining the exploration waypoints. The task assignments are briefly described in the following paragraphs.

T1a-ctrl—Open-loop locomotion control—First, students get familiarized with the CoppeliaSim environment and control the robot in an open-loop fashion. The task is to implement a `goto` function that steers the robot towards the desired goal position using a velocity command consisting of the desired forward and angular speed of the robot that is passed to the provided controller. The robot is supposed to reach a sequence of positions as depicted in Fig. 4a.

T1b-react—Reactive obstacle avoidance—Next, students are requested to improve the navigation capabilities of the robot through sensory-motor feedback. Again, students implement the `goto` function that determines the robot steering command based on the sensory input. The accompanied labs guide students with Bug and Bug2 algorithms and AI models of Braitenberg vehicles [7]. The expected behavior of the robot is as depicted in Fig. 4b.

T1c-plan—Path planning—The third task is grid-based path planning, where students are requested to deploy some graph (or grid) based search technique such as A*, which they already know from the AI courses. However, within the robotics context, the obstacles need to be grown to consider the embodiment of the robot; see Fig. 5a. Besides, the grid path represented as a sequence of neighboring grid cells can be too dense to be smoothly followed. Therefore, students are tasked to simplify the path to a sequence of waypoints; see Fig. 5b. Such a sequence of waypoints can be utilized in the reactive controller from task T1b. Students implement the functions `grow_obstacles`, `plan_path`, and `simplify_path` that are validated in the BRUTE for defined planning scenarios. The validation procedure is available with the reference solutions bundled in Python's `pickle` object serialization.

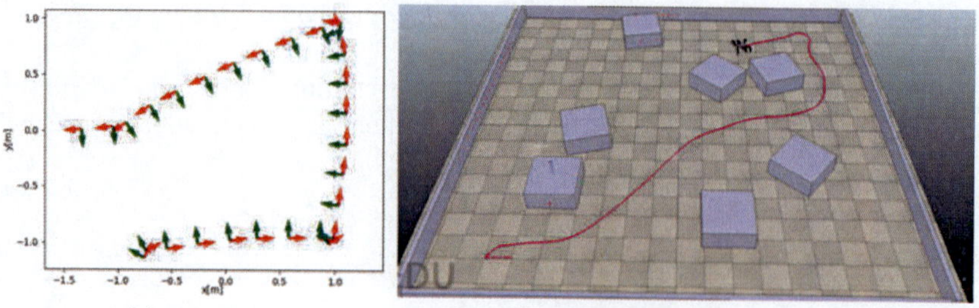

(a)T1a-ctrl (b)T1b-react

Fig. 4 Expected robot path in the robot control tasks T1a-ctrl and T1b-react

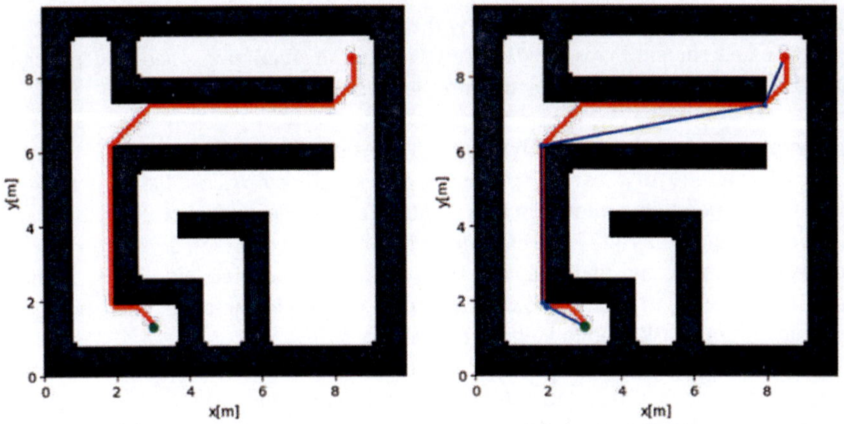

(a) Grown obstacles (b) Path simplification

Fig. 5 Grid-based path planning and path simplification. The red curve is the planned grid path. The simplified path with waypoints where the robot changes its orientation is in the blue

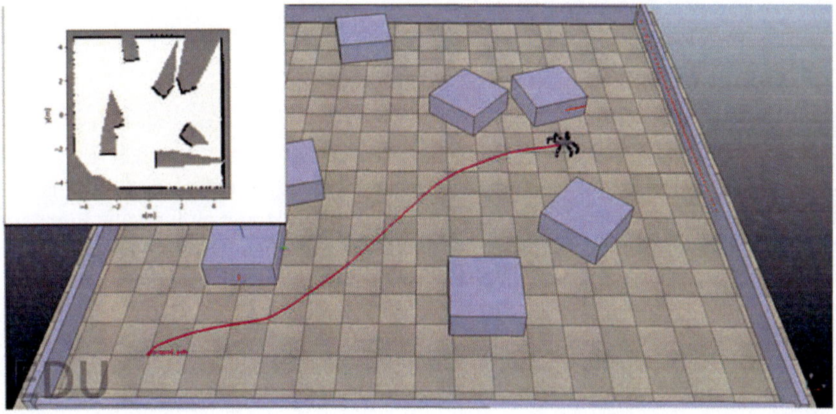

Fig. 6 Example of the robot path in the CoppeliaSim with visualization of the grid map created by thresholding the occupancy grid map. The gray cells denote unknown parts; the obstacles are black, and the free space is white

T1d-map—Mapping—In the mapping task, students are to fuse laser scanner measurements by implementing the `fuse_laser_scan` function. For the validation, students can utilize the reactive controller from T1b-react to capture scans along a defined path as depicted in Fig. 6. Besides, similar to the previous task, testing scenarios are available. The automated task evaluation is based on computing an accumulated difference of the map fused by the student's implementation and the reference map. In real deployments, a small difference in a map built over multiple trials can be expected because of numerical issues and uncertainty in the pose estimation that is, however, provided by the CoppeliaSim. Differences between the reference and student maps are expected even in the simulation since students can experiment with different parametrizations of the Bayesian map update.

T1e-expl—Exploration—Finally, students are tasked to implement the function `find_free_edge_frontiers` that is supposed to provide a list of positions representing possible exploration waypoints. Here, students can exploit advanced functions of the SciPy [27] and use only a few lines of code to determine frontiers cells, cluster them, and find free edge representatives as depicted in Fig. 3. To that end, students need to reason about the problem, proper formulation, and design of the convolution mask used to identify the frontier cells.

As the result of the overviewed tasks, students have prepared the individual functions to be integrated into the exploration framework consisting of three threads working in parallel as described in Sect. 2.1. Besides, when implementing the tasks, students are provided with a support code in Python that handles communication with the simulator, low-level robot motion control given the velocity command, and data types for sensory inputs.

4 Evaluation of the Students' Achievements

Students submit tasks' implementations and project into BRUTE, where they can report time spent on the particular task. The time reporting is voluntary, but students fill reasonable values in most cases; therefore, we only filter out outlier values longer than 100 h, which is unrealistic, specifically as they have one to two weeks per each small task. The filtered reported time spent is depicted as the five-number summary in Fig. 7.

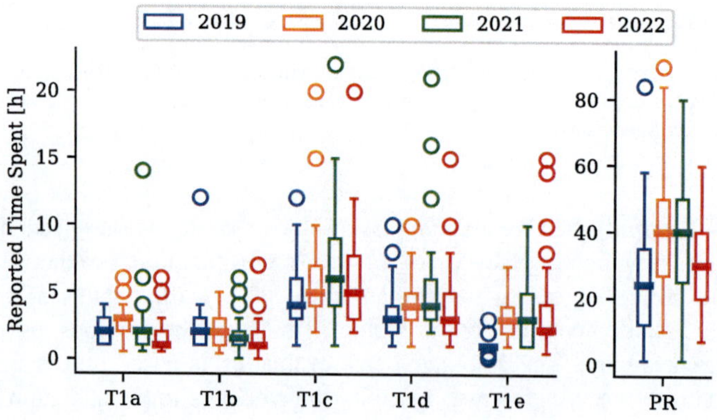

Fig. 7 Reported students times spent on the tasks T1a–e and project PR

The maximal reported values support the suitability of outliers being thresholding at 100 h. Besides, in the authors' opinion, it might be the case that students often round the reported hours to tens for the project. The maximum hours spent on the project correspond to more than ten working days, which might be realistic if a student struggles with programming. However, most of the values are around half of that, which is our expected value. The success of our teaching mission can be further evaluated by an in-depth analysis of how students spent the time budget and what features they implemented within the autonomous mobile robot exploration project, detailed in the rest of this section.

4.1 Selected Students' Implementations

We can imagine various extensions of the exploration framework to improve mission performance. For students' convenience, we list possible extensions with expected scoring as the project can be up to 30% of the course grading. However, students are not limited to the list and are encouraged to discuss the viability of other extensions with lab instructors. The mandatory implementation of the project represents about 10 % corresponding to 10 points. Seven selected extensions are further discussed as follows.

E1-dmap—Dynamic map size (+2 points)—Since a fixed size, a sufficiently large map is assumed in the T1d-map task, the extension is to implement a dynamically resizable map representation.

E2-clust—Multiple-representative free-edge clusters (+2 points)—A single representative is determined even for a long free-edge in the T1e-expl task; hence, students can implement splitting long free-edges using [16] as depicted in Fig. 8a.

E3-mi—Mutual information (+7 points)—Multiple representatives can be ranked by considering the information gained by observing their surroundings [10], where each cell yields information based on the entropy of its obstacle-probability and cells are considered independent. Since assessing the expected information gain can be demanding, students can implement a simplified solution that approximates the information gain as the sum over all cells within the sensor range, e.g., using raycasting to determine the visible cells as depicted in Fig. 8b.

E4-tsp—Non-myopic planning (+4 points)—Non-myopic decision-making can be implemented as a solution of an open-ended TSP instance, as illustrated in Fig. 8c. Students are expected to utilize the LKH solver [20], already used during the courses. However, they can also use alternative solvers.

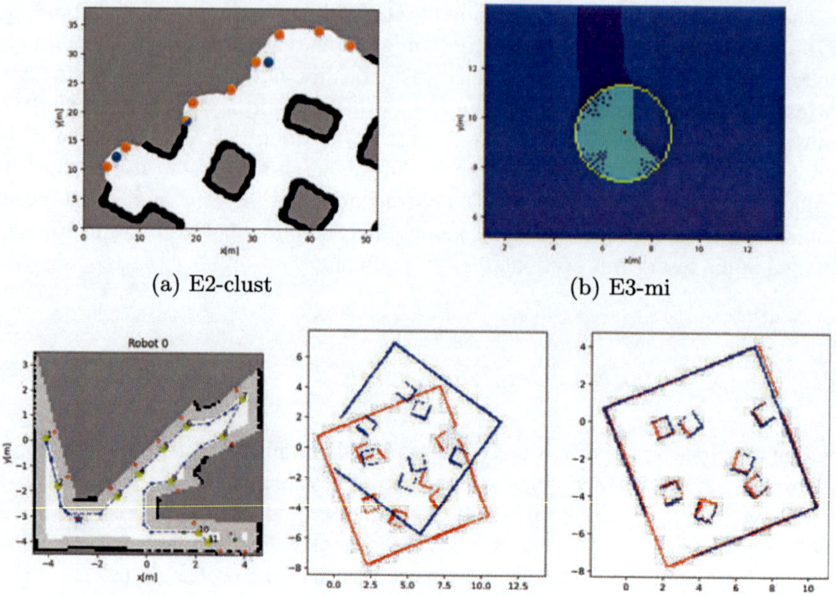

(a) E2-clust (b) E3-mi

(c) E4-tsp (d) Individual frames (e) E7-icp

Fig. 8 Illustrations of the suggested project extensions. **a** single representative of the free edge is in blue, while multiple representatives are in orange; **b** raycasting for determining visible cells in E3-mi; **c** TSP-based selection of the exploration waypoint in E4-tsp; **d** and **e** maps without and with the ICP-based transformation of the coordinate frames, respectively, in E7-icp

E5-mre—Multi-robot exploration (+5 points)—Using multiple exploring robots can improve exploration performance, and students can use various technical solutions, such as inter-process communication.

E6-dec—Multi-robot exploration with decentralized task allocation (+8 points)—Further extension of multi-robot exploration towards decentralized decision-making is suggested to be based on the MinPos [4], where each robot ranks possible waypoints based on its and other robots' distance to the waypoint being ranked.

E7-icp—Multi-robot exploration with individual coordinate frames (+10 points)—In the CoppeliaSim, the robots' positions are reported in a common coordinate frame. Students can assign each robot its coordinate frame based on its respective initial position; see Fig. 8d. Then, the transformation between the coordinate frames can be found by the Iterative Closest Point (ICP) algorithm [11] to get a joint coordinate frame as depicted in Fig. 8e.

We collected students' implementations from 2021 and 2022, where 42 and 33 students implemented at least one extension of the exploration project among 55 and

Table 1 No. of students implementing the individual project extensions

Year	PR	E1-dmap	E2-clust	E3-mi	E4-tsp	E5-mre	E6-dec	E7-icp
2021	42	31	39	30	19	21	12	1
2022	33	21	30	26	13	18	14	0

66 enrolled students in the course, respectively. The distribution of the implemented extensions is reported in Table 1.

The distribution suggests that students prefer to implement single-robot extensions over multi-robot options. Overall, the dynamic map size (E1-dmap), multiple-representative free-edge clusters (E2-clust), and mutual information (E3-mi) extensions are the most popular. The popularity of E3-mi is somewhat surprising to the course instructors since mutual information computation is relatively complex compared to the other popular options; however, closer inspection reveals that more than one-third of the students opted for the simplified version each year that omits raycasting, making the computation significantly easier. Among the single robot extensions, the TSP-based planning (E4-tsp) is the least popular, likely due to their choice to work with an external library, for which students report issues when using macOS and Windows.

About half of the students opted to use multiple robots (E5-mre). The data suggest that after implementing the multi-robot exploration, the students are motivated to add the advanced task allocation (E6-dec), with about two-thirds doing so. The exploration without a common coordinate frame (E7-icp) is selected rarely, likely because it requires an extensive modification of the provided supporting code. The only student who implemented the extension in 2021 noted his interest in mobile robot localization.

One more student has implemented ICP-based matching of the robot scan to a priori prepared maps of the environment; however, although similar, the extension was part of a single robot exploration and evaluated as a custom extension. Other custom extensions include D* Lite path planning [21], an obstacle distance field [5], or a ROS2-based implementation.

Overall, the project is popular with students, showing a considerable participation rate through non-trivial and custom extensions.

5 Conclusion

In this paper, we share our experience on autonomous mobile robot exploration in a robotic course; within it, we aim to prepare the students for the final project—an exploration framework—through a sequence of small tasks that are then integrated into the framework. Based on the students' progress and reported feedback, the students enjoy the way of small incremental steps to build a solution for a relatively complex behavior of autonomous exploration, which most students would not imagine at the beginning of the course. Although the small tasks have been fixed for several years, and our submission system automates evaluating the students' solutions, we do not detect significant plagiarism. The students acknowledge the tasks as steps to understand the topics and the incremental way of building a complex solution to avoid frustration from being overwhelmed. Based on our course experience, we prepared a dedicated, short, a few days long course[1] with four introductory lectures and a series of tasks leading to a robotic exploration framework. The supporting files can be used for further similar courses.

Acknowledgements The work on mobile robot exploration has been supported by the OP VVV-funded project CZ.02.1.01/0.0/0.0/16_019/0000765 "Research Center for Informatics." The effort of Petr Čížek and Jan Bayer on tasks implementation and Daniel Večerka on BRUTE design are also gratefully acknowledged.

References

1. Amigoni, F., Caglioti, V.: An information-based exploration strategy for environment mapping with mobile robots. Robot. Auton. Syst. **58**(5), 684–699 (2010)
2. Course Web Page—B4M36UIR and BE4M36UIR—Artificial Intelligence in Robotics. https://cw.fel.cvut.cz/wiki/courses/uir/start. Accessed 26 Jan 2023
3. Basilico, N., Amigoni, F.: Exploration strategies based on multi-criteria decision making for searching environments in rescue operations. Auton. Robot. **31**(4), 401–417 (2011)
4. Bautin, A., Simonin, O., Charpillet, F.: Minpos : A novel frontier allocation algorithm for multi-robot exploration. In: International Conference on Intelligent Robotics and Applications (ICIRA), pp. 496–508 (2012)
5. Bayer, J., Faigl, J.: On autonomous spatial exploration with small hexapod walking robot using tracking camera intel realsense t265. In: European Conference on Mobile Robots (ECMR), pp. 1–6 (2019)
6. Bouman, A., Ginting, M.F., Alatur, N., Palieri, M., Fan, D.D., Touma, T., Pailevanian, T., Kim, S.K., Otsu, K., Burdick, J., Agha-Mohammadi, A.: Autonomous spot: Long-range autonomous exploration of extreme environments with legged locomotion. In: IEEE/RSJ International Conference on Intelligent Robots and Systems (IROS), pp. 2518–2525 (2020)
7. Braitenberg, V.: Vehicles, Experiments in Synthetic Psychology. MIT Press (1984)
8. Bresenham, J.E.: Algorithm for computer control of a digital plotter. IBM Syst. J. **4**(1), 25–30 (1965)
9. Burgard, W., Moors, M., Stachniss, C., Schneider, F.E.: Coordinated multi-robot exploration. IEEE Trans. Robot. **21**(3), 376–386 (2005)
10. Charrow, B., Liu, S., Kumar, V., Michael, N.: Information-theoretic mapping using Cauchy-Schwarz quadratic mutual information. In: IEEE International Conference on Robotics and Automation (ICRA), pp. 4791–4798 (2015)
11. Chen, Y., Medioni, G.: Object modeling by registration of multiple range images. In: IEEE International Conference on Robotics and Automation (ICRA), pp. 2724–2729 (1991)
12. CoppeliaSim. https://www.coppeliarobotics.com/. Accessed 26 Jan 2023
13. Daniel, K., Nash, A., Koenig, S., Felner, A.: Theta*: Any-angle path planning on grids. J. Artif. Intell. Res. **39**(1), 533–579 (2010)

[1] https://cw.fel.cvut.cz/wiki/courses/crl-courses/redcp/start.

14. Faigl, J., Kulich, M.: On determination of goal candidates in frontier-based multi-robot exploration. In: European Conference on Mobile Robots (ECMR), pp. 210–215 (2013)
15. Faigl, J., Kulich, M.: On benchmarking of frontier-based multi-robot exploration strategies. In: European Conference on Mobile Robots (ECMR), pp. 1–8 (2015)
16. Faigl, J., Kulich, M., Přeučil, L.: Goal assignment using distance cost in multi-robot exploration. In: IEEE/RSJ International Conference on Intelligent Robots and Systems (IROS), pp. 3741–3746 (2012)
17. Faigl, J., Čížek, P.: Adaptive locomotion control of hexapod walking robot for traversing rough terrains with position feedback only. Robot. Auton. Syst. **116**, 136–147 (2019)
18. Gerkey, B.P., Mataric, M.J.: A formal analysis and taxonomy of task allocation in multi-robot systems. Int. J. Robot. Res. **23**(9), 939–954 (2004)
19. Harabor, D., Grastien, A.: Improving jump point search. Proc. Int. Conf. Autom. Plan. Sched. **24**(1), 128–135 (2014)
20. Helsgaun, K.: An effective implementation of the lin-kernighan traveling salesman heuristic. Eur. J. Oper. Res. **126**(1), 106–130 (2000)
21. Koenig, S., Likhachev, M.: Fast replanning for navigation in unknown terrain. IEEE Trans. Robot. **21**(3), 354–363 (2005)
22. Kulich, M., Faigl, J., Přeučil, L.: On distance utility in the exploration task. In: IEEE International Conference on Robotics and Automation (ICRA), pp. 4455–4460 (2011)
23. M. Wurm, K., Stachniss, C., Burgard, W.: Coordinated multi-robot exploration using a segmentation of the environment. In: IEEE/RSJ International Conference on Intelligent Robots and Systems (IROS), pp. 1160–1165 (2008)
24. Moravec, H.: Sensor fusion in certainty grids for mobile robots. AI Mag. **9**(2), 61–74 (1988)
25. Quigley, M., Conley, K., Gerkey, B., Faust, J., Foote, T., Leibs, J., Wheeler, R., Ng, A.Y., et al.: Ros: an open-source robot operating system. In: ICRA Workshop on Open Source Software, vol. 3, p. 5 (2009)
26. ROS—Robot Operating System. https://www.ros.org/. Accessed 26 Jan 2023
27. SciPy—Fundamental algorithms for scientific computing in Python. https://scipy.org/. Accessed 26 Jan 2023
28. Yamauchi, B.: A frontier-based approach for autonomous exploration. In: IEEE International Symposium on Computational Intelligence in Robotics and Automation (CIRA), pp. 146–151 (1997)
29. Yamauchi, B.: Decentralized coordination for multirobot exploration. Robot. Auton. Syst. **29**, 111–118 (1999)
30. Zlot, R., Stentz, A., Dias, M., Thayer, S.: Multi-robot exploration controlled by a market economy. In: IEEE International Conference on Robotics and Automation (ICRA), pp. 3016–3023 (2002)

Introducing Coding and Robotics: Prospective Mathematics Teachers' Metacognitive Thinking in Scratch

Marietjie Havenga and Tertia Jordaan

Abstract This paper reports on prospective mathematics teachers' metacognitive thinking when introduced to coding and robotics in the Scratch programming environment. A convenience sample of 61 Intermediate Phase second-year mathematics students participated in this qualitative study. Students were randomly assigned to groups of 5–6 members and worked together on several activities. Data were gathered by means of two Scratch programming tasks, a PowerPoint presentation, and a video as well as group reflections on their thought processes. Individual students also completed twelve reflective questions using Google Forms. Data were manually analyzed, and the findings revealed that students' individual and group metacognitive abilities were crucial in supporting their thinking in coding tasks. In addition, we suggested meta-construction as an important skill to be successful in coding and robotics activities.

Keywords Coding and robotics · Group collaboration · Mathematics education students · Metacognitive thinking

1 Introduction

According to Hitchens [1], "[t]he essence of the independent mind lies not in what it thinks, but in how it thinks." This notion is related to metacognitive inquiry, namely thinking about knowledge, beliefs about one's abilities, thinking about task requirements and knowledge about relevant strategies to aid in cognitive processing [2].

Metacognitive knowledge and control of one's thinking are crucial in addressing challenges and solving ill-structured problems [3]. Havenga et al. [4] emphasize the interplay and coherence between metacognition and problem-solving regarding

M. Havenga (✉) · T. Jordaan
Research Unit Self-Directed Learning, North-West University, Potchefstroom, South Africa
e-mail: marietjie.havenga@nwu.ac.za

T. Jordaan
e-mail: tertia.jordaan@nwu.ac.za

R. Balogh et al. (eds.), *Robotics in Education*, Lecture Notes in Networks and Systems 747, https://doi.org/10.1007/978-3-031-38454-7_2

computer science teachers' programming activities. Accordingly, such thinking is essential in coding and robotics. Socratous and Ioannou [5] highlight metacognition and problem solving as interrelated and profound to encourage good thinking practices when solving problems related to educational robotics (ER). Coding in Scratch involves the use of block-based instructions in a web-based environment to program the output as performed by visual objects, known as sprites [6]. Scratch also allows for the use of Raspberry Pi and Arduino to integrate robotics [7]. Such coding and robotics require high-level thinking as well as frequent monitoring and reflection on one's thinking to achieve the desired outcomes.

Although mathematics students are used to the practices of applying formulas and solving problems, knowledge regarding coding and robotics is unknown to most. Development of such skills is imperative, as most schools in South Africa consider implementing the draft coding and robotics curriculum in the near future to address challenges in the digital age [8]. Consequently, it is important to introduce future educators to the knowledge and skills involved in coding and robotics, as well as the importance of metacognition to support higher-order thinking.

This paper aims to report on prospective mathematics teachers' metacognitive thinking while working on several coding activities.

2 Theoretical Framework

2.1 Constructivism and Constructionism

Scholars highlight constructivism and constructionism as important philosophical views for robotics. The theory of constructivism originated from Jean Piaget's views on mental development. According to Piaget, individuals are actively involved in the construction of new knowledge [9]. Papert [10] regards the learner as the 'builder' or constructor of own knowledge (constructivism) and emphasizes 'objects to think with' as powerful means of providing for the development of several thinking skills. Papert consequently developed the theory of constructionism, which involves learners constructing new knowledge as a result of creating physical objects or artefacts [10, 11]. He also viewed objects as a means of providing for reflection 'on one's own actions and thinking' [10], clearly emphasizing metacognition as a crucial skill for coding and robotics.

2.2 Metacognitive Thinking in Coding and Robotics

The integration of higher-order thinking (e.g. problem solving, decision making, critical and creative thinking, innovation as well as computational thinking) and metacognitive skills are important for coding and robotics. Flavell distinguishes between metacognitive knowledge (personal knowledge, knowledge about the task, and knowledge about strategies) and metacognitive thinking (planning, monitoring, evaluation and reflection) as metacognitive processes that aid in cognitive inquiry [2]. Although metacognition is related to an individual's thinking, Socratous and Ioannou [5] point out the importance and benefits of metacognitive group thinking when students collaborate in robotics activities and problem solving. They refer to group metacognition as the ability to reflect on members' cognitive thinking, which involves skills such as awareness of members' abilities, management of information, revision, and the ability to make decisions collectively [5]. Stewart, Baek, Kwid and Taylor [12] emphasize the practice of metacognition when students identify erroneous thinking while engaging in coding and robotics. In addition, metacognitive thinking is also considered an essential ability of self-directed learners when solving problems [4]. Furthermore, Lopez-Caudana et al. [13] highlight that ER provides for the development of key competences and fosters curiosity, group work, decision-making as well as active and responsible learning.

Since metacognitive thinking is essential in coding and robotics, we would like to propose *meta-construction* as an important skill where students (as individuals and groups) are actively involved in purposeful collaboration when planning their code, creating robots (online or physical), monitoring group activities, reflecting on their thinking and evaluating robot performance according to their initial planning—thus, constructing new knowledge and creating robot artefacts as objects for cognitive and metacognitive thinking. In other words, students learn from working with open-ended problems where they are challenged with coding and robotics. In Scratch, the meta-construction may involve creating a new sprite, planning the trajectory, monitoring the movement of the sprite according to their coding of instructions, reflecting on the movement, reviewing and evaluating their thinking, and determining what they could do 'better' to optimize the coding.

2.3 Prospective Teachers and the Learning and Teaching of Coding and Robotics

It is imperative that prospective teachers be exposed to the knowledge and skills associated with coding and robotics. Yüksel [14] emphasizes the importance of training prospective teachers regarding coding and robotics, as it allows for meaningful learning and knowledge construction, enhances problem solving, creativity, computational thinking and group work, and enriches reflective thinking. However, such activities must be connected to students' real-world context to be motivating [15]. Chahine et al. [16] reported on an exploratory case study regarding middle-level science and mathematics teachers' engagement in robotics and engineering design activities. It is evident that these teachers made connections between mathematical thought and real-life aspects. Furthermore, they were curious and interested in mathematics, and their shared involvement and inquiry resulted in the development of new ideas while engaging in the above-mentioned activities [16]. For these teachers, it was valuable to understand the importance of problem solving in coding and robotics, to think logically, to be actively involved in meaningful learning, and to develop confidence when engaged in such activities.

Yüksel [14] further noted that pre-service science teachers highlighted the practice-oriented aspect of ER, linking science concepts to coding, emphasizing enjoyment as well as active learning, and participating in solving problems. Initially, these teachers experienced the coding aspect as the most difficult part of ER, as they did not have previous knowledge about programming [14].

Prospective teachers must be able to apply coding and robotics in their teaching and they have to be confident to facilitate skill development in learners [17]. Future teachers must therefore create a learning environment that triggers learners' curiosity about ER [8]. In addition, they should facilitate learners' creativity, promote their confidence in coding and robotics, and assist them to become responsible, lifelong and self-directed learners [4, 15]. Prospective teachers must also provide an environment where learners have to set clear and realistic goals, plan the coding task, monitor their own progress and the execution of the robot, participate in critical discussions, reflect on their thinking, and evaluate their performance.

3 Research Methodology

A general qualitative methodology was followed in this study. Since the mathematics students had no prior experience of coding, the lecturer decided to introduce Scratch as a block-based visual programming environment and educational tool. The first phase involved introducing students to coding, while the second phase comprised robotics activities. In this paper, we only discuss results based on phase 1.

3.1 Participants

The participants were a cohort of 61 second-year education students in an Intermediate Phase Mathematics course. Students worked mainly online due to the COVID pandemic. The second researcher randomly assigned students to groups of 5–6 members where they completed the coding tasks. Each group leader coordinated coding activities and uploaded the assignment on the learning management system (LMS). The researcher employed delayed informed consent after completing the assignment, consisting of two activities as detailed below. The research was approved by the Ethics Committee of the Faculty of Education at the relevant university.

3.2 *Structuring of Students' Activities*

As part of phase 1, students completed Scratch programming tasks consisting of two activities.

Activity 1—Draw a Shape

To introduce students to Scratch, they had to draw a specific shape. In the first activity, students had to program the sprite to draw a red decagram using five images of block codes with related output (program execution) sequentially. With each attempt, group members were required (1) to show the coding blocks that they used, (2) display the related output of the sprite, and (3) reflect on their thinking. Figure 1 shows an example of Group 3's first attempt. The quotes are unedited and are reproduced verbatim.

Activity 2—Formulate a Problem Related to a Mathematics Topic

In the second activity, prospective teachers had to formulate their own 'Scratch problem', related to a specific topic in the curriculum, to introduce future Grade 4 to 6 learners to coding and robotics (Fig. 2, topic: 'space and shapes'). This was followed by activities where group members had to display their coding, consecutively including five Scratch images with output, as well as their individual and group reflections (similar to Activity 1). As part of didactics, the groups submitted a lesson plan for coding and robotics and indicated active teaching–learning strategies relevant to this problem. For example, they could use problem-based learning or cooperative learning to assist in active learning. In addition, they had to clearly specify the learning objectives taking Bloom's revised taxonomy into consideration [18]. For example, students had to design and produce their own original shape for the sprite to draw (create—level 6 in Bloom's taxonomy).

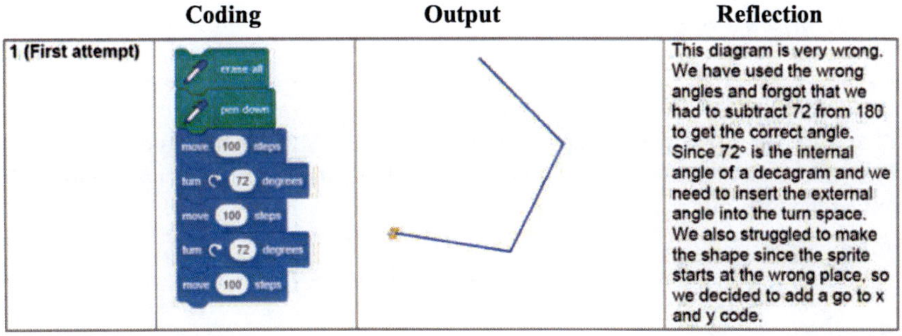

Fig. 1 Example of group 3's first coding attempt

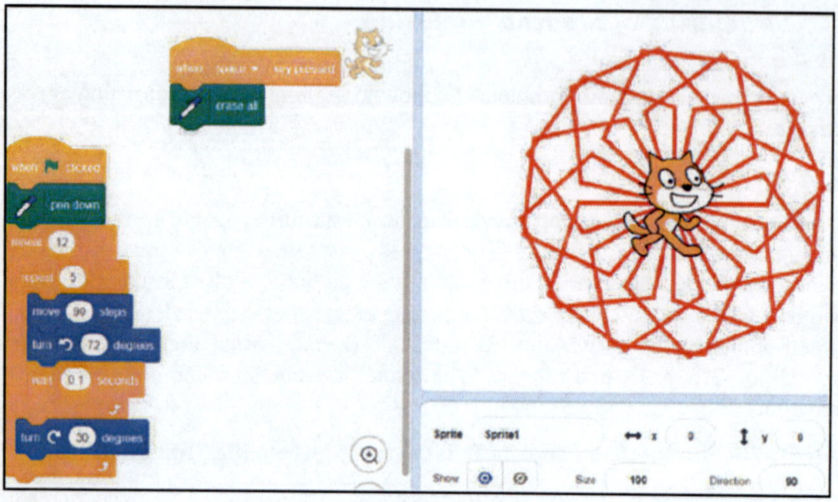

Fig. 2 Group 10's formulation and execution of their own problem aimed at grade 4–6 learners

Table 1 Data collection methods

Data gathered	
Group's data	• Scratch programming tasks (Activity 1 and 2); • POWERPOINT presentation (the group's experiences and collaboration on Scratch activities); • Video of each group's Teams meeting (during which they reflected on their metacognitive thinking processes in completing Activity 1 and 2)
Student's individual data	• Completion of a questionnaire comprising twelve reflective questions on Google Forms

3.3 Data Collection

Data were collected from the twelve groups as well as from individual students as outlined in Table 1. Each group submitted their two Scratch programming tasks, a PowerPoint presentation, and a video. Thereafter, individual students completed twelve reflective questions using Google Forms.

3.4 Data Analysis

The second author used descriptive coding (short phrase or word) [19] to analyse the data. Findings are presented in Table 2 according to aspects regarding metacognitive knowledge and thinking (Sect. 2.2).

Table 2 Students' responses regarding their thinking while doing coding activities

Metacognitive knowledge	
Personal knowledge	*I stressed a little … I am not good with … coding* *I needed more resources to understand the tasks*
Knowledge about the task	*Never heard of Scratch before. I was uncertain … did not know what to do* *Difficult to understand … read it a few times to realize what was expected of us* *I was confused about what exactly had to be done* *Challenging to understand what to do … had to practice how to use the Scratch app first*
Knowledge about strategies	*We used extra resources … attended the lecturer's meeting … asked questions … this helped a lot*
Metacognitive thinking	
Planning	*We initially struggled … watched YouTube and things became easier* *Initially … play around … had to assist each other and collaborate to figure it out* *[Sprite] needed to move a certain number of steps forward … turning 12 to the left … repeated 9 times before reaching [start]*
Monitoring	*Our coding wasn't going to work but we managed to … correct them [mistakes]* *I had to re-do everything for about 5 times before I could get the actual shape right* *We changed the angles until we finally found the correct angles to make the shape perfect* *Shape was double … getting closer … main problem … lines did not connect … at the end one is too long and going past the end point … if we get that right it should work* *The degrees were decreased and set to the original … steps were increased by another 50 steps … the shape was better … still too many corners* *We [deleted] codes outside the repeat code and it resulted in repeating squares at angle of 45° forming the unique shape that we wanted*
Reflection	*I could see what I needed to do and when I needed to do that part* *One challenge … difficulty of agreeing on a mutual time to do group work … planning* *Once we understood how Scratch works, we started enjoying playing around to make the best drawings. It was interesting to see how all the codes made a difference to the product. To change something small, add or take something away, influenced the product and changed it* *I enjoyed the coding activities … to learn a new program and play around … see all the things I can do* *I used more brainpower in Scratch than in any other mathematics problems*

(continued)

Table 2 (continued)

Evaluation	*To be honest, I did not believe that we could do such a great coding task, being the first time doing such an activity on the Scratch app ... and we were satisfied* *We can give ourselves a pat on the shoulder. After all the struggling ... we have reached a successful product. Everything works as it should and looks quite cool* *I really enjoyed the fact that the activity allowed me to think out of the box* *We had enough time to work on the task ... let our creativity go to submit a good task. The group worked together well ... we completed the task* *Repeated the code ten times ... lines met each other ... sprite had to turn 108° ... sides met each other ... decagram was successfully sketched*

4 Findings and Discussion

Table 2 displays selected examples of students' responses on their metacognitive knowledge and thinking.

Metacognitive knowledge and thinking, as suggested by Flavell [2], play an important role in problem solving in general and in coding and robotics specifically [5]. Initially, students were honest about their lack of knowledge and skills, as they had little or no background of information technology (IT), coding and robotics. However, the mathematics students were positive that they would be successful in completing the activities in collaboration with their peers.

Students *planned* activities by meeting online to share ideas and discuss their *modus operandi*. Initially, they struggled; however, they used additional resources, watched YouTube videos, attended the lecturer's meeting, played around with Scratch, and assisted each other.

The mathematics students *monitored* their progress while working on the tasks. They experienced several challenges when drawing the decagram (Activity 1) and formulating and designing their own activity (Activity 2). These challenges were related to the following: incorrect logical thinking and integration of block codes, which resulted in incorrect angles; lines passing or not meeting each other; incorrect number of repetitions; incorrect pen size; incorrect color (it had to be red); and the edges of the shapes were too thick or too thin, which resulted in inappropriate shapes. However, students persisted, they 'constantly' checked and managed to draw the shapes as requested (see Table 2). Some set reminders of what to do and when to do it while others used a checklist to keep them on track.

In addition, students *reflected* on the coding tasks, their group collaboration, and addressing the challenges. Using Scratch, they discovered previously unknown applications of shapes. Students were motivated, made connections to the real-world and applied their knowledge about mathematics (e.g. degrees, internal and external angles) to create interesting sprite patterns (Activity 2). One student mentioned: *once we understood how Scratch works, we started enjoying playing around to make the best drawings.* Another noted: *I used more brainpower in Scratch than in any other mathematics problems.* In general, they enjoyed learning the Scratch programming environment, and 'playing around' to see what they could achieve.

Evaluation of activities showed that, despite their challenges, students were satisfied with the quality of their coding tasks as a first introduction to Scratch. They had enough time to complete the tasks and they valued the opportunity to be creative and to think outside the box. In addition, mathematics students indicated that they worked together well, and they were dependent on each other while working on the tasks in terms of their assistance and collaborative decision-making. They sometimes had to redo the coding several times before they were successful. Ultimately, most of the students were satisfied with the development of their own Scratch activity and enjoyed the tasks despite their challenges.

Metacognitive thinking was essential in students' coding activities, as highlighted by Stewart et al. [12]. Despite their initial lack of metacognitive knowledge, they were successful in creating the decagram and their own shape. Several examples of *metaconstruction* were evident where students were actively involved. They planned and figured out what needed to be done, designed their own program and structured the coding blocks in such a way to ensure correct logic that would consequently lead to the expected output and movement of the sprite.

Future teachers' higher-order thinking skills and metacognitive thinking should be deliberately fostered during their training to enable them to teach these skills to their learners [17]. The activities that students completed can be seen as a start to their exposure to coding and robotics to empower them, as future teachers, to introduce the 'new' subject, to their Intermediate Phase learners in future. One student emphasized, *what I enjoyed most about the assignment was the opportunity it offered me to learn about coding.* Another student commented, *I valued the fact that I was forced to think outside the box and be creative.*

5 Conclusion

The goal of the study described in this paper, was to investigate prospective mathematics teachers' metacognitive thinking in completing Scratch tasks. Findings showed that students were aware of their metacognitive knowledge—or the lack thereof—and implemented metacognitive thinking while working on the tasks. They planned their activities, monitored their progress, reflected on their thinking, and evaluated the quality of the output. It was evident that metacognitive thinking plays an important role in coding activities. Future research on the development of metacognitive thinking in a face-to-face environment, and specifically in coding and robotics, is suggested. Furthermore, effective ways to teach coding and robotics in the mathematics classroom also need to be explored further.

References

1. Hitchens, C.: Letters to a young contrarian. Art of Mentoring, pp. 1–160. Basic Books, New York (2001)
2. Flavell, J.H.: Metacognition and cognitive monitoring: a new area of cognitive–developmental inquiry. Am. Psychol. **34**(10), 906–911 (1979)
3. Yorulmaz, A., Uysal, H., Çokçalışkan, H.: Pre-service primary school teachers' metacognitive awareness and beliefs about mathematical problem solving. J. Res. Adv. Math. Educ. **6**(3), 239–259 (2021)
4. Havenga, M., Breed, B., Mentz, E., Govender, D., Govender, I., Dignum, F., Dignum, V.: Metacognitive and problem-solving skills to promote self-directed learning in computer programming: teachers' Experiences. SA-eDUC J. **10**(2), 1–14 (2013)
5. Socratous, C., Ioannou, A.: Evaluating the impact of the curriculum structure on group metacognition during collaborative problem-solving using educational robotics. TechTrends **66**, 771–783 (2022)
6. Dickins, R., Melmoth, J.: Coding for beginners: using scratch. Usborne, London (2015)
7. Dhakulkar, A., Olivier, J.: Exploring microworlds as supporting environments for self-directed multimodal learning. In: Mentz, E., Laubscher, D., Olivier, J. (Eds.) Self-directed learning: an Imperative for Education in a Complex Society. NWU Self-Directed Learning Series, vol. 6, pp. 71–106. AOSIS, Cape Town (2021)
8. Department of Basic Education (DBE) Republic of South Africa: Curriculum and Assessment Policy Statement. Proposed Amendments for the Curriculum Assessment Policy Statement (CAPS) to Make Provision for Coding and Robotics Grades R–9. Pretoria (2021)
9. Piaget, J.: Piaget's theory. In: Mussen, P. (ed.) Carmichael's Manual of Child Psychology, pp. 703–832. Wiley, New York (1970)
10. Papert, S.: Mindstorms, Children, Computers, and Powerful Ideas. Basic Books, New York (1980)
11. Ackermann, E.: Piaget's constructivism, Papert's constructionism: What's the difference? (2001). https://learning.media.mit.edu/content/publications/EA.Piaget%20_%20Papert.pdf. Accessed 12 Jan 2023
12. Stewart, W.H., Baek, Y., Kwid, G., Taylor, K.: Exploring factors that influence computational thinking skills in elementary students' collaborative robotics. J. Educ. Comput. Res. **59**(6), 1208–1239 (2021)
13. Lopez-Caudana, E., Ramirez-Montoya, M.S., Martínez-Pérez, S., Rodríguez-Abitia, G.: Using robotics to enhance active learning in mathematics: a multi-scenario study. Mathematics **8**(12), 1–21 (2020)
14. Yüksel, A.O.: Investigation of pre-service Science teachers' learning experiences on educational robotics applications. J. Comput. Educ. Res. **10**(19), 50–72 (2022)
15. Rodríguez-Martínez, J.A., González-Calero, J.A., Sáez-López, J.M.: Computational thinking and mathematics using Scratch: an experiment with sixth-grade students. Interact. Learn. Environ. **28**(3), 316–327 (2020)
16. Chahine, I.C., Robinson, N., Mansion, K.: Using robotics and engineering design inquiries to optimize mathematics learning for middle level teachers: a case study. J. Math. Educ. **11**(2), 319–332 (2020)
17. Schina, D., Esteve-Gonzalez, V., Usart, M.: Teachers' perceptions of Bee-Bot robotic toy and their ability to integrate it in their teaching. In: Lepuschitz, W., Merdan, M., Koppensteiner, G., Balogh, R., Obdržálek, D. (eds.) Robotics in Education (RiE). Advances in Intelligent Systems and Computing, vol. 1316, pp. 121–132. Springer, Cham (2021)
18. Krathwohl, D.R.: A revision of Bloom's taxonomy: an overview. Theory Pract. **41**(4), 212–218 (2002)
19. Saldaña, J.: The Coding Manual for Qualitative Researchers, 3rd edn. Sage, Los Angeles (2016)

Validation of Teachers' Competencies for Applying Robotics in Science Education

Doaa Saad, Igor Verner, and Rinat B. Rosenberg-Kima

Abstract This research aims to identify the required competencies needed to integrate robotics activities into science education, particularly due to the lack of comprehensive models that address the necessary competencies for teachers to effectively teach science with robotics. A list of competencies needed by middle school teachers to integrate robotics activities into science classrooms was developed in the following steps. First, an initial list of competencies was developed based on a literature review in the field of robotics education that focused on various aspects of the TPACK model and open observations during a teachers' development program. Second, experts and experienced robotics teachers were interviewed regarded the competencies needed to develop and implement robotics activities suitable for science education, which resulted in an updated list of competencies and adding additional competencies in the context of 21st-century skills. Third, Fifty-five teachers rated the items on a scale from 1-not necessary to 5-very necessary. Factor analysis was performed and the items were examined concerning the rating they received. Understanding how robotics can be coordinated with pedagogy and scientific knowledge for effective teaching, the present study adapted a TPACK instrument for use of robotics to teach classroom science and reformulated it with 21st-century skills.

Keywords Science-education · Educational robotics · Competencies · TPACK model · 21st-century skills

D. Saad (✉) · I. Verner · R. B. Rosenberg-Kima (✉)
Technion—Israel Institute of Technology, Haifa, Israel
e-mail: sdoaa14@campus.technion.ac.il

R. B. Rosenberg-Kima
e-mail: rinatros@technion.ac.il

I. Verner
e-mail: ttrigor@technion.ac.il

R. Balogh et al. (eds.), *Robotics in Education*, Lecture Notes in Networks and Systems 747, https://doi.org/10.1007/978-3-031-38454-7_3

1 Introduction

Despite the aspirations of educational systems and the research findings on the positive effects of integrating technologies like robotics into science classrooms, many teachers avoid incorporating technological accessories into their lessons, preferring to teach using traditional methods [1]. The literature reports a lack of STEM teachers' competencies for integrating robotics activities in science education and teachers' low self-efficacy in using a robotics environment in their teaching [2, 3].

While the literature points to the importance of guiding science teachers to integrate robotics activities into their teaching, few studies present a comprehensive model of the competencies or the extent of teacher guidance needed to support the development of the teaching competencies to use robotics in the classroom [4].

The identification and development of competencies needed to integrate robotics activities into science education can support the implantation of robotics in science education. Compared to learning that focuses on one discipline, integrating robotics activities in science education requires a synthesis of different disciplines. The TPACK (Technological Pedagogical and Content Knowledge) framework leverages the interactions between content knowledge, pedagogical knowledge, and technological knowledge. According to the TPACK model, teachers need to know not only how to operate robots but also how to integrate them into effective instruction. Teachers need to be confident in the use of technology, know the content they want to teach, and the pedagogical strategies needed to teach the content with the technology [2, 5]. Previous studies (e.g., [3, 6]) consider the TPACK model as necessary to train teachers to integrate educational technologies in general and robotics in particular into their teaching. *Technological Knowledge*, e.g., Learning new technologies such as LEGO robotics. *Technological Content Knowledge,* e.g., Knowing of technology to use for understanding and doing science/math. *Technological Pedagogical Knowledge*, e.g., Select technology to enhance teaching and learning. *Technological Pedagogical Content Knowledge*, e.g., Use of LEGO robotics to illustrate science/ math topics.

The TPACK model may be necessary to prepare science teachers to integrate robotics activities into their teaching, but it may not be sufficient. 21st-century skills may also be required. The ABET (Accreditation Board for Engineering and Technology) defined a list of essential professional skills for teachers, e.g., an ability to function on multi-disciplinary teams, an ability to communicate effectively, a recognition of the need for, and an ability to engage in lifelong learning [7]. In addition, other studies (e.g., [8, 9]) identified creativity as one of the important skills of the 21st-century skills.

Therefore, this research aims to identify the competencies required for integrating robotics activities into science education for middle school teachers based on the conceptual framework of the TPACK model in addition to 21st-century skills such as ABET recommendations [10], NRC 21st-century skills [9], and OECD 21st-century competencies and skills [8].

2 Methodology

2.1 Participants

Fifty-five teachers (22 female and 33 male) responded to the questionnaire on competencies for teaching science with robotics. Thirty-five teachers (~64%) had a previous background in robotics and experience teaching robotics. Seventeen teachers had up to 10 years of teaching experience, 23 teachers had 11–25 years of teaching experience, and 15 teachers had more than 25 years of teaching experience.

In addition, sixteen middle school teachers participated in a robotics professional development for science teachers and were observed with respect to the competencies they demonstrated. Furthermore, five experts in the field of educational robotics and five teachers experienced in robotics were interviewed by one of the researchers about robotics and the skills needed to teach with robotics.

2.2 Research Tools

Literature Review. Based on a literature review in the field of robotics education, we defined an initial list of competencies needed to integrate robotics activities into the science classroom (e.g., [1–3, 6, 11–14]). The full list of competencies can be seen in Table 2.

Interviews. In parallel with the literature review, we interviewed experts in the field of educational robotics and experienced robotics teachers to update the list of competencies needed to integrate robotics activities into science classrooms. The interview questions were based on research tools developed in other studies [2] on this topic and adapted for this study and were validated by experienced experts in the field of educational robotics.

Open Observations. We designed a 30-h professional development program for teachers based on the Task-Centered Instructional Strategy [15]. Three tasks were developed that included technological, pedagogical, and scientific knowledge at increasing levels of difficulty and decreasing levels of support. During the teachers' professional development program, one of the researchers observed the teachers' learning processes in solving science problems, including the development of robotics models (building and programming with EV3 LEGO Mindstorms software), experiences with the models, analysis of the results, and conclusions.

Demographics questionnaire. The demographic questions included gender, seniority, teaching subject, previous background in robotics, and prior experience in teaching robotics.

Competencies for teaching science with robotics questionnaire. This questionnaire aimed to identify competencies for integrating robotics activities into middle

Table 1 Overview of the validation process

Stage	Description
Initial identification of the competencies	The competencies were initially extracted from a literature review and open observations during a teachers' development program
Expert validation	Interviews were conducted with five experts and five experienced robotics teachers, who were presented with the initial competencies list and could comment on the items and suggest additional items
Statistical validation of the questionnaire	Fifty-five teachers rated the items on a scale from 1—not necessary, 2—slightly necessary, 3—moderately necessary, 4—necessary, and 5—very necessary Factor analysis was performed and the items were examined with respect to the rating they received

school science classes. The questionnaire contained closed and open-ended questions. The closed-ended questions were statements that were rated on a 5-point Likert scale of 1—not necessary, 2—slightly necessary, 3—moderately necessary, 4—necessary, and 5—very necessary. Teachers rated the competencies that might be needed to integrate robotics activities into science lessons. In addition, teachers were asked to add other relevant competencies to the list and provide feedback. Table 1 provides an overview of the validation procedure of the questionnaire.

3 Results

3.1 Initial Identification of the Competencies

Following a literature review of conceptual frameworks for teacher professional development and, in particular, teacher training in educational robotics, along with open observations of the study's researchers in robotics education, we developed a list of competencies for teachers to integrate robotics activities into science instruction. The initial list included about forty-two competencies based primarily on the TPACK model.

3.2 Expert Validation

As a next step, we interviewed experts in the field of robotics to validate the questionnaire. After the interviews, we removed some of the competencies that were not well formulated, or that were too general (i.e., we did not include items that refer

Table 2 Factor Analysis results for competencies for integrating robotics into science lessons questionnaire (N = 55)

	Competencies	Factors			Ratings	
		TPK	21st skills	TPACK	M	SD
1	Basic ability to solve faults in robot operation	0.79			4.00	1.09
2	Basic ability to program an educational robot	0.76			3.96	1.15
3	Ability to teach students to build a robot	0.75			3.89	1.18
4	Ability to guide students to a robotics project	0.72			4.04	1.12
5	Ability to teach students to program a robot	0.70			3.96	1.23
6	Basic ability to build an educational robot	0.67			4.07	1.05
7	Ability and motivation to learn to operate educational robots	0.67			4.16	0.96
8	Ability to use robotics activities in class to increase educational motivation	0.62			4.35	0.82
9	Ability to develop and manage educational environments to experiment with robotics	0.63			4.00	0.96
10	Ability to improve study results in robotics classes based on evaluation of the previous experience	0.61			4.09	0.95
11	Ability to cooperate		0.82		4.27	0.95
12	Ability to solve problems		0.82		4.44	0.94
13	Teamwork ability		0.79		4.25	1.00
14	Creativity		0.77		4.33	0.94
15	Ability to think critically		0.74		4.20	0.95
16	Self-regulated ability		0.74		4.35	1.04
17	Communication ability		0.73		4.07	1.05

(continued)

Table 2 (continued)

18	Ability to make decisions		0.71		4.25	1.02
19	Ability to plan and perform an experiment with a robot to expand the scientific content			0.70	4.07	0.92
20	Ability to direct robotics activities in class to develop higher-order thinking skills			0.67	4.15	1.03
21	Ability to enrich an explanation of a scientific concept to deepen its understanding and illustrate it using a robot			0.65	4.11	0.99
22	Ability to define an applied problem in robotics and solve it based on mathematical and scientific methods			0.60	3.60	1.13
23	Ability to model natural phenomena with the help of a robotic system			0.60	3.80	1.01
F1	Overall perceived necessity of TPK				4.05	1.05
F2	Overall perceived necessity of 21st-century skills				4.27	0.98
F3	Overall perceived necessity of TPACK				3.95	1.01

only to pedagogy or only to the content knowledge, as we assume these are general competencies that are required for any science teacher). Other competencies were added or reworded, based on the interview.

Another finding that emerged from the interviews was the need to expand the list of competencies to include additional competencies for the effective integration of robotics into science education, such as self-regulation, teamwork, creativity, etc. For example, "I think there are other important competencies that do not appear in the list, such as the ability to work in a team, the ability to persevere, and self-regulation. "Another expert: "What about creativity? Without the ability to be creative, how can a teacher integrate robotics activities into the science classroom?" It was possible to determine that all of the additional competencies suggested by the researchers were 21st-century skills. Therefore, the next step was to conduct another literature review on 21st-century skills. Following this process, we arrived at thirty-four competencies. We then distributed these to fifty-five teachers.

In the following, the results of the statistical analysis are presented. The Factor Analysis test of competencies is presented, and the Means and Standard Deviations of all competencies. The final version of the list contains 23 competencies.

3.3 *Factors of the Competencies for Teaching Science with Robotics Questionnaire*

Fifty-five teachers were asked to rate the necessity of the competencies on a scale of 1—not necessary to 5—very necessary. The means of the items range from 3.44 to 4.44 (see Table 2). To identify the factors of the questionnaire, we conducted an exploratory factor analysis using principal axis factorization with varimax rotation. First, we evaluated the scree plot to determine the number of factors. We found a flattening point that indicated the importance of the first three factors, where each item loaded on one factor. Nine items were deleted because their factor loading was below the threshold of 0.60. Another two items were deleted, although they belonged to the factors, because of a low rating of the items in terms of their perceived necessity (mean less than 3.5). Overall, 23 competencies were identified, belonging to one of three factors (see Table 2). The three factors we identified were: (1) Technological Pedagogical Knowledge (TPK) for integrating robotics activities (10 items, α = 0.95), (2) 21st-century skills (8 items, $\alpha = 0.95$), and (3) Technological Pedagogical Content Knowledge (TPACK) for integrating robotics activities in science education (5 items, α = 0.85). All three factors received high ratings; the factor with the highest necessity rating was 21st-century skills ($M = 4.27$, $SD = 0.98$), followed by the factor TPK ($M = 4.05$, $SD = 1.05$), and the factor TPACK, which received the lowest necessity rating ($M = 3.95$, $SD = 1.01$) (see Table 2).

4 Discussion and Conclusions

In this study, we provide an overview of the competencies required to integrate robotics activities into science classrooms, particularly given the lack of comprehensive models that address the necessary competencies for teachers to effectively integrate robotics into science education [2, 3]. Consistent with previous studies (e.g., [2, 5]), this study demonstrates the importance of the interactions between science knowledge, pedagogical knowledge, and technological knowledge, which in this case involved robotics. In addition, this study highlights the importance of developing 21st-century soft skills to incorporate robotics activities into science teaching. This is consistent with the recommendation from ABET [7], the OECD [8], and NRC [9].

We believe that the list of competencies, which we validated within the current study, can be used to design effective professional development for science teachers to enrich their teaching and learning activities in robotics. Effective professional development can develop competencies in teachers, which in turn can positively influence their sense of self-efficacy [16], reduce anxiety about integrating robotics activities into their teaching, and develop positive attitudes towards robotics [17].

In summary, this study identified the set of competencies required to integrate robotics into science education. Our findings highlight the importance of 21st-century skills for integrating innovative technologies such as robotics into the classroom in addition to technological, pedagogical, and scientific knowledge. Future studies can utilize this set of competencies to assess the effectiveness of professional development programs and evaluate teachers' self-efficacy with respect to these competencies.

References

1. Castro, E., Cecchi, F., Salvini, P., Valente, M., Buselli, E., Menichetti, L., et al.: Design and impact of a teacher training course, and attitude change concerning educational robotics. Int. J. Soc. Robot. **10**(5), 669–685 (2018)
2. Rahman, S.M., Krishnan, V.J., Kapila, V.: Exploring the dynamic nature of TPACK framework in teaching STEM using robotics in middle school classrooms. In: Proceedings of ASEE Annual Conference and Exposition (2017)
3. You, H.S., Chacko, S.M., Kapila, V.: Examining the effectiveness of a professional development program: integration of educational robotics into science and mathematics curricula. J. Sci. Educ. Technol. https://doi.org/10.1007/s10956-021-09903-6
4. Atmatzidou, S., Demetriadis, S., Nika, P.: How does the degree of guidance support students' metacognitive and problem solving skills in educational robotics? J. Sci. Educ. Technol. **27**(1), 70–85 (2018)
5. Mishra, P., Koehler, M.J.: Technological pedagogical content knowledge: a framework for teacher knowledge. Teachers College Rec. **108**(6), 1017–54 (2006)
6. Schmidt, D.A., Baran, E., Thompson, A.D., Mishra, P., Koehler, M.J., Shin, T.S.: Technological pedagogical content knowledge (TPACK): the development and validation of an assessment instrument for preservice teachers. J. Res. Technol. Educ. **42**(2), 27 (2009)
7. Chidthachack, S., Schulte, M.A., Ntow, F.D., Lin, J.L., Moore, T.J.: Engineering students learn ABET professional skills: a comparative study of project-based-learning (PBL) versus traditional students. In: 2013 North Midwest Section Meeting (2021)
8. Ananiadou, K., Claro, M.: 21st century skills and competences for new millennium learners in OECD countries [Internet]. OECD Education Working Papers; vol. 41. Report No.: 41 (2009). https://www.oecd-ilibrary.org/education/21st-century-skills-and-competences-for-new-millennium-learners-in-oecd-countries_218525261154
9. NRC. Education for Life and Work: Developing Transferable Knowledge and Skills in the 21st Century [Internet]. National Academies Press, Washington, D.C.(2012). http://www.nap.edu/catalog/13398
10. ABET. Criteria for Accrediting Engineering Programs, 2018–2019 | ABET [Internet] (2019). https://www.abet.org/accreditation/accreditation-criteria/criteria-for-accrediting-engineering-programs-2018-2019/
11. Arikan, S., Erktin, E., Pesen, M.: Development and validation of a STEM competencies assessment framework. Int. J. Sci. Math. Educ. 1–24 (2020)

12. Atmatzidou, S., Demetriadis, S.: Advancing students' computational thinking skills through educational robotics: a study on age and gender relevant differences. Robot. Auton. Syst. **75**, 661–670 (2016)
13. Balyk, N., Barna, O., Shmyger, G., Oleksiuk, V.: Model of professional retraining of teachers based on the development of STEM competencies (2018)
14. Chalmers, C.: Preparing teachers to teach STEM through robotics. Int. J. Innov. Sci. Math. Educ. **25**(4), 17–31 (2017)
15. Merrill, M.D.: A task-centered instructional strategy. J. Res. Technol. Educ. **40**(1), 5–22 (2007)
16. Mallik, A., Rajguru, S.B., Kapila, V.: Fundamental: analyzing the effects of a robotics training workshop on the self-efficacy of high school teachers. In: American Society for Engineering Education Annual Conference & Exposition, June 2018, pp. 24–7
17. Papadakis, S., Vaiopoulou, J., Sifaki, E., Stamovlasis, D., Kalogiannakis, M.: Attitudes towards the use of educational robotics: exploring pre-service and in-service early childhood teacher profiles. Educ. Sci. **11**(5), 204 (2021)

How to Draw Cardioids with LEGO Robots: A Technical-Mathematical Project in Higher Education

Attila Körei and Szilvia Szilágyi

Abstract Drawing robots have been made for decades, but there are only a few designs that can be used directly to teach mathematics in higher technical and IT education. Our goal is to present the case of using LEGO robots for drawing cardioid curves. Two different derivations of the cardioid curve were used to design the robot mechanism. In the first case, the curve is drawn by means of gears rolling on each other, and in the second case by means of circular movements carried out by two motors. This paper examines the details of the construction of both types, together with the theoretical background. Finally, we present a STEAM learning guided project based on the use of cardioid drawing robots.

Keywords Educational robotics · LEGO robots · Parametric curves · Cardioid · STEAM education

1 Introduction

Nowadays educational robotics plays an important role in mathematics education because it is a powerful and flexible educational tool with a great motivational factor. The use of robotics and programming has a long history in mathematics. In 2020 Zhong and Xia conducted a systematic review to explore the potential of educational robotics in mathematics education [1]. They reviewed the empirical evidence on the use of robotics in mathematics education and identified future research perspectives

Prepared in the "National Laboratory for Social Innovation" project (RRF-2.3.1-21-2022-00013), within the framework of Hungary's Recovery and Resilience Plan, with the support of the Recovery and Resilience Facility of the European Union.

A. Körei (✉)
Department of Applied Mathematics, University of Miskolc, Institute of Mathematics, Miskolc, Hungary
e-mail: attila.korei@uni-miskolc.hu

S. Szilágyi
Department of Analysis, University of Miskolc, Institute of Mathematics, Miskolc, Hungary
e-mail: szilvia.szilagyi@uni-miskolc.hu

R. Balogh et al. (eds.), *Robotics in Education*, Lecture Notes in Networks and Systems 747, https://doi.org/10.1007/978-3-031-38454-7_4

for robot-assisted mathematics education. The authors analyzed 20 empirical studies on how to teach and learn mathematical knowledge through robotics. The results indicate that most studies were conducted with a small sample size, the largest research groups were elementary and secondary school students, most studies used LEGO robots, robots were primarily applied to teach and/or learn graphics, geometry, and algebra, and almost half of the studies taught mathematics by engaging students in game-like interactions with robots. The use of educational robotics in higher education is also growing in popularity as confirmed by Sánchez et al. [2]. There are already many examples of robots being used in higher education during the teaching of maths, with impressive results [3–6].

All engineering and computer science BSc students encounter the cardioid curve during their first year of studies because calculus and mathematical analysis courses invariably discuss this well-known curve. It is important that students develop the skills of understanding the cardioid, i.e. they have to know them by heart. By using STEAM teaching techniques, we can create opportunities for students to encounter the cardioid curve in a new way, gain first-hand experience and acquire practical knowledge. We chose to work on projects with LEGO robots because we have already developed good practices in discussing hypotrochoid curves. To draw these kinds of curves and to study their parametric equations, we built the Spikograph 1.0 robot using a LEGO SPIKE Prime Core Set, which models the mechanism of the Spirograph game [6]. The Spirograph is a geometric drawing toy that produces mathematical curves known as hypotrochoids and epitrochoids [7, 8]. There are several robots on the web that use the principle of Spirograph or Hoot-Nanny the Magic Designer toy, and the creators usually present them by sketching spectacular plane curves. In some cases, the drawn curve has been given a fancy name by the creators to distinguish between the drawn curves [9]. If the mechanism for building the robot follows the principle of the original Spirograph game, then there is no challenge in mathematically naming the curve and writing down its parametric equations, as in [6] for members of the hypotrochoid family.

In this paper drawing the cardioid curve is the main goal we want to achieve using two different approaches based on LEGO parts. In this context, we present a three-step project in which educational robotics plays an important role.

2 The Most Important Facts About Cardioids

The cardioid was developed by the Danish astronomer Ole Christensen Roemer in 1674. It was discovered during an investigation to find the best design for gear teeth [10]. The cardioid curve is a special case of the epicycloid and the limaçon of Pascal [11]. The name comes from the Greek word for heart. Interestingly, the cardioid was given its evocative name relatively late, only in 1741, when Johann Castillon named it in his treatise in the Philosophical Transactions of the Royal Society. The heart-shaped cardioid curve has fascinated mathematicians for centuries because of its mathematical properties, graphical beauty, and practical applications [12, 13].

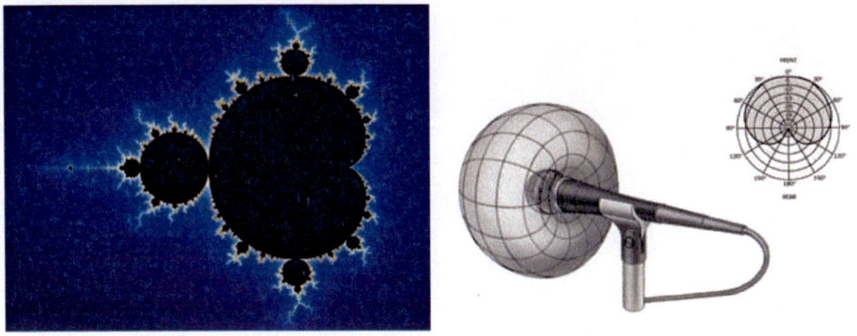

(a) The Mandelbrot set. (b) Cardioid microphone.

Fig. 1 Cardioids in mathematics and physics

The cardioid appears in many seemingly different areas of mathematics, playing an important role in both optics and fractal geometry, for example as the central figure of the famous Mandelbrot set (Fig. 1a) [13]. Cardioid shapes appear in the cams that direct the even layering of thread on bobbins and reels in the textile industry, and in the signal-strength pattern of certain radio antennas [14]. The cardioid pattern antenna (or the cardioid antenna) derives its name from the shape of its radiation lobe pattern. It is typically a mono-band antenna that is used most often for terrestrial communication [15]. Cardioids even show up in audio engineering. The most common unidirectional microphone is a cardioid microphone (Fig. 1b), so named because the sensitivity pattern is a cardioid. The cardioid family of microphones is commonly used as vocal or speech microphones [16].

3 Theoretical Framework of Cardioid Drawing Robot Design

Cardioids are curves that can be classified into several families of curves. We now consider these as special epitrochoids. An epitrochoid is a plane curve produced by tracing the path of a point at a given distance from the center of a circle which rolls without slipping around a fixed circle [17]. This general definition allows the two circles to have different radii, and the point under consideration can be inside the circle, on its circumference or even outside the circle. By varying the radius of the circles and the location of the point, a very wide variety of curves can be created. If the radius of the fixed and moving circles are the same, then it is a limaçon, and if the point is on the moving circle, then it is a cardioid. By the most commonly accepted definition, a cardioid is a plane curve traced by a point on the perimeter of a circle that is rolling around a fixed circle of the same radius. The cardioid can be determined by various mathematical formulas, see e.g. [18]. For our purposes, parametric equations

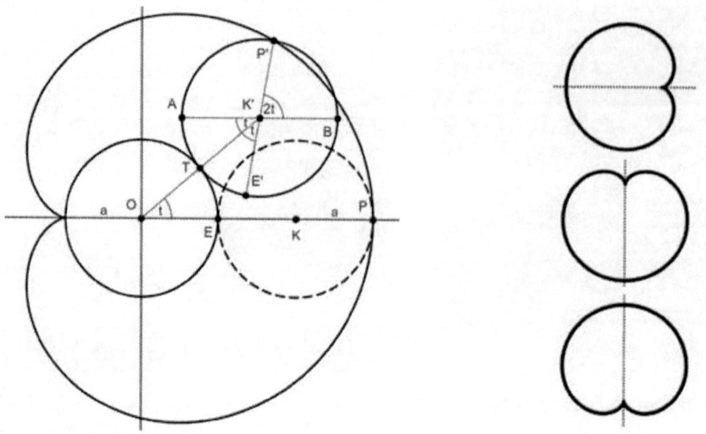

(a) Forming a horizontal cardioid. (b) A horizontal and two vertical cardioids.

Fig. 2 Different orientations of cardioids

are the most appropriate, meaning that the coordinates of the points on the curve are given by different functions, but both the x- and y-coordinates depend on the same variable, called the parameter. When deriving the parametric equations of the cardioid, this parameter is usually the angle between a line through the center of both circles and the x-axis. See Fig. 2a, where this angle is denoted by t. Here the centre of the fixed circle is at the origin, denoted by O, and let's suppose that the radius of the circles is a. Consider the motion of the point P, the initial position of which is at $t = 0$ is $(3a, 0)$. After the angular rotation t, the new position of P is denoted by P', the new center of the moving circle is K', and the point denoted by E is placed in E'. Because of the non-slip rolling, the length of arc TE is equal to the length of arc TE', and since the circles have the same radius, it follows that angle t is equal to angle $TK'E'$. Clearly, angle $TK'A$ also equals t, implying that angle $P'K'B$ is $2t$. The centre of the moving circle traces on a circle of radius $2a$, so the coordinates of point K' is $(2a \cos t, 2a \sin t)$ for angular rotation t. Using the observation obtained for angle $P'K'B$, point P' has coordinates $(a \cos 2t, a \sin 2t)$ with respect to the centre K' as an origin. Summing these results we obtain the coordinates of vector $\overline{OP'}$, which are the parametric equations of the cardioid:

$$\begin{aligned} x(t) &= 2a \cos t + a \cos 2t = a(2 \cos t + \cos 2t) \\ y(t) &= 2a \sin t + a \sin 2t = a(2 \sin t + \sin 2t), \end{aligned} \tag{1}$$

where $t \in [0, 2\pi]$.

Depending on its axis of symmetry the cardioid has 4 different orientations, of which only one of the horizontal cardioids' equations have been derived. The other three possibilities are shown in Fig. 2b and the parametric equations of them are as follows:

$$x(t) = a(2\cos t - \cos 2t), \quad y(t) = a(2\sin t - \sin 2t), \quad t \in [0, 2\pi], \text{(up)},$$
$$x(t) = a(2\sin t - \sin 2t), \quad y(t) = a(2\cos t - \cos 2t), \quad t \in [0, 2\pi], \text{(middle)},$$
$$x(t) = a(2\sin t + \sin 2t), \quad y(t) = a(2\cos t + \cos 2t), \quad t \in [0, 2\pi], \text{(down)}.$$

4 LEGO Robots Drawing Cardioids

Two types of robots were designed and built to draw cardioid curves using mainly the parts of a LEGO SPIKE Prime Set. The first model is based on the Spirograph principle, using gears to model the fixed and the moving circle (Fig. 3). In the second model, no gears are used and the circular motions are carried out by two motors (Fig. 5). Using the right elements and settings, both models are suitable for drawing cardioids.

4.1 Drawing Robot with Gears-Spikograph 2.0

The operating principle of our first drawing robot is based on the definition of epitrochoids. Outside the circumference of a fixed circle, another circle is rolled along to which we attach a pen at a given distance from its centre. In the case of cardiods, this distance is equal to the common radius of the moving and fixed circles. The circles are modelled with gears that ensure rolling without slipping. One of the two gears is fixed and the other is rotated around it by a motor. The simplest construction is to pass the motor-driven axle through the centre of the fixed gear and from there, with the help of a lever, it moves the other gear.

Standard LEGO gears have an axle hole instead of a standard pinhole in the middle, which means that the gear, once locked in place, would not allow the rotation of the arm to turn the moving circle. Therefore a gear with a hole in the middle was needed.

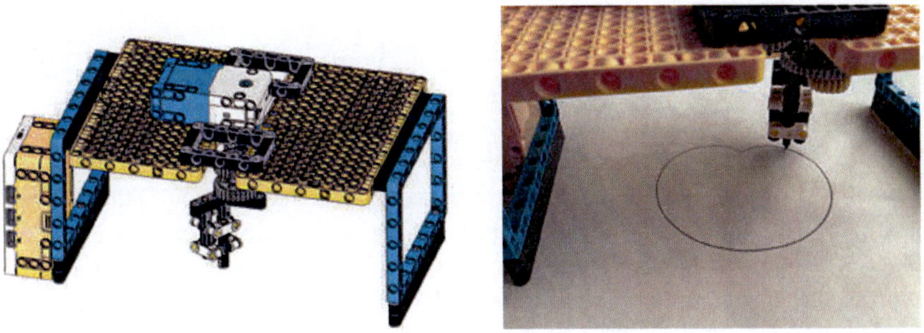

(a) The model. (b) Spikograph 2.0 in action.

Fig. 3 Drawing device with gears

(a) Positioning with the 28-tooth gear. (b) Positioning with the 40-tooth gear.

Fig. 4 Finding the best position of the drawing head

Such an element is included in the expansion set, which is not really a gear but the upper part of the so-called turntable, but it served our purposes perfectly. In addition, its diameter is 28 mm and there is also a normal gear in this size, so we had an immediate solution for the moving circle. The next task was to place the drawing pen on the circumference of the moving gear. The pen holder we built is fixed to the moving gear by a beam, but the spacing of the holes in the beam also determines the position of the pen. The distance between two adjacent holes is 8 mm on a straight LEGO Technic beam, which means that the pen can be placed at a distance of a multiple of 8 mm from the centre of the gear. In order to make the robot draw a curve that most closely resembles a cardioid, the pen holder should be fixed at a distance of 16 mm from the centre of the 14 mm radius gear, using a straight beam. In fact, the drawn curve is a looped limaçon.

A slightly better result can be achieved if a beam bent at 53° is used to fix the pen. In Fig. 4a the light brown pin denoted by A is in the centre of the moving circle, while the intended position of the pen is in point C and indicated by the arrow. Using trigonometry to find distance d between the points A and C we have

$$d = | \overline{AC} | = 2 \cdot 0.8 \cdot \sin \left(\frac{127}{2} \right)^{\circ} . \tag{2}$$

The result is approximately 14.3 mm, which in theory still gives a looped limaçon, but the difference is barely visible on the curve drawn by the robot.

Finally, with a little trick, we got our robot to draw a perfect cardioid. The idea was inspired by the fact that pin-holed, 3D-printed Lego-compatible gears are available on the Internet. We took a Technic gear with a diameter of 40 mm and simply drilled it through the middle, turning the axle hole into a pinhole, so it could now be incorporated into the model as a fixed circle. We chose this size because the geometry of the gear allowed us to install the drawing pen in the desired position on the rolling gear. The 40-tooth Technic gear features alternating axle and standard pinholes placed 4 mm apart, allowing for a shift of this size when placing the other components. Figure 4b shows our solution which required two additional gears. The light brown 20 mm diameter gear is only used to stabilize the position of the other 24 mm diameter gear, which is fixed to the large, 40-tooth rotating gear by a pin. This pin is located 8 mm from the centre and the hole marked with the arrow is 12 mm from the pin. Thus the pen, when fixed to this hole, will be exactly 20 mm from the centre. The cardioid made by this drawing head is shown in Fig. 3b.

4.2 Drawing Robot with Two Motors

Now we introduce a robot construction that can also be used to draw a cardioid, but its operating principle differs from the previously presented one. Studying the movement of the drawing pen mounted on the Spikograph robot, we can notice that the beam to which the head is attached rotates around a point, which also describes a circular path. This planetary-like motion can be easily modelled with a robot equipped with two motors. One motor rotates the whole construction around a point, while the other rotates the drawing head. We built a robot that performs the two circular movements, the model is shown in Fig. 5a.

Next we derive the general parametric equations of the curves which can be drawn by this robot. Let us call the motor that rotates the entire robot the main motor, and the other motor the arm motor. Assume that the speed of the arm motor is c times that of the main motor. In practice, this means that while the main motor makes one turn, the arm motor makes c turns. Let b denote the distance between the rotation axes of the two motors, and let the pen rotate in a circle of radius a.

In Fig. 6b, the point around which the robot rotates is denoted by O, and the point around which the pen-holder beam rotates is denoted by K. P indicates the position of the pen. It can be read from the figure that while the main motor makes t turns, the robot arm closes an angle $ct + t$ compared to its original position if both motors rotate in the same direction. Denote $(x(t), y(t))$ the coordinates of the moving point P, after

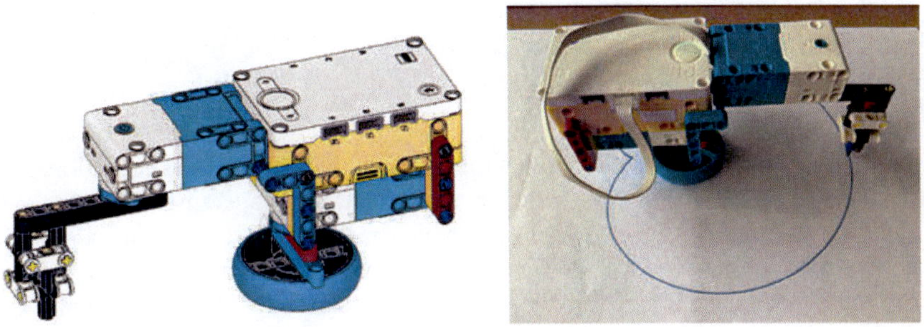

(a) The model. (b) Drawing a perfect cardioid.

Fig. 5 Drawing robot with two motors

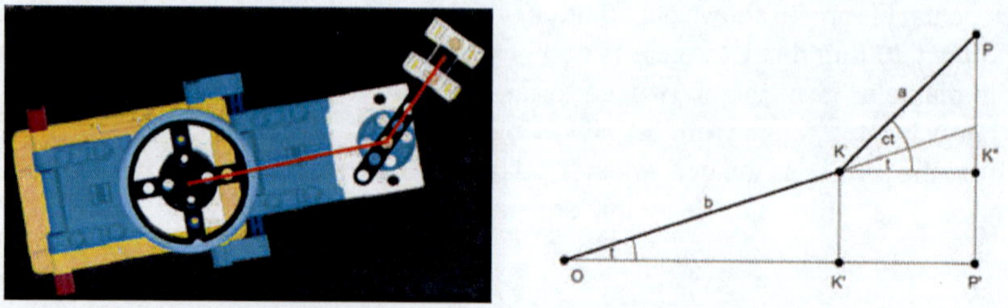

(a) The robot from bottom view. (b) Determining the coordinates of point P.

Fig. 6 Position of the drawing head after the main motor has turned at angle t

rotating the main motor by an angle t. Then $x(t) = OK' + K'P', y(t) = P'K'' + K''P$. Expressing the length of these line segments by the variable t, we get the parametric equations of the curve traced by the point P:

$$\begin{aligned} x(t) &= b\cos t + a\cos((c+1)t) \\ y(t) &= b\sin t + a\sin((c+1)t), \end{aligned} \tag{3}$$

where $t \in [0, 2\pi]$, if we follow the motion for one turn. Note that if $c = 1$ and $b = 2a$, then Eq. (3) takes the form of Eq. (1) i.e. we get a cardioid. This means that if we want to draw a cardioid with our robot, the two motors must be operated at the same speed and the point around which the pen-holder beam rotates describes a circle twice the radius of the circle traced by the pen itself. The construction on Fig. 5a meets the geometrical requirements, since $a = 40$ mm and $b = 80$ mm. In the program we operate the motors parallel with the same speed, making one rotation with both motors. Thus, the two motors run for exactly the same amount of time, and the pen returns to its starting position, completing the drawing of the cardioid curve (Fig. 5b).

5 A Student Project on Cardioids

In this section, we suggest some exercises that can be used by students to practice and understand cardioids better. Following the principles of STEAM education, the project is structured in a way that engages students in cooperative learning, and teamwork with LEGO robots, they can share their ideas and apply old and new knowledge to gain a deeper understanding of the problems. The tasks to be solved are rather practical, so LEGO drawing robot and Desmos graphing calculator play an important role in solving them. The whole project consists of 3 steps. The steps of the project can be carried out once a week over a period of three weeks, in parallel with the curriculum. The steps build on each other, solving increasingly complex tasks. Each step is accompanied by an electronic task sheet with links to the construction guides and the path to the Desmos graphing calculator. Students also have the opportunity to document their work via the task sheet. Steps 2 and 3 contain more tasks, which provides differentiation. Considering that for the cardioid, besides the parametric equations, the implicit and polar equations can be easily written, students can choose between several solution paths. Added to this is the diversity of variation provided by horizontal and vertical cardioids that are equivalent to each other in many test aspects. The hints and solutions to the tasks are not included on the worksheet, they continue their work on the basis of verbal instructions for those who need the teacher's help and intervention. The interesting thing about the project is that we draw the cardioid with a different robot in each three-step. Before starting the project work, the theoretical background of the robot's operation is clarified.

Tools needed: LEGO SPIKE Prime Core Set (No. 45678), LEGO SPIKE Prime Expansion Set (No. 45680), 3 pcs LEGO Technic gears 40 Tooth (No. 3649), 1 pcs LEGO Technic gear 24 Tooth (No. 3648), scissors, ruler, compass, pen, markers, paper.

Step 1-Checking the symmetry property
Verify that the cardioid is an axially symmetric curve.

(a) Build the Spikograph 2.0 robot (Fig. 3a) using the building instructions.
(b) Put a pen in the writing head, then make and start a program that moves the motor for one rotation.
(c) Create cardioid templates. Draw at least three curves on a piece of paper with the robot, then cut them out.
(d) Fold the cut-out shape in half to test the axial symmetry of the cardioid.
(e) Place the template in the Cartesian-coordinate system as shown in Fig. 2 and draw it around. Give the parametric equation of the drawn curve for all four cases. Check the written parametric equations with Desmos graphing calculator. Hint: Using theoretical results in Sect. 3 we can give the parametric equations because $a = 1.4$ due to the use of a 28-tooth gear.

Step 2-Finding the Equation of a Tangent Line
Give the equation of the tangent line for the given point P of the cardioid. To solve this problem, you need to know the relationship between the rectangular and polar coordinates.

(a) Build the Spikograph 2.0 robot using LEGO Technic gears with 40 Teeth. Follow the steps in the building guide.
(b) Put a pen in the writing head, then make and start a program that moves the motor for one rotation. Create a new cardioid template from paper.
(c) Place the template in the Cartesian-coordinate system as shown in Fig. 7a and draw it around. Find the circles to get this cardioid and draw them both with a compass.
(d) Give the polar and implicit equation of the cardioid curve.
Hint: Consideration should be given to the cusp of the cardioid is at the origin and $a = 2$ due to the use of a 40-tooth gear.

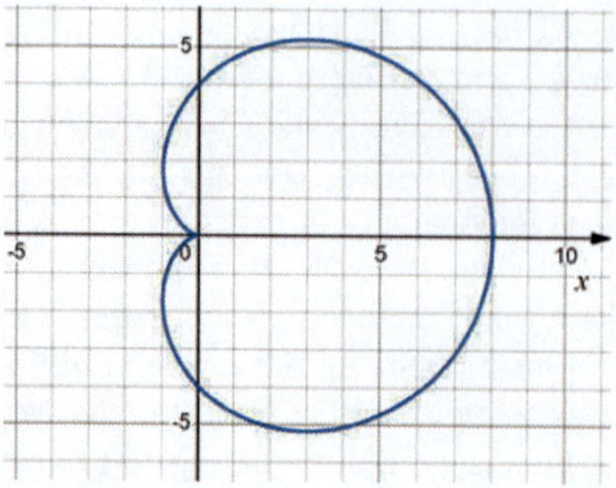

(a) Cardioid curve for Step 2.

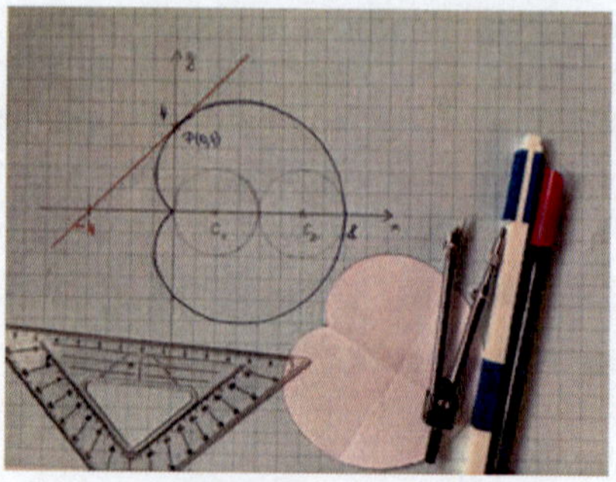

(b) The tangent line.

Fig. 7 Exercises with cardioids in Step 2

Solution: The polar equation of the curve:

$$r(\theta) = 4(1 + \cos\theta), \qquad \theta \in [0, 2\pi]$$

and the implicit equation:

$$x^2 + y^2 = 4\sqrt{x^2 + y^2} + 4x.$$

(e) Find $\frac{dy}{dx}$ for the given cardioid.
Hint: There are two alternative ways to find the derivative. You can start from both the polar equation and the implicit equation.
(f) Evaluate the derivative at $P(0, 4)$.
(g) Write the equation of the tangent line to the cardioid at point $P(0, 4)$. Graph the tangent line and the cardioid together (Fig. 7b).
(h) Find the horizontal and vertical tangent lines of the cardioid.

Step 3-Finding Area and Length
Find the area of the region in the plane enclosed by the cardioid and the length of the curve.

(a) Build the cardioid drawing robot with two motors (see Fig. 5a). Follow the steps in the building instructions.
(b) Put a pen in the writing head, then make and start a program that moves one rotation with the main motor and two rotations with the arm motor at the same power.
(c) Determine the area of the region in the plane enclosed by the cardioid.
Hint: In Steps 1 and 2, we can see that the curve is symmetric, and the curve can be placed in the Cartesian coordinate system when the cusp is at the origin and the axis of symmetry is the x-axis. Of course in this case the area above and below the x-axis is equal. There are two possible placements.
Solution: In the first case the polar equation $r(\theta) = 2a(1 + \cos(\theta))$, in the second

$r(\theta) = 2a(1 - \cos(\theta))$. Now $a = 4$, so the total area can be calculated in the first case as follows:

$$A = \int_0^{\pi} r^2(\theta)d\theta = \int_0^{\pi} a^2(1 + \cos(\theta))^2 d\theta = \frac{3}{2}\pi a^2 = 24\pi.$$

(d) Draw with a compass the biggest circle you can find in the cardioid. Determine the area of the region that lies inside the cardioid and outside the circle.
(e) Find the area outside the cardioid $r(\theta) = 8 + 8\sin(\theta)$ and inside the circle $r(\theta) = 24\sin(\theta)$.
(f) Find the length of the curve.
Hint: It is useful to calculate the arc length for the curve $r(\theta) = 8 + 8\cos(\theta)$. Because of symmetry, it is sufficient to integrate over the interval $[0, \pi]$.
Solution:

$$s = 2\int_0^{\pi} \sqrt{r^2(\theta) + \left(\frac{dr(\theta)}{d\theta}\right)^2} d\theta = 2\int_0^{\pi} \sqrt{128(1 + \cos(\theta))} d\theta = 64.$$

(g) Find the length of any chord through the cusp point.

6 Conclusions and Perspectives

At the university level, our teaching goals are to support engineer and IT students to study and retain relevant information, encourage their creative and critical thinking, and teach them how to study effectively through experimental learning. Consequently, we have modified the traditional coursework by using new tools and STEAM techniques to attract and motivate the students to actively participate in the learning process. Therefore, we use educational robotics and a project-based learning approach to support students to learn parametric curves in a meaningful way. In this paper, we present two LEGO-based designs for cardioid curve drawing. They have in common that their principle of operation follows the mathematical method of curve derivation. Considering that the cardioid can be derived by several methods, we built two types of robots with different mechanisms: one using gears to model slip-free rolling, while the other incorporating two motors. The theoretical foundations of building robots and the technical aspects required to design them are reviewed in detail. In addition, we have created a three-step project to integrate cardioid drawing robots into education. This project-based learning programme is currently a work in progress. It is planned to go into live testing during the spring semester at the University of Miskolc.

Although in this article we have focused specifically on the study and drawing of cardioids, it is clear that other types of curves can be plotted using the robots we have built. With Spikograph 2.0, it is possible to draw additional epitrochoids by varying the available gears and by placing the drawing head in different positions. The double-motor version also offers many possibilities for further investigations depending on the position of the drawing head and the speed of the motors relative to each other. Our future work will be aimed at being able to draw and study additional curves by developing existing robot prototypes and designing new constructions.

References

1. Zhong, B., Xia, L.: Systematic Review on Exploring the Potential of Educational Robotics in Mathematics Education. Int. J. Sci. Math. Educ. **18**, 79–101 (2020). https://doi.org/10.1007/s10763-018-09939-y
2. Sánchez, H., Martínez, L.S., González, J. D.: Educational robotics as a teaching tool in higher education institutions: a bibliographical analysis. J. Phys.: Conf. Ser. (8th International Conference on Mathematical Modeling in Physical Science) **1391**, 1–7 (2019). https://doi.org/10.1088/1742-6596/1391/1/012128
3. Nerantzi, C., James, A.: LEGO for University Learning: Inspiring academic practice in higher education (2019). https://doi.org/10.5281/zenodo.2813448
4. Francis, K., Caron, F., Khan, S.: Robots in Mathematics Education. In: Holm, J., Mathieu-Soucy, S. (eds.) Proceedings of the 2018 Annual Meeting of the Canadian Mathematics Education Study Group/Groupe Canadien d'Étude en Didactique des Mathématiques, pp. 63–75 (2018)
5. Zaldivar, X., Niebla-Zatarain, J., Zaldivar, U., Marmolejo-Rivas, C., Bernal-Guadiana, R.: Learning with robotics, new technologies in higher education to learn programming and mathematical functions. In: Proceedings of EDULEARN17 Conference. Barcelona, Spain, pp. 10447–10454 (2017). https://doi.org/10.21125/edulearn.2017.0979
6. Körei, A., Szilágyi, Sz.: Displaying parametric curves with virtual and physical tools, The Teaching of Mathematics, **XXV/2.**, pp.61–73. (2022) https://doi.org/10.57016/TM-EHGC7743
7. Hall, L.M.: Trochoids, roses, and thorns—beyond the spirograph. Coll. Math. J. **23**(1), 20–35 (1992). https://doi.org/10.2307/2686194
8. Phan-Yamada, T., Gwin, E.: Hypocycloids and Hypotrochoids. MathAMATYC Educator **6**(1), 1–13 (2014)
9. Simple Drawing Machine, https://jkbrickworks.com/simple-drawing-machine/. Last accessed 15 Jan 2023
10. Larson S.: Introducing the Cardioid, https://mse.redwoods.edu/darnold/math50c/CalcProj/Sp99/SeanL/cardioidf.htm. Last accessed 11 Jan 2023
11. Lawrence, J.D.: A Catalog of Special Plane Curves, pp. 118–121. Dover, New York (1972)
12. Weisstein, E.W.: Cardioid. From MathWorld-A Wolfram Web Resource, https://mathworld.wolfram.com/Cardioid.html. Last accessed 11 Jan 2023
13. Pickover, C.A.: The Math Book: From Pythagoras to the 57th Dimension, 250 Milestones in the History of Mathematics. Sterling Publishing Co., Inc., New York City (2012)
14. Thomas, G.B., Weir, M.D., Giordano, F.R.: Thomas' Calculus, 11th edn. Pearson Education Inc., Addison-Wesely (2005)
15. Miscellaneous Antennas, https://vu2nsb.com/antenna/miscellaneous-antennas/. Last accessed 11 Jan 2023
16. What Is a Cardioid Microphone? https://mynewmicrophone.com/what-is-a-cardioid-microphone-polar-pattern-mic-examples/. Last accessed 11 Jan 2023
17. Yates, R.C.: Curves and Their Properties, Classics in Mathematics Education, vol.4. National Council of Teachers of Mathematics (1974)
18. Lockwood, H.: A Book of Curves. Cambridge U.P, Cambridge (1967)

Evaluation of a Robotic System in Secondary School

Christopher Bongert and Reinhard Gerndt

Abstract This paper addresses a need for basic education in the field of computational thinking skills to prepare students for ever-faster digitization. We present the underlying considerations and a teaching approach for secondary school students with educational robots and detail seven teaching units. We show the results of evaluating the approach in a case study with nine 12th-grade vocational school students, using an adapted Callysto test [7]. The results suggest an increasing students' motivation and interest in computational thinking.

Keywords Educational robots · Computational thinking · Teaching approach

1 Introduction

Competencies in digitization are increasingly crucial for modern education. Students need to gain technological literacy on top of classical subjects. Computational thinking is a process for solving complex problems that involves formulating problems such that they can be solved with the help of a computer, organizing and analyzing data logically, representing data through abstractions, automating solutions through algorithmic thinking, identifying, analyzing, and implementing solutions to find the most efficient and effective combination of steps and resources, and generalizing and transferring this problem-solving process to a wide range of problems [8]. With educational robots and suitable teaching modules, students can learn computational thinking and programming. With our work we primarily want to evaluate the following questions:

1. Do students' computational thinking skills improve as a result of robotics lessons?

C. Bongert (✉) · R. Gerndt
Ostfalia University of Applied Science, Wolfenbuttel, Germany
e-mail: christopher@bongert.eu

R. Gerndt
e-mail: r.gerndt@ostfalia.de

R. Balogh et al. (eds.), *Robotics in Education*, Lecture Notes in Networks and Systems 747, https://doi.org/10.1007/978-3-031-38454-7_5

2. Is the use of educational robots helpful in the classroom?

In the following we will present our approach to teach computational thinking with robots. After sketching the underlying considerations, we will detail the approach and then, prior to our conclusions, describe and evaluate our real-world experiments with the Edison robot from the Australian company Microbric [5] in a vocational high school in the state of Lower Saxony in Germany.

2 State of the Art

There is no uniform nationwide model for teaching STEM (Science Technology Engineering Mathematics) skills in Germany. Due to the federal system, every state creates its own curriculum. In Lower Saxony, basic computer science will be introduced as a compulsory subject in grades 8–10 in all types of general education schools from 2023 on [3]. Vocational high schools ('Berufliche Gymnasien') offer full-time courses of secondary education spanning classes 11–13 [1]. In the technology track, programming and image processing skills are taught in grade 12. This vocational orientation distinguishes the lessons from computer science lessons at a general education school, where primarily the laws and principles of information processing are taught [4].

2.1 The Teaching Approach

The teaching approach, though independent of a specific robot type, builds on a number of conditions: The robots are easy to use for different age groups to focus on the learning process and come at a price of less than €100 to equip as many students as possible with a robot. The Edison robots used in this approach come with a pedagogical concept, a development environment and teaching materials, in the form of a guide for teachers and worksheets for students, freely available under a Creative Commons license so they can be easily adapted by teachers [6].

This work foresees 2 use cases scenarios: Exploration and programming. For exploration, several pre-programmed applications can be called via barcodes, scanned by the robots. Programs can be developed either via the very simple block language, "EdBlocks", the more advanced "EdScratch" based on Scratch [2], or the text-based, Python-like language "EdPy", which was used for this work. A browser-based development environment (Fig. 1) showed to be advantageous. The teaching material was derived from the teaching materials provided and translated to German.

Teaching was divided into 4 lessons of 90 min and one or two units each, with one robot for every group of 2–3 students:

In unit 1, students explore the robot's capabilities using pre-made programs.
In unit 2, they learn to write and download simple programs to the robot.

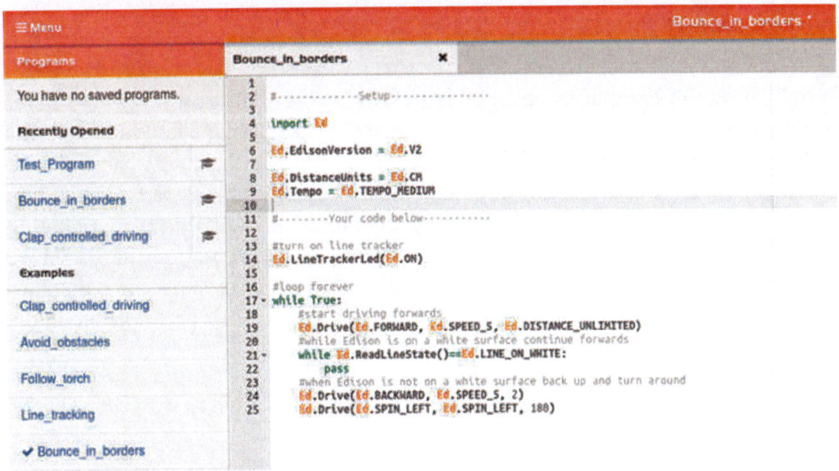

Fig. 1 Screenshot of the "EdPyApp" development environment

Unit 3 challenges students to program the robot to navigate a maze.
Unit 4 introduces expressions and variables for angle calculations and geometric shapes.
In unit 5, students use loops and variables to program the robot to dance.
Unit 6 teaches students how to write programs that respond to audio inputs from the environment.

Finally, in unit 7, students learn about decision-making and conditional program flow using infrared sensors for obstacle detection. Throughout the curriculum, students learn programming concepts such as while and for loops, comment signs, motor control functions, and event-based programming. All documents for the units as well as the test can be downloaded in the edited versions in German on GitHub [9].

3 Evaluation

The teaching approach was evaluated in a 12th-grade class with 9 students from a vocational high school with a focus on technology. Following a technical track, the students receive a much better STEM education than students at a general education school. The evaluation was started with a pre-test to determine the students' entry-level capabilities of computational thinking. Then classroom sessions were conducted as detailed above. Finally, a post-test was carried out.

3.1 The Callystotest

To determine the computational thinking skills in pre- and post-test and answer the research questions, the Canadian "Callysto Computational Thinking test" was used [7]. The test was translated into German to eliminate language barriers. The adapted test takes about 20 min to complete and consists of two parts. In the first part, the participants' handling of technology, complex problems, and programming knowledge is evaluated. The second part asks for typical algorithmic skills. There are 11 programming exercises in the form of text tasks with supporting graphics (Fig. 2). At the end, students provide a self assessment on their performance in the second part.

3.2 Results

In the first part of the pre-test, all students showed to be enthusiastic about technology. Most could solve complex problems but not explain the steps. Six had basic programming experience. Seven were not excited by programming challenges. All

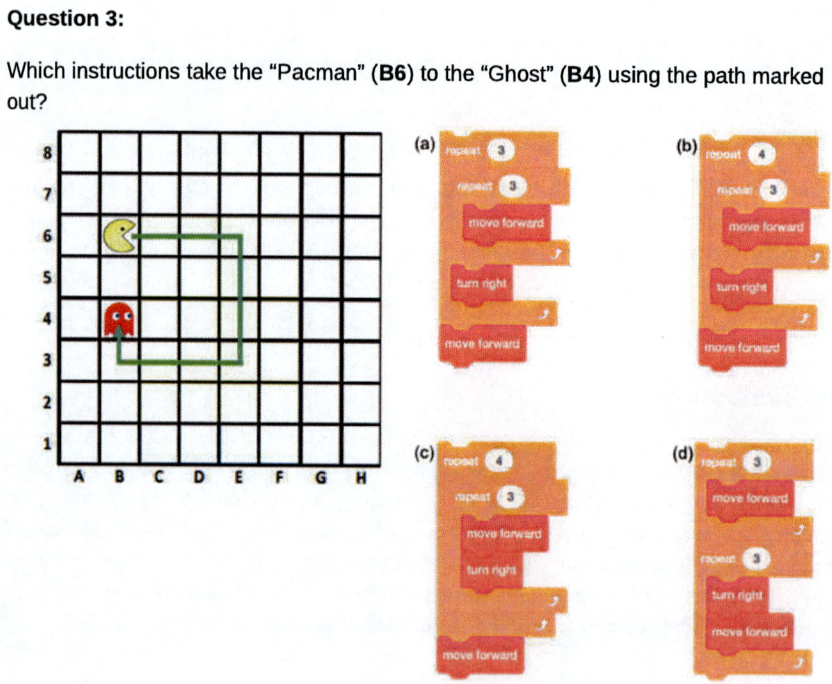

Fig. 2 Example task from the Callystotest [7], answer **a** is correct

Fig. 3 Two Edisonrobots battle each other

felt frustrated by errors and wanted to fix them before asking for help. The majority was fond of computational thinking (Table 1 'pre'). In the second part of the pre-test, students performed well in applying computational thinking with 50% scoring 8 or 9 out of 10 points and gave an average self-assessment feedback of 7 out of 10 (Table 2 'pre').

During the teaching units, the students have been highly motivated and committed to working on the worksheets and with the robots. They also repeatedly designed their own programs, tested their own ideas and had robots compete against each other (Fig. 3). These very playful breaks did not disturb the flow of the teaching units. On the contrary, the students' playful engagement with the robots even promoted mutual exchange and learning progress.

In the post-test, after the teaching units, all students still were enthusiastic about technology and liked using technical devices. They could solve and explain complex problems. Six were excited by programming challenges and all were frustrated by errors but wanted to fix them before asking for help. Eight out of nine were enthusiastic about computational thinking (Table 1 'post'). In the second part of the post-test, half scored 7–9 out of 11 with a median of 8 and one student scored 11, and self-assessed an average of 7 out of 10 (Table 2 'post').

3.3 *Discussion*

The comparison of the first parts of the post- and pre-tests shows a positive development of the student's motivation and interest in the robot-based lessons, visible in the clusters "use of computational thinking", "solving of complex problems" and "working with code". Only in the cluster "use of technologies" the development is

Table 1 Results of the test part 1 from pre- and post-test with median before and after (++ strongly agree, + agree, – disagree, – – strongly disagree)

Cluster	Question	Pre	Post
Use of technologies	I enjoy using technology	++	++
	I find it easy to use new technology	+	+
	I am confident I can fix my computer myself when it is not working	–	–
	People ask me for help with their computer	+	–
Solving of complex problems	I can figure out the steps to solve a complex problem	+	+
	When I am solving a complex problem, I try to break it up into smaller or simpler problems	+	–
	When I am solving a complex problem, I think about other problems I've solved before to see if I can solve this problem in a similar way	+	+
	I can explain the steps of how I solved a complex problem	–	+
Working with code	The challenge of coding appeals to me	–	+
	I am comfortable writing code to solve problems	–	–
	I feel frustrated and want to give up when I encounter an error in my code	–	–
	When my code has a bug, I try to fix it myself rather than ask someone else to fix it	+	+
Use of computational thinking	It is important to develop computational thinking	+	+
	I have the skills to teach others about computational thinking	–	+
	I know how to make learning about computational thinking interesting	–	–
	I am excited by the idea of learning and/or using computational thinking in school	+	+

Table 2 Results of part 2 from pre- and post-test

	Pre	Post
Points	9 out of 11	8 out of 11
Feedback	7 out of 10	7 out of 10

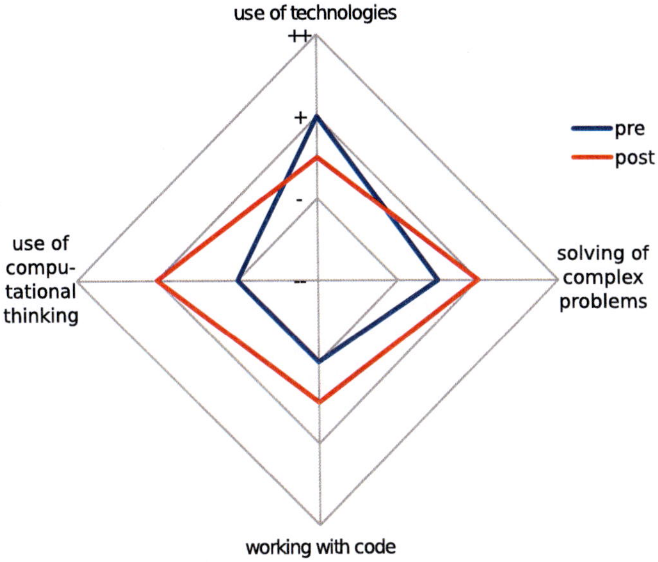

Fig. 4 Results of the test part 1, pre- and post-test in comparison, grouped by cluster, ++ strongly agree, + agree, – disagree, – – strongly disagree, own figure

negative (Fig. 4). All other clusters show a positive development. In the second part, testing the knowledge and competencies of the students, a negative development is visible in question 9 only [7, 9]. The study design allows for high evaluation objectivity, as the tests are evaluated anonymously. However, the low number of students, lack of a control group and randomization limit reliability of results. Still, this study indicates general viability of approach and shows necessity for further research on this subject.

4 Conclusions

Our case study showed an interest in computational thinking and an increase in motivation to acquire these skills. Thus, it is definitely worth it to do lessons with the robotic system. Notably, the robots alone do not make for good teaching, a good lesson plan is equally important. The materials provided by the manufacturer and adapted by the authors allow easy up-take and further adjustments to specific requirements by further teachers. With a teaching approach based on general assumptions and a study design based on a general framework, we expect our approach and our findings to be generally applicable and indicate the benefits of educational robotics systems in schools in general.

References

1. Lower Saxony Ministry of Education and Cultural Affairs: Das Berufliche Gymnasium (The vocational high school), Brochure, 4–8 (2020)
2. Scratch Foundation Homepage, https://www.scratchfoundation.org. Last accessed 22 Dec 2022
3. Federal state of Lower Saxony, Press information, Informatik wird ab dem Schuljahr 2023/2024 Pflichtfach (Computer science will be a compulsory subject as of the 2023/2024 school year), mk.niedersachsen.de/startseite/aktuelles/presseinformationen/informatik-wird-ab-dem-schuljahr-2023-2024-pflichtfach-weitere-qualifizierungskurse-fur-lehrkrafte-starten-184807.html. Last accessed 22 Dec 2022
4. Lower Saxony Ministry of Education and Cultural Affairs (2022), Rahmenrichtlinien für das Profilfach Berufliche Informatik im Beruflichen Gymnasium (general teaching approach for the vocational computer science profile subject in the vocational Gymnasium), 3
5. Edison Homepage, https://meetedison.com/. Last accessed 23 Dec 2022
6. The EdPy Lesson Plans Set by Brenton O'Brien, Kat Kennewell and Dr Sarah Boyd is licensed under a Creative Commons Attribution-ShareAlike 4.0 International License, https://creativecommons.org/licenses/by-sa/4.0/
7. Cutumisu, M., Adams, C., Yuen, C., Hackman, L., Lu, C., Samuel, M.: Callysto Computational Thinking Test (CCTt) Student Version [Measurement instrument] (2019). Available: https://www.callysto.ca/
8. Barr, D., Harrison, J., Conery, L.: Computational thinking: a digital age skill for everyone. Learn. Lead. Technol. **38**(6), 20–23 (2011)
9. C. Bongert (2023), Edited teaching material and Callystotest in German, https://www.github.com/ruebe5w/Evaluation-of-a-robotic-system-in-secondary-school-material
10. Bocconi, S., Chioccariello, A., Kampylis, P., Dagienė, V., Wastiau, P., Engelhardt, K., Earp, J., Horvath, M.A., Jasutė, E., Malagoli, C., Masiulionytė-Dagienė, V. and Stupurienė, G.: Reviewing computational thinking in compulsory education. In: Inamorato Dos Santos, A., Cachia, R., Giannoutsou, N., Punie, Y. (eds.) Publications Office of the European Union. Luxembourg (2022). ISBN 978-92-76-47208-7, https://doi.org/10.2760/126955, JRC128347

Single Session Walking Robot Workshop for High School Students

Martin Zoula and Filip Kučera

Abstract This paper exposes our single-session Walking Robot Workshop for high school students. Students program a walking pattern for real hardware from scratch in four hours. We aim to motivate the students to have a deeper appreciation and understanding of science by emphasizing the interdisciplinary context of the solved task. Hence, a discussion of the direct kinematic task with a three-degrees-of-freedom robot leg is included in the Workshop. The Workshop design and recent experience are described so others can reproduce our results.

Keywords Legged robot · Kinematics · Secondary education

1 Introduction

Contemporary educational robotics conveys soft skills and technical knowledge [1] to students of various ages [2]. In this paper, we emphasize its importance as a supplementary discipline connecting other fields of study. Thus, we address students' long-term and short-term goal awareness while demystifying the technology to enhance their future flow experiences [3]. The more the students understand the reason and connection behind partial pieces of knowledge, the easier they can apply them, becoming self-motivated and self-appreciated. Our Walking Robot Workshop aims to provide such linking.

This paper provides insight into the Workshop launched in October 2022. Students (primarily high school, 16–19 years of age) are tasked to make an assembled legged robot walk during a single 4 h session. So far, there have been 17 installments with a wide variety of students; from broadly-focused students with a solid science

The work has been supported by the National Center for Informational Support of Research, Development and Innovation, Ministry of Education Youth and Sports of the Czech Republic under research project No. MS2101.

M. Zoula (✉) · F. Kučera
National Library of Technology, Prague, Czech Republic
e-mail: martin.zoula@techlib.cz
URL: https://ror.org/028txef36

R. Balogh et al. (eds.), *Robotics in Education*, Lecture Notes in Networks and Systems 747, https://doi.org/10.1007/978-3-031-38454-7_6

background to absolute programming beginners. Positive feedback from students and their teachers was obtained in all cases.

The rest of this paper comprises a course specification in Sect. 2. The summary of our experience from so-far performed Workshop installments can be found in Sect. 3. The paper is concluded in Sect. 4.

2 Course Design

The Walking Robot Workshop is held in a classroom at the National Library of Technology, Prague, Czech Republic. We admit organized classes of up to 12 students, regardless of prior study focus; the Workshop is free of charge to such groups. The Workshop lasts 4 h, breaks included, and is delivered by one lecturer. Although the students are tasked with making an assembled legged robot walk from scratch, our didactic ambition lies elsewhere: We aim to emphasize and explain mathematical and physical principles involved in intermediate tasks. Thus, we hope to connect the partial pieces of knowledge the students already have learned, motivating and encouraging the belief that the knowledge "painstakingly" attained in school is worth learning.

Four robots are handed out to the students, who are divided into groups of at most 3 per robot. We use the Robotis Premium[1] kits assembled into three "King Spider" (hexapod) configurations and one "Humanoid A Type" configuration to introduce diversity (see Fig. 1). An external tethered power adapter powers the robots as the battery would deplete during the four hours. The programming is done on ordinary desktop computers running *Robo Plus Task 1.0* [2] software. We opted for an earlier software version as it provides a cleaner interface without visual distractions. Programming over the wireless channel was rejected due to possible delayed responses or random disconnections. Thus we aim to reduce the cognitive load, which is critical, especially with inexperienced students.

After introducing the hardware and the software to students, a repetitive movement using only one leg is set as the first task. Then, the students implement a single-step movement for a single leg. We continue with a discussion of leg kinematics and involved trigonometry. As the students cooperate, they can implement the walk before the end of the Workshop. In the end, the students exploit their work while using preprogrammed motions and sensors in free play. The course syllabus is detailed below and follows the course slideshow. [3] The Workshop also draws from [4], where robotics-related details may be found.

Preparation. Before the Workshop starts, approximately 30 min is required to prepare the classroom. Four "stations" are selected in the classroom, each with three

[1] https://www.emanual.robotis.com/docs/en/edu/bioloid/premium/.

[2] https://www.emanual.robotis.com/docs/en/software/rplus1/task/getting_started/.

[3] https://www.github.com/zoulamar/walking_robot_workshop.

Fig. 1 Used robotis premium (https://www.emanual.robotis.com/docs/en/edu/bioloid/premium/) robots: Humanoid in the middle, hexapods by sides

neighboring desktop computers as seen in Fig. 2a. The lecturer also plugs the power and USB cables and ensures the robot is operational.

Course Introduction. The course starts with approx. 30 min block where we explain the rules of the Workshop and describe the robots to be used. The rules are set explicitly free:

(1) Investigate, inquire and ask when not sure.
(2) Cooperate at will.
(3) Exploit the opportunity to work with unusual robots fully.
(4) Try not to destroy the robots by sudden moves or mechanical shocks, as the gearbox may break.

The students are encouraged to move robot limbs while the robot is powered off such that they understand the kinematics intuitively. Further, we mention minimal technical details, namely, how the angle reckoning works with used *Dynamixel AX-12A* servomotors. Also, we declare the servomotors are uniquely identified, as seen in the diagrams included in the slideshow, which the students may use freely.

First Moves. The first task is to create and deploy a code that moves a single servo. To this end, we explain the basic program structure and describe how to operate the *Robotis CM-530* control unit. Then, students need to move a single limb periodically. Here, we focus on teaching basic programming concepts of looping and conditional execution or verifying students understand them already. As the servomotors have their own motion controllers, the control unit is independent of them and runs commands immediately as they come by default. The students now need to understand the necessity to block the program until a move-command finishes.

Step Exercise. The students are now tasked to develop a prerequisite for the walk: A code that raises a leg, moves it forward and lowers it in front of the robot. To this end, the students must find the correct angles for the leg movement. We suggest finding the angles by direct computation, trial-and-error or using the *RoboPlus Manager* tool, part of the *RoboPlus 1.0*, which interactively shows the current servomotor angle in a tabular view.

Servo Kinematics. Around 90 min into the Workshop, we dive into the mathematics describing the leg motion. We start by analyzing single servo kinematics, described as a right triangle with known hypotenuse length l and one angle ϕ. I.e.,

$$x = l\cos(\phi) \qquad\qquad y = l\sin(\phi). \tag{1}$$

We discuss two angle-reckoning systems. The servomotor uses the "technical angle" ψ, whereas our triangle reckons with the "math angle" $\phi = \psi - 150°$. This part of the Workshop is the most important with regard to the interdisciplinary contextual view we aim to provide to the students; the students should see that the "dull" math theory is, in fact, very useful because it can determine, e.g., the height of the robot body based on servomotor angle.

Leg Kinematics. Discussion and analysis of forward leg kinematics follow. The discussion concerns only the hexapod leg, which has three controllable degrees of freedom (CDOF) in a yaw-pitch-pitch configuration. The joints are named *Coxa*, *Femur* and *Tibia*, from the body to the foottip. We motivate the forward kinematics with a question: *Would it not be great if the robot knew beforehand whether a collision with some known obstacle would occur given some set servomotor angle?* Students are thus hinted toward a path planning idea.

We start solving the kinematics by exposing the planar task, which considers fixed Coxa (the first joint). Two right triangles where the angles are determined by the *Femur* angle β and the sum of *Femur* and *Tibia* angle $\gamma + \beta$ are analyzed. We expose the trigonometry similar to the one servo case. If time and mental strength allow, we even extend the planar task to 3D. We draw the situations on a whiteboard and invite the students to discuss and ask questions frequently.

Gait Exercise. Around the Workshop halftime, programming the walk itself commences. For the students with the hexapod, we explain the *pentapod* gait, when only one leg swings forward, while the other legs support the robot. When all legs are placed forward, they move backward simultaneously, propelling the robot. We discuss the best of the leg swing order: As legs may clash, the pentapod gait has to start from the front legs to make space for the latter legs. To support intuition, we liken the gait to swimming. Legs reach forward to subsequently flush the water/ground behind, thrusting the body.

The humanoid robot is suggested to use ankles and hips for simplicity. The robot starts squatting with the center of mass as low as possible. By turning one ankle, the robot leans slightly sideways, stably standing on one leg thanks to wide "shoes". The hip movement then places the airborne leg before the supporting leg, and the ankle straightens. Alternating the legs makes the walk. We discuss the stability of both robots in terms of the center of mass and support polygon.

Implementation-wise, we explain the beautiful trick of wrapping user code into named function blocks and reusing them later in the main part of the program. Thus, the students are encouraged to create clean-cut reusable motion primitives. We note that the legs slide against the floor during the walk. We only mention that inverse kinematics would help this issue, hoping that some students could investigate the problem at home. Once the robot starts walking, the main objective for the students is accomplished.

Sensors and Preprogrammed Patterns. The Workshop's end is dedicated to free creation. Students are revealed existing motion primitives which they can start using. Further, we briefly explain available sound-pulse sensors and infrared distance and proximity sensors, which the students can start using. The lecturer now advises and counsels the students. We suggest creating a "dance" sequence of preprogrammed movements for students with no ideas.

3 Experience

Our experience until March 2023 is summarized in this section. Among 17 installments so far, we met student groups of different backgrounds. Moreover, the actual age span from around 14 (initially expected 16) years to 19 years of age. Hence, students' initial knowledge varied highly, from fluent programmers to absolute beginners, moreover without suggested trigonometry knowledge. Thus, our initial assumptions were violated, and each installment needed individual adjustments. Despite this, the students were fully invested in the course in each installment. Moreover, all student groups successfully completed the titular task—making the robots walk. The main adjustments regarded the depth of the mathematics discussion. Advanced students were curious even about inverse kinematics and its possible uses. On the contrary, programming beginners were content with basic single-joint kinematics, further focusing on baseline tasks. Summarized, we discovered a beneficial inherent course modularity.

We observed no serious issues with students using the *RoboPlus Task 1.0* environment, not even with non-programmers. The students required technical explanation in the scope of the accompanying slideshow. The boss-worker model was understandable for the students, although younger students struggled with realizing where exactly to put the program blocking. Students generally appreciated technical details and insights like a side note about the angle-measurement principle in the AX-12A servomotor.

The three students per robot scheme (see Fig. 2b) worked well; the students cooperated, developing their custom experiments or distributing computations of the leg angles. Although the humanoid robot provides an observable psychological bonus as the students relate to the robot better, we saw high interest in the hexapod robots as well. Students regarded both types of robots as high technical level; it was generally their first programming such complex machines. The work was challenging for the students. However, given a proper initial explanation, all students were able to work on the programs meaningfully until the end. Especially the younger students were extremely self-motivated, even rejecting the lecturer's help as they wanted to tackle problems themselves.

The challenge in tasks served as a positive motivation; we showed the students that an initial black box (robot itself or the programming) can be delved into. We consider

(a) Student group with hexapod.

(b) Classroom Overview.

(c) Humanoid robot exerscising.

(d) Humanoid robot walking.

Fig. 2 Photographs from the workshop

using heterogeneous robots beneficial as the student collective gains a richer experience while learning the same principles. So far, the reported statements are based on our observations and ad-hoc interviews. We consider conveying a detailed survey among students as we see interesting potential relations regarding the appreciation of the in-depth understanding of the problem.

We are proud that the students came up with and implemented clever ideas at the end of the Workshops. Some students tuned the walk, improving locomotion speed, by implementing the *tripod gait* when three legs are always swinging, the rest supporting. Proximity sensors were used to stop the robot's motion in several cases. We also saw several "dances" when students invoked available motion primitives in a meaningful order. Namely, the students from pedagogics-focused class made the robot perform push-ups and head-stand (see Fig. 2c, d), then waving the hands and then clapping to raise viewer's applause; once viewers clapped hands in response, the robot recognized it and bowed.

We conducted a voluntary feedback query among teachers who were present with their classes. So far, among 6 responses (some teachers came with multiple student groups, some did not participate) one suggested that hexapods are more suitable for first touch with complex robotics than humanoids. Further, the emphasis on interdisciplinary context with the goal of motivating the students in further education was appreciated. The teachers endorsed the practicality; the students did not dully compute on paper. Moreover, our approach of only hinting at the students instead of spoiling the correct answer was also endorsed. Finally, the teachers saw the length and difficulty of the Workshop as appropriate. Altogether, the Workshop turned out to be a good education supplement.

4 Conclusion

We conclude that our Workshop has been successful so far. The students and their teachers acknowledged our Workshop even afterward when we had a chance to talk to them retrospectively. The Workshop fulfills the goal of linking the theoretical knowledge students gain during ordinary education with practical tasks related to walking robots. We are keen to continue and further improve and systematically evaluate the Workshop in the future.

References

1. Zeaiter, S., Heinsch, P.: Robotikum. In: Robotics in Education, pp. 3–15. Springer, Berlin (2021)
2. Eguchi, A.: Key factors for keeping students engaged in robotics learning. In: Robotics in Education, pp. 38–48. Springer, Berlin (2021)
3. Csikszentmihalyi, M.: Applications of Flow in Human Development and Education. Springer, Berlin (2014)
4. Čížek, P., Zoula, M., Faigl, J.: Design, construction, and rough-terrain locomotion control of novel hexapod walking robot with four degrees of freedom per leg. IEEE Access **9**, 17866–17881 (2021)

Integrating Secondary School and Primary School Learners to Grasp Robotics in Namibia Through Collaborative Learning

Annastasia Shipepe, Lannie Uwu-Khaeb, David Vuyerwa Ruwodo, Ilkka Jormanainen, and Erkki Sutinen

Abstract Learning by applying motivates learners to put the theory into practice. However, at least in the Global South, it seems that the concept of applying seems to be not common yet in K-12 education, even for technology related subjects. We conducted an educational robotics (ER) workshop with the aim of exploring how collaborative learning can aid secondary school and primary school learners grasp robotics. The qualitative analysis indicates that learners were exposed to robotics and coding knowledge through this ER workshop, and they further wanted to learn more about these technologies. The results can be applied by practitioners in Namibia when considering incorporating robotics in the formal school curriculum, after-school robotics programs and robotics boot-camps.

Keywords Robotics · Coding · Collaborative learning · Primary school · Secondary school · Global south

A. Shipepe (✉) · I. Jormanainen
School of Computing, University of Eastern Finland, 80100 Joensuu, Finland
e-mail: ashipepe@students.uef.fi

I. Jormanainen
e-mail: ilkka.jormanainen@uef.fi

L. Uwu-Khaeb · D. V. Ruwodo · E. Sutinen
Department of Computing, University of Turku, 20014 Turku, Finland
e-mail: Lannie.Uwu-Khaeb@utu.fi

D. V. Ruwodo
e-mail: DavidVuyerwa.Ruwodo@utu.fi

E. Sutinen
e-mail: Erkki.Sutinen@utu.fi

R. Balogh et al. (eds.), *Robotics in Education*, Lecture Notes in Networks and Systems 747, https://doi.org/10.1007/978-3-031-38454-7_7

1 Introduction

Walvis Bay is a harbor town in western Namibia with an estimated population of 60000+ people. Most activities in the town are related to the fishing industry and tourism industry. It is a working town. There are both private and public schools in Walvis Bay but with little collaboration between them. They offer a wide variety of subjects at the schools from the basic subjects like languages and mathematics to accounting, design and technology, however there are no robotics offerings in any school in the town to the authors' knowledge.

Educational robotics (ER) is picking up speed in the Namibian context mostly driven by non-formal educational activities through after-school programs. More and more private and public schools in Namibia have shown interest in having their learners equipped with robotics knowledge and the application of it. The principals of two schools in Walvis Bay were interested in introducing such a program to their schools, as they saw the importance of it in the fast-changing world. ER enables learners to apply the robotics technologies by designing and developing robotics systems using technologies such as Arduino and LEGO alongside programming behind the functionality of the systems [4, 10]. Educational robotics is widely known for motivating learners in learning STEAM related subjects. Furthermore, learning with robotics is believed to also equip learners with the 4Cs of the 21st century skills namely creativity, collaboration, critical thinking and communication. These skills can also be achieved through collaborative learning where learners work in small groups to grasp robotics.

The integration of robotics in K-12 education is a well investigated research field [1, 7]. However, analyses of collaborative learning of robotics are scarce, especially in the Namibian or more widely African context and even less in scenarios where learners from diverse socioeconomic backgrounds teach and learn from each other. Therefore, we hosted a robotics workshop with the two schools: the peer teachers were learners from a public secondary school with students from mid-income or poor backgrounds and the peer learners were from a private primary school with students from more affluent backgrounds. The aim of this educational robotics workshop was to explore how collaborative learning can aid secondary school and primary school learners grasp robotics. In our approach of peer learning between the two schools, we chose to follow the qualitative approach for discovering key themes in the learning process for further elaboration, interpretation, and application by practitioners.

The rest of the paper is structured as follows. Section 2 focuses on background and related work while Sect. 3 is on research design. The paper further continues with Sect. 4 which focuses on the results and thereafter Sect. 5 with the discussion and finally, Sect. 6 has the conclusion.

2 Background and Related Work

2.1 Robotics

The use of robotics technologies is reported to be commonly used in non-formal educational activities [3]. In a study carried out in Barcelona, Spain, a robotics workshop was carried out aiming to motivate secondary school students to take up degrees related to Science, Technology, Engineering, Arts and Mathematics (STEAM) field [3]. The workshop enabled students to learn the basics of robotics and inspired them to provide creative solutions to real-world problems. Similarly, a robotics workshop was carried out in Namibia to equip participants with technologies behind the robotics systems [4].

The use of robotics tools in education aid learners and students adopt easily to problem-based learning and improve in areas of learning such as electronics, programming as well as mechanics [2]. Learning with robotics technologies also motivates learners and students during their times at schools by allowing them to practically apply and witness the theoretical knowledge in STEAM [2]. Robotics is currently not in most of the primary and secondary school curricula although it has started to dominate the educational sector [2]. Having said that, it is important to equip learners with robotics technologies through non-formal educational activities, putting the sustainability of it as well as the teachers' perspectives into consideration. According to [7], teachers find learning with educational robotics helpful particularly in equipping students with not only computational thinking skills but also with transferable skills.

2.2 Gender Aspects in Educational Robotics

It is evident that gender representation is important in almost any given project to ensure that both masculine and feminine viewpoints are taken into consideration [6]. In an education robotics (ER) summer camp carried out by Pedersen et al. at Teknologiskolen in Denmark, there was gender imbalance because only few girls initially showed up for their ER summer camp. They then strategized to have a 'girls' only team' among their teams for the 2020 summer camp [5].

2.3 Collaborative Learning

Collaborative learning is a learning approach in which a group of learners gather to learn a concept together to ensure participation [11]. Collaborative learning was used in a study conducted in [11] where the main objective was to equip students with advanced robotics skills to develop a garden robot. The results of this study indicate that the students collaboratively worked in teams to investigate and provide the solution to the problem [11]. Various studies have shown that collaborative learning creates a safe, welcoming, and healthy learning environment using icebreakers [10]. Icebreakers are defined as simple activities usually at the beginning of an event that get the learners' attention and create a bond among the team(s) as well as help teachers understanding the learners' expectations [10]. Icebreakers were used in a study conducted in the UK to enhance students' collaborative skills while learning robotics with LEGO EV3 [12].

There have been debates on how computer science and robotics should be integrated in the curriculum, especially in the primary school curriculum [9]. Although a lot of initiative have been created in the Global South to tackle the challenge of integrating robotics in the school curriculum [9], this still remains a challenge in the Namibian context and the aim of our study is contributing to filling this gap in Namibia by using a collaborative learning approach to teach the secondary school learners in order for them to teach the primary school learners the robotics technologies.

3 Research Design

3.1 Research Problem

The study presented in this paper is part of a bigger PhD project that is looking at the impacts of educational robotics and sensor technologies in Namibia. The bigger PhD project aims at creating digital solutions that answer the expectations of extending the role of ER in Namibian schools and 4IR in the whole society. Therefore, the study presented in this manuscript is contributing to the bigger project by answering the research question in Sect. 3.2.

3.2 Research Question

How can we recommend organizing ER workshops that would develop the connection of educational robotics and theoretical knowledge from other subjects among learners from different groups using collaborative learning in Namibia?

3.3 Research Approach and Methodology

This study followed a collaborative learning approach where secondary school learners were trained to train the primary school learners to learn robotics technologies.

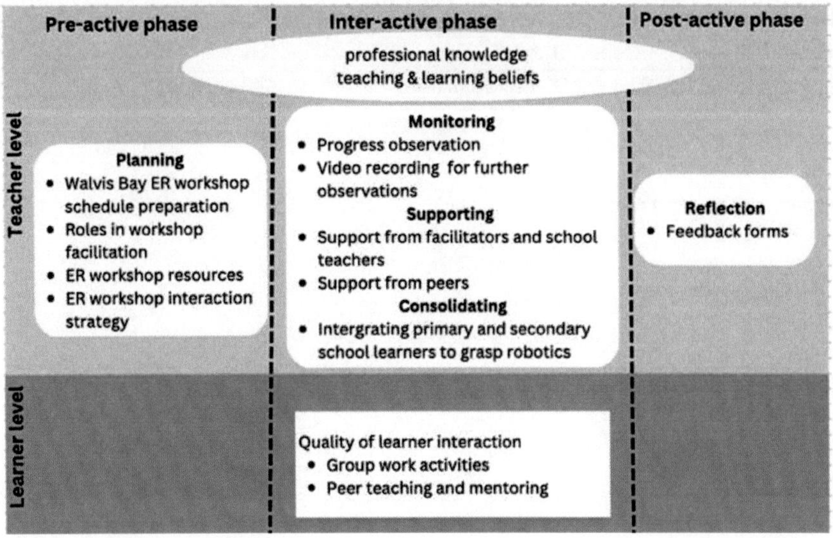

Fig. 1 ICLC Framework for collaborative learning adapted from [13]

The study further adapted the ICLC framework for collaborative learning by Kaendler et al. [13] as illustrated in Fig. 1. As mentioned in 3.1 the study presented in this paper is part of a bigger Ph.D. project. The bigger project follows a Design Science Research methodology. The study in this paper uses a qualitative research method, following the inductive qualitative coding explained later in this section.

3.4 Context Description

This study was carried out for three days, from 11 May 2022 to 13 May 2022. It involved two schools, i.e. De Duine Secondary School and Walvis Bay Private School, both located in Walvis Bay, in the coastal side of Namibia. De Duine is a secondary school with grades 8 to 11 as well as Advanced Subsidiary (AS) level (grade 12). The school is situated in the Narraville suburb of Walvis Bay and it admits learners from its main feeding school, Narraville Primary School, as well as from other local primary and secondary schools. There are no specific preference criteria to the school's selection process.

Walvis Bay Private School on the other hand has grades 1–12. Their grades are separated into four phases i.e Junior Primary Phase for grades 1–3, Senior Primary Phase for grades 4–7, Junior Secondary Phase for grades 8 and 9 and Senior Secondary Phase for grades 10, 11 and 12 . The school is located in the town center of Walvis Bay and its admission rules include a language and mathematics readiness test for new applicants.

Table 1 Workshop schedule

Day 1	Day 2	Day 3
Introduction and Informed consent	Reflection of day 1	Introduction and Informed consent
Group formation and Icebreaker	Installing Arduino IDE	Group formation and Icebreaker
Explain the Tinkercad platform	Block coding	Explanation of Tinkercad platform
– Why it was developed	Blinking LEDs	Build simple LED circuits
– How it works	– Blinking 1 LED	Blinking LEDs
Explain Components in the Arduino Keyestudio kits	– Blinking 2 LEDs	– Blinking 1 LED
Resistors (types of resistors based on colors)	– Blinking 3 LEDs	– Blinking 2 LEDs
Simple LED circuit	Push buttons	– Blinking 3 LEDs

Keeping in mind that the main aim of the study was to explore how collaborative learning can aid secondary school and primary school learners grasp robotics, we first trained the secondary school learners for the first two days and the secondary school learners then trained the primary school learners with the supervision of the main facilitators as shown in Table 1.

3.5 Workshop Participants

The workshop was attended by 29 participants in total, 14 participants were the secondary school learners (aged between 13 and 16) and 15 participants were primary school learners (aged between 12 and 13). The secondary school learners were from grades 8, 9, 10 and 11, while the primary school learners were all from grade 7. The workshop participants were selected by the school teachers. Five groups were formed for both secondary learners and primary school learners as summarized in Table 2. As it can be seen in the second column of Table 2, the primary school learners were assigned two secondary school learners per group as their mentors. Furthermore, the workshop involved five workshop facilitators of which two of them were researchers. The workshop further involved 2 school teachers and a school principal.

3.6 Data Collection and Analysis

The data was collected for a qualitative analysis through a feedback form where learners provided written feedback on what they learned, what they want to learn next and

Table 2 Workshop participant groups with gender representation

Secondary school day 1 and 2	Primary school day 3
Group 1: 3 participants (1 F: 2 M)	Groups 1: 3 participants (2 F: 1 M) 2 mentors (1 F: 2 M)
Group 2: 3 participants (1 F: 2 M)	Groups 2: 3 participants (1 F: 2 M) 2 mentors (1 F: 1 M)
Group 3: 3 participants (2 F: 1 M)	Groups 3: 3 participants (3 F: 0 M) 2 mentors (2 F: 0 M)
Group 4: 3 participants (2 F: 1 M)	Groups 4: 3 participants (1 F: 2 M) 2 mentors (2 F: 0 M)
Group 5: 2 participants (0 F: 2 M)	Groups 5: 3 participants (0 F: 3 M) 2 mentors (0 F: 2 M)

their teaching experience. Data was further collected through informal interviews, field notes as well as video recordings. The feedback provided were later captured to the code book on Google sheet for data analysis using an inductive qualitative coding approach [14] where themes were derived from the data. The data was coding following the steps below:

Step 1: Data preparation—In this step we captured the data to the code book i.e. Google sheet in preparation for data analysis ; Step 2: Create the first sets of codes—Here were gave descriptive codes based on the feedback data provided by the learners—The codes were allocated different colors, however the codes that were similar in description were allocated the same color ; Step 3: Three rounds of regrouping codes—In this step, we regrouped the codes by merging the codes with a common interest ; Step 4: Categorize codes into themes—Here the group of themes were allocated a theme ; Step 5: Themes evaluation—In this step, we revised the themes and merged themes that were similar.

3.7 Ethical Consideration

The ethical guidelines of the University of Eastern Finland (UEF) were followed in the workshop. Permission was granted by the school principals of De Duine Secondary School and Walvis Bay Private School. All participants were informed in all honesty about the purpose of the workshop, and they were further informed that participant was voluntary. The autonomy of the participants was respected by the researchers throughout the workshop.

Fig. 2 Two arduino super learning kits allocated to a group

3.8 Robotics Technologies Used

Arduino robotics technologies were used in the workshop of which 11 new Super Learning Kits for Arduino by keystudio were available to use throughout the workshop. Each group was allocated 2 kits as it can be seen in Fig. 2.

4 Results

4.1 Introduction and Groups Formation

The workshop introduction was made in a form of an icebreaker of which learners introducing themselves and further mentioning their fun facts. It was noted that the learners were freely interacting after the icebreaker. After the icebreaker five groups were formed and the workshop activities began.

4.2 Activities

ER workshop day 1 at De Duine secondary school focused on the introduction of robotics components and platforms used to design and develop robotics systems. The workshop facilitators first introduced the learners to the Tinkercad platform by not only explaining what Tinkercad is but also explaining why it was developed and further showing how it is used. This was done through designing simple circuits which involved LEDs and resistors. It was observed that the learners were openly interacting with each other and with the facilitators, asking questions on anything that they needed clarity on. This was helpful in the context of the study approach because the secondary school learners needed to grasp the robotics knowledge well to transfer what they learned to the primary school learners.

ER workshop day 2 picked from where we stopped on day 1 and it involved three main activities, i.e. installing Arduino IDE, block coding, blinking LEDs and push buttons. Block coding and blinking LEDs involves coding. It is important to mention here that the learners had no prior knowledge on coding, hence the facilitators had to explain coding to the learners. Coding a push button added up being one of the activities only for one group which was way ahead of the other groups because they insisted that they wanted to code a push button. It was coded in such a way that once you press it, an LED light turns on. Although coding was new to the learners, it was observed that they were interested in knowing more about coding.

ER workshop day 3 with the primary school learners at Walvis Bay Private School was a repetition of day 1 and 2. Like with the secondary school learners, the workshop started off with group formations. Five groups were formed and every group had two secondary school learners to teach or train the primary school learners on what they have learned in the previous days. This was done through a chronological order from the Tinkercad activity to turning on LED lights. Although the primary school learners were only trained for one day due to the limited time, it was observed that they were excited about learning the robotics technology and showed interest in continuing learning with the secondary school learners through a robotics club which was birthed from this study.

4.3 Learners' Feedback: Qualitative Analysis

The data was analyzed and four themes were derived from the feedback data as presented in Table 3.

Table 3 Themes derived from the data

Codes	Themes
T1	Exposure to robotics and coding
T2	Educational robotics motivation
T3	More time needed to learn robotics
T4	Robotics knowledge transmission through collaborative learning

Theme 1—Exposure to robotics and coding: This theme was derived based on the three words that were trending in the set of codes of the feedback data provided by the learners. The three words are exposure, robotics, and coding.

Almost every learner mentioned that it was their first time being exposed to both robotics and coding. It is noted that the learners managed to apply the basics of robotics. For example, a secondary school learner allocated ID number DRW2 is as quoted in the next sentence from the feedback form. "*I really enjoyed the lessons. It was great trying out something new. For the past two days, we lit up LED lightbulb and did a bit of coding.*" The idea of having a robotics club at the two schools was also discussed during this workshop and learners indicated that they were excited to become members of the robotics club. A secondary school learner allocated ID number DRW5 is quoted saying "*I just want to say I am very excited to be a member of robotics clubs, because the things I have experienced were very magnificant because I have experienced a lot more than I thought I have experience. On the next workshop I would want to make a timer on the breadboard (connecting it). And in the mean time I want to do something that is more enough than my thoughts*". The learners further indicated that they want to do more coding as indicated by a primary school learner allocated ID number WRW7 saying "*It was amazing and I want to do it again. I want to learn more about coding and how to make buttons. I am definitely going to google how to make a light work with a button.*"

Theme 2—Educational robotics motivation: This theme was derived from the data that had its focus on how this educational robotics inspired them.

It was noted in the feedback form that this educational robotics workshop helped them to understand the application of the theory they learn in science subjects and allowing them to be as creative at possible. A secondary school learners with ID number DRW9 and a primary school learner with ID number WRW14 is as quoted below.

DRW9 "It was an amazing experience. I never knew that I would find robotics interesting but after this 2 days I want to learn more . I never knew I would be this excited to light up an LED bulb."

WRW14 "It was very exciting and fun. I hope to learn more about robotics because I feel like it is a subject where I can enhance creativity. Next time I hope to emphasize more on coding because I struggled a bit to understand it."

Theme 3—More time needed to learn robotics: It was noted that in the feedback of primary schools learners (WRW1 and WRW6) that they felt that the journey fun but stressful, hence the sets of code related to stressful emotions were allocated theme 3. This could be due to the limited time that was allocated to primary school learners to grasp the basics of robotics.

WRW1 "We have never had a chance to try robotics, so it was quiet fun but a stressful journey"

WRW6 "It was very fun and a little bit stressful. I learned a lot of how to switch on a light and codes"

This theme may help after school robotics programs and schools offering robotics classes when planning robotics teaching schedules to keep in mind that learning robotics require more time and patience.

Theme 4—Robotics knowledge transmission through collaborative learning: This theme is guided by the feedback the secondary school learners gave based on how they transferred the robotics to the primary school learners. Learners DRW2 and DRW13 are as quoted below.

DRW2 "I have never really tried teaching before but it was an eventful journey to meet other students and to teach them what I previously learned. I enjoyed it. The learners were fast learners and I would love to teach robotics again". DRW13 "My teaching experience was awesome. They all listened and understood very well. I hope they will also teach others. We taught them how turn on LED lights which was easy and coding."

4.4 Results Contribution to the Namibia Context

The reported results may not be new elsewhere but new and crucial for the Namibia context where the concept of educational robotics is picking up speed but still at an emergent stage. Although the results presented in this paper may be different from the regular teaching methods, these results would be transferable in aiding with the integration of robotics into the school curriculum, the after school curriculum as well as robotics holiday boot-camps in Namibia.

5 Discussion

In this study we explored how collaborative learning can aid secondary school and primary school learners grasp robotics. It can be seen in the results section that the three days educational robotics workshop did not only enable the secondary school learners to grasp elementary robotics to train the primary school learners but this workshop also opened doors to opportunities and collaborations.

A discussion to have a robotics club for the learners from both schools took place at this workshop and both learners and teachers were excited of the idea.

The Sustainable Development Goals (SDG) 4 talks about quality education for all [8]. Many scholars have argued that there is need for a promotion of guided learning to boost learning and enable learners to apply robotics [10]. While putting guided learning into consideration, it is also imperative for the educators to put sustainability of learning ER technologies into consideration especially with novice learners given the fact that ER is yet to be integrate in the K-12 Namibian curriculum.

Keeping in mind SDG 4 "Quality education", this initiative opened doors for learners to continue learning robotics and coding among themselves under the supervision of the teachers. It can also be seen in the feedback section that the secondary school learners wish the primary school learners transfer what they grasped to other learners at their school. This is an indication that learners want to learn these technologies every day. However, this will be made possible by the integration of robotics and coding in the school curriculum.

In the feedback given by learners, they mentioned that they had no clue of what was going on in day 1 because things were new to them but they could at least relate some technologies to movies. They further mentioned that they theoretically knew how electricity flows but never practically, however after the "turning on LEDs" activity which also involves the basics of electronics the learners got a broader picture of how circuits work. With this, we recommend educational practitioners to incorporate educational robotics in the public and private school curriculum to help the learners with the STEAM related subjects as well as to motivate them to take up the STEAM degrees at institutions of higher learning as argued by other researchers as noted in the literature review [4].

It was further noted in the learners' feedback that it was their first time for them as secondary school learners to encounter coding. Frankly, this is a bit of a pity given the fact that the world is moving at a fast pace with artificial intelligence (AI) and robotics technologies and yet our learners are not equipped with the basic knowledge of these technologies. The latter is however not only a problem in Namibia but a challenge in the Global South. There are however few after-school initiatives for learning coding and robotics in Namibia and more in the Global South however we recommend that AI, robotics and coding become part of the K-12 curriculum for both public and private schools in Namibia.

Although scholars have it that girls seem to have gender stereotypes towards robotics technologies [7], this was however not the case with the educational robotics workshop we had at Walvis bay: both females and males learners were observed to be equally interested in learning the robotics technologies.

The results presented in this study is important in the Namibian context to contribute to the integration of robotics and coding in the school curriculum. The four themes identified from the qualitative data contribute to answering the research question in a way that learners can grasp robotics through exposure of applying the robotics technologies in a hands-on way. The study integrated secondary school learners from a public school with primary school learner from a private school to learn robotics, which is rare in the Namibian context where private schools are more advanced as opposed to public school. Furthermore, the results of this study is crucial in guiding the educational practitioners on how to design a meaningful curriculum that exceeds the outdated limitations or barriers between schools of different kinds.

6 Conclusion

The study explored how primary school learners and secondary school learners grasp robotics through collaborative learning. The results show that both primary and secondary school learners successfully grasped robotics. However, neither the primary school learners from the private school nor the secondary school learners from the public school had robotics and coding skills. Therefore, it can be concluded that it is important to integrate robotics in the school curriculum of both public and private school to enable learners grasp and apply cutting edge technologies from the grass-root level.

Acknowledgements We thank the school principals of De Duine Secondary School and Walvis Bay Private School for inviting us to carryout the robotics workshop at their schools. We further thank Leevi Seesjärvi and Luke Arendse for assisting in facilitating the workshop.

References

1. Alves-Oliveira, P., Arriaga, P., Paiva, A., Hoffman, G.: Children as robot designers. In: Proceedings of the 2021 ACM/IEEE International Conference on Human-Robot Interaction, pp. 399–408 (2021)
2. Novák, M., Pech, J., Kalová, J.: An arduino-based robot with a sensor as an educational project. In: International Conference on Robotics in Education (RiE), pp. 64–71. Springer, Cham (2021)
3. Cárdenas, M.I., Campos, J., Puertas, E.: How to promote learning and creativity through visual cards and robotics at summer academic project Ítaca. In: International Conference on Robotics in Education (RiE), pp. 52–63. Springer, Cham (2021)
4. Shipepe, A., Jormanainen, I., Duveskog, M., Sutinen, E.: Screams of joy yield creative projects at the educational robotics workshop in Namibia. In: 2020 IEEE 20th International Conference on Advanced Learning Technologies (ICALT), pp. 103–105. IEEE (2020)
5. Pedersen, B.K.M.K., Larsen, J.C., Nielsen, J.: Girls and technology-insights from a girls-only team at a reengineered educational robotics summer camp. In: International Conference on Robotics in Education (RiE), pp. 119–133. Springer, Cham (2021)
6. Dray, S.M., Peer, A., Brock, A.M., Peters, A., Bardzell, S., Burnett, M., Churchill, E., Poole, E., Busse, D.K. (2013) Exploring the representation of women perspectives in technologies. In: CHI'13 Extended Abstracts on Human Factors in Computing Systems, pp. 2447–2454
7. Chevalier, M., El-Hamamsy, L., Giang, C., Bruno, B., Mondada, F.: Teachers' perspective on fostering computational thinking through educational robotics. In: International Conference on Robotics in Education (RiE), pp. 177–185. Springer, Cham (2022)
8. SDG, U.: Sustainable Development Goals, p. 7. The energy progress report, Tracking SDG (2019)
9. El-Hamamsy, L., Chessel-Lazzarotto, F., Bruno, B., Roy, D., Cahlikova, T., Chevalier, M., Parriaux, G., Pellet, J.P., Lanarès, J., Zufferey, J.D., Mondada, F.: A computer science and robotics integration model for primary school: evaluation of a large-scale in-service K-4 teacher-training program. Educ. Inf. Technol. **26**(3), 2445–2475 (2021)
10. Sapounidis, T., Alimisis, D.: Educational robotics curricula: current trends and shortcomings. In: Educational Robotics International Conference, pp. 127–138. Springer, Cham (2021)
11. Correll, N., Rus, D.: Peer-to-peer learning in robotics education: lessons from a challenge project class. Comput. Educ. J. **1**(3), 60–66 (2010)
12. Zarb, M., Scott, M.: Laughter over dread: early collaborative problem solving through an extended induction using robots. In: Proceedings of the 2019 ACM Conference on Innovation and Technology in Computer Science Education, pp. 249–250 (2019)
13. Kaendler, C., Wiedmann, M., Rummel, N., Spada, H.: Teacher competencies for the implementation of collaborative learning in the classroom: A framework and research review. Educ. Psychol. Rev. **27**, 505–536 (2015)
14. Linneberg, M.S., Korsgaard, S.: Coding qualitative data: A synthesis guiding the novice. Qual. Res. J. **19**(3), 259–270 (2019)

Methodology and Pedagogical Aspects

Revisiting the Pedagogy of Educational Robotics

Amy Eguchi

Abstract In 2012, I published a book chapter called "Educational Robotics Theories and Practice: tips for how to do it right," focusing on the theoretical foundation of educational robotics and pedagogical approaches to enhance student learning through educational robotics as a learning tool. The pedagogical approaches include the "learner-centered approach," "project-/inquiry-based approach," "supporting student learning with good scaffolding," and "promoting documentation." The chapter was written even before Computing/Computer Science (CS) Education was officially introduced to schools (i.e., The U.K. made computing teaching compulsory from the age of 5 in 2014). The paper revisits the chapter and reexamines the pedagogical approaches, including those in CS education and engineering education, to update the knowledge base to deepen the understanding of the power of educational robotics (ER) in constructionist learning environments. It aims to support educators who are new to ER learning but interested in integrating constructionist ER learning practices in their teacher education or/and teacher professional development programs as well as in their classrooms.

Keywords Educational robotics · Constructionism · Robotics as a learning tool · Pedagogy · Project-based learning

1 Introduction

Educational robotics (ER) or Robotics in Education (RiE) is the term broadly used in describing the use of robotics as a learning tool in the classroom [1]. ER has become a familiar term in classrooms in the last ten years or so. The concept of ER was initiated by Seymour Papert [2] and has its roots in the Constructionist theory. Papert developed Logo, a computer programming language for children, and Turtle robot, which in turn became the foundation for the development of the *Programmable Brick* for the LEGO Mindstorms [3]. LEGO Mindstorms was first released in September

A. Eguchi (✉)
University of California, San Diego, CA, USA
e-mail: a2eguchi@ucsd.edu

R. Balogh et al. (eds.), *Robotics in Education*, Lecture Notes in Networks and Systems 747, https://doi.org/10.1007/978-3-031-38454-7_8

1998 and has been one of the popular robotics tools in education. The development of robotics components and tools appropriate for school-age children, once accessible only to experts including robotics scientists and researchers, made learning with robotics accessible for teachers and students of any age [4–6]. There has been a variety of robotics tools made available for ER in classrooms [1, 7].

Constructionist theory, developed by Papert, focuses on children's active knowledge construction and reconstruction through their experience, and direct interaction with the world [8]. It provides powerful learning experiences enforces children's retention of constructed knowledge (following direct instruction on a fact or knowledge does not *stick* to children). Furthermore, children's learning experiences should be supported by their interaction with manipulatives—*objects to think with*, which contributed to the development of robotics tools. To ensure meaningful knowledge construction among children, they should be encouraged to explore, express, and exchange their views with others [9]. Constructionist educators highlight that constructionist learning is more effective when learning is supported by making an object that is sharable [10], which helps children's externalization of their feelings and ideas, further contributing to their understanding of knowledge [9]. The constructivist learning activities ensure that children continue to build upon, revisit, or reconstruct previously constructed knowledge [11]. Robotic tools can create tangible learning experiences for children with which children shape, reshape, and sharpen their knowledge and/or ideas.

Another theory that supports exploratory learning that ER promotes is the pedagogy of critical exploration by Eleanor Duckworth [12]. The critical explanation focuses on children's exploration of their *wonderful ideas* through inquiry and interaction with complex materials and/or confusions, investigating with trial and error approaches, which leads to the construction of new knowledge [13]. It discourages direct instructions by teachers/adults and emphasizes children's personal process of learning facilitated by adults around them. Duckworth strongly suggests that having of wonderful ideas is essential to children's intellectual development [12]. With critical exploration, learners explore their journey of learning while facing unknowns, taking risks, unexpected discoveries while teachers join the children's exploration together [13]. Duckworth emphasizes that teachers must focus on understanding children's experiences and construction of their knowledge by encouraging children to think about, reflect on, and share their experiences. In a sense, critical exploration supports two-way exploration – exploration of materials/subjects by children/learners and exploration of children's thinking by adults/teachers [12].

Robots naturally excite children and spark children's interests and curiosity, a perfect learning tool, to promote children's exploration of their ideas through inquiries and trying out their hypotheses. Children can explore their wonderful ideas, find new discoveries, and build new knowledge through their interactions with the real world. Robots provide instantaneous feedback when children's ideas are tested on a robotic tool, which can challenge and inspire the further exploration of their ideas.

This paper reexamines pedagogical approaches that support constructionist learning among children and further explores instructional strategies that effectively support constructionist learning. In the following section, pedagogical approaches that enhance ER learning among students are presented.

2 Pedagogical Approaches Enhance ER Learning

Educational robotics is an effective tool for promoting students' active learning. It becomes effective when students engage in making or creating using hands-on robotic tools, which provides a fun and engaging hands-on learning environment for students. With ER, students engage in multiple attempts to perfect their robotics creation by manipulating, assembling, disassembling, and reassembling materials while going through design learning and engineering iteration processes. In the process, students go through problem-solving steps through trial and error while improving their codes and/or robot constructions.

The pedagogical approaches that make learning with robotics successful are project-based, inquiry-based, learning by design, and maker education approaches where the focus is on the *process of learning*, rather than the final product [1]. All those approaches could be considered as a sub-set of constructivist learning since they promote students' construction of knowledge through students' experience. With ER, those approaches support constructionist learning through hands-on robotics manipulatives.

Project-based learning and learning by design are similar approaches focusing on student learning through an iterated process of designing and problem-solving. Project-based learning engages students in solving real-world problems/challenges while learning by design is centered around the construction of knowledge through the conscious effort for creating an artifact which is functional and aesthetically meaningful to students [14]. Inquiry-based learning can be considered as a subset of project-based learning since throughout the iterative process of project-based learning, students intend to solve problems that they face through their inquiries.

Maker education focuses specifically on the process of making and students' knowledge construction through the act of making. *Making* is one of the fundamental elements for us as a human being and for our lives [15]. The maker movement gained increasing attention from the education community in the early to mid-2010s when Maker Faires started to be held in various communities. ER naturally fit in maker education since it utilizes students' making activities to enhance their learning and knowledge construction. Maker education is also considered to have its foundation in constructionism [10] providing opportunities for learning by making.

Maker education contributed us to rethink the definition of a learner and a learning environment [16], and how to promote new types of *learning* among students. Maker education fosters students' active engagement in their own learning by assembling, disassembling, reassembling, tinkering, evaluating, and sharing. Moreover, digital making, such as robotics making, is a powerful learning experience with which students gain the power to realize their wonderful ideas in the real world [17]. Digital fabrication tools, such as 3D printing and laser printing/cutting, have become more and more accessible to primary and secondary students in recent years, adding more value to implementing maker education in classrooms [17].

All approaches require open-ended and authentic projects and/or environments to foster student agency, creativity, authorship of their project, and reflective learning. However, such learning experiences do not happen without careful planning and scaffolding by teachers. Prescribed teacher-centered lessons with instructions for students to follow do not effectively create such learning experiences. The following section discusses various tips and steps necessary to promote effective ER learning for students.

3 Tips to Promote Effective ER Learning in Classrooms

Making with robotics is effective because it creates a *fun* and engaging hands-on learning environment for students. It provides students with opportunities to explore, mess around, test, and evaluate new ideas. In many ways, ER disrupts the traditional practice of teaching still prevalent in schools. A traditional classroom practice where students sit, and a teacher instructs in a more disciplined and structured way has very limited space in ER maker classroom. Rather, students engage in manipulating, assembling, and reassembling materials while going through the design learning process and problem-solving program through trial-and-error approaches.

3.1 Structural Change

With ER maker activities, it is highly recommended that students work in small groups of 2–3 students [7, 18–20]. Working in groups encourages students to learn to collaborate and obtain the skills needed for effective collaboration. The *pair programming* technique widely used in Computer Science (CS) classes can be implemented to ensure desired collaborations happening with ER learning. With pair programming, two students sit next to each other while using one computer to work collaboratively to develop, test, and refine codes. One student called a "driver" is responsible for developing and typing the code, while the other called a "navigator", who is responsible for observing the driver's work, detects errors and provides suggestions or ideas for solving problems they face [21–24]. Throughout the pair programming activity, the pair alternate their roles to make sure that their work is distributed fairly, and collaboration is enhanced. The technique can be used with multi-member groups in ER. With robotics making, there are various roles, especially with construction and coding. Students can share tasks and work on the task in pairs using the pair programming technique. Working in groups, students are excited and motivated to share their ideas, engage in collaborative decision-making, and provide constructive feedback, which helps them to acquire communication skills [25–27] while exploring and solving real-world problems with their peers.

In ER making classroom, students are not sitting quietly in their chairs. Students are allowed to move around, interact, and talk with their peers. Teachers have to learn to teach in *chaos* [28]. In typical ER classrooms, levels of groups/students gradually start to diverse as they progress. This adds to the chaos in classrooms. It is not chaos in the sense that teachers lose control of the students. Rather, it is a *controlled chaos* [19] where students are engaged in the development of their project by working on tasks, discussing, and sharing their ideas, making decisions, and solving problems. The level of noise and movements in the classroom is expected to be medium to high with a buzz in the classroom throughout a session. If you pay close attention to the students, they are all *busy* doing something related to their robotics projects [19].

The classroom should be set up in a way that sparks students' creativity to support robotics creations. Teachers are encouraged to create a maker space type of environment where a variety of maker materials are available for students to explore and select for their robotics projects including craft materials, such as markers, crayons, colored pencils, paints, various types of papers, yarns, ribbons, glues/glue guns, cardboard boxes. We collect recycled materials such as egg cartons, tissue boxes, cereal boxes, milk cartons, plastic bottles, etc. Anything that inspires students' ideas is welcomed. Digital fabrication tools such as 3D printers and laser printers/cutters are powerful tools to make students' ideas into reality.

3.2 *Paradigm Shift*

With the constructionist approach, children are at the center of the learning and are the agent to *program* the computer and robots. Traditional ways of *teaching*—a teacher is a provider of information and knowledge to students following a rigidly structured curriculum, do not support effective ER learning. With that, teachers have to face a paradigm shift—from teachers to facilitators, mentors, or scaffolders [18, 29]. Teachers are required to transfer from familiar lecture-type instructions or the teacher's own agenda of a lesson to focusing on providing technical and subject knowledge and responding to students' needs (on-demand instructions). Constructionist teachers create a learning community where peer instructions are encouraged and supported and the shift of the roles of teachers and students is required [30].

With constructionist ER learning, students also shift from passive learners to active participants in their own learning [18]. The author has witnessed, in two decades of experience teaching in higher education classrooms, that many students come to a classroom expecting an instructor to *lecture* and their role is just to listen and take notes of the lecture (passive). They usually wait for specific instructions to follow. They expect the instructor to have the right answers to deliver to them. In other words, they firmly believe that they must know the *right* and the *only* answer. They tend to ask the instructor if their code is correct even before testing it on a robot coming from a fear of making mistakes. All of these are because of the type of education that their primary and secondary schooling provided, which focuses on standardized testing. For them, ER learning is very confusing since they are asked to think and create and there tends to be no one right answer [31].

The fact that there are multiple ways to solve problems and challenges, and there is rarely one and the only answer with ER projects confuse both teachers and students who are used to having one solution and one correct answer. Again, schooling, which has not changed much since it was instituted, *educated* people that teachers are the answers (master of content knowledge), and they *teach* students (direct instructions). However, with ER lessons, as the teacher's role shifts, teachers become masters of scaffolding and facilitating who can guide students to stay on the right track [28].

The biggest challenge for most teachers is to tell their students, "I don't know" when they do not have an answer. A common question I receive after an ER workshop for teachers is if they know enough to teach ER. They have the pressure of knowing all the answers. The mindset is quite similar to those of students who ask if their code is correct before testing it. Not knowing the *right* answer makes both teachers and students uncomfortable. However, we should embrace this as a wonderful learning opportunity for both teachers and students. And this is what professional scientists and engineers do—inquiry and problem-solving. Teachers should use this opportunity to implement inquiry-based and critical exploration approaches. Students benefit and learn the most from discussion and teacher inquiries penetrating their creativity and curiosity [31]. Moreover, this is a great opportunity for teachers to learn *with* their students. In this sense, effective classrooms can be created with learner-centered approaches where both teachers and students learn together/from each other. Teachers should not be afraid of telling their students, "Oh, I don't know. Let's figure it out together." Occasionally (with some classrooms, quite often), it is students who know or find the answers and help teachers to acquire new knowledge and teachers should be excited about the opportunities.

Another important paradigm shift happens with students' roles in ER classrooms. When the level of mastery started to be diverse and each group need different support from teachers, from basic coding questions, robot construction issues, to more complicated coding questions, the teacher needs to multi-task to help all students. To turn the classroom into an effective ER learning environment, teachers need to accept that they need more teachers in the class. Here, we suggest that advanced students take on the teacher's facilitator of learning role and provide support to other students in need. Teachers need to initiate the shift by asking those capable students to provide support to other students. Most likely, those students are delighted to help. This experience helps advanced students to reach the highest level of knowledge/skill mastery since teaching is the highest level of learning [18, 20, 26].

Keeping it challenging, but not too challenging is another skill that teachers need to master for providing successful ER learning experience. Successful educational robotics learning occurs when students are challenged and engaged, sometime to the point of failure. But determining how much challenge is needed to enhance their learning to the maximum point is always a challenge for teachers. Csíkszentmihályi's *flow* [32] explains the state that maximize students' ER learning. The flow state occurs when people's body or mind is stretched to its limits while voluntarily trying to accomplish something difficult or worthwhile. Prensky explains how game designers create the flow state so that players continue playing games [33]. This narrow zone is between too hard (I am giving up) and too easy (I'm not challenged enough). Teachers use differentiated instructions to create such learning experiences. It is another balancing act that teachers play in classroom by providing additional challenges for advanced learners or inviting them to help other students (explained in detail below).

3.3 *Making Everyone Accountable*

In learner-centered ER classrooms, especially with group work, it is expected that everyone is accountable and completes tasks at hand in a collaborative manner. In undergraduate classrooms, I have recognized that some students hated group work since they had unpleasant experiences in secondary education where they did not receive support from other group members or others who enjoyed an easy ride where their group members completed the projects for them. In recent years, possibly because of the pandemic, many students do not know how to efficiently do group projects and they clam everything at the last minute, which usually results in poor project executions. One strategy that has been adopted for ER projects is for each group to create a timeline at the beginning of the project, aiming to keep everyone accountable (which was rather a common practice before the pandemic.) With more capable students, Figma (www.figma.com) is a tool that provides flexibility for students to create a timeline in the way that is best for their project (Fig. 1). I usually provide a template to start with; however, each group can decide to or not to use the template. Google sheets or docs are ideal for younger students.

Each group should assign tasks to the members expecting them to complete the task by the deadline, which helps everyone to be accountable. Teachers should have access to all the timelines and check on each group before the deadlines to help them keep themselves accountable.

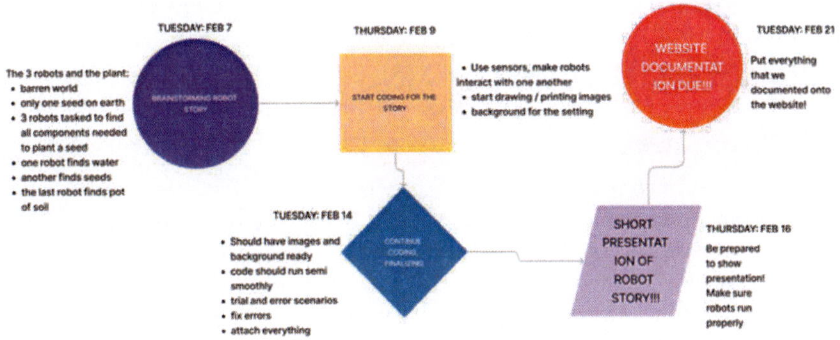

Fig. 1 Figma timetable

In ER learning environments, projects that students tackle should be open-ended and authentic [18, 29, 34, 35] which encourages students to sit in the driver's seat of their own learning. They are responsible for their own learning and have various decision-making opportunities to accomplish the goals of the project that they set by themselves. This provides them authorship of their project, adding excitement and enthusiasm to complete the project successfully. This makes teachers' role as facilitators of their learning more complex requiring teachers' balancing acts—when to step in or to let go; how much structure or freedom; when to listen or when to provide support, etc. [36]. To accommodate the needs of various projects that students decide to work on, teachers need to provide on-demand/on-the-spot instructions/support. Deciding when and how to intervene and/or support is another challenge when each group has different needs from coding, robot assembly, to different technical support. It is important for teachers to understand each group's process and problems that they are facing. To *make their learning visible* [37], documentation of their learning process provides critical information for teachers [18, 19, 26, 35, 38]. The documentation process is used to promote self-reflective practices. Students evaluating their own work encourage them to learn from their experiences, including both successful achievements and mistakes and/or failures, nurturing their ability to improve themselves [30, 39]. Moreover, it puts students responsible for their own learning [30, 38]. Sharing their own learning with others makes learning *visible* to others. However, most importantly, it makes their learning *visible* to themselves [12].

By making documentation as a part of their everyday practice, students develop skills in evaluation and reflection, promoting them to become reflective learners. Teachers can use different journaling techniques; however, an engineering journal, either digital using Google sites/docs (or similar online tools) or analog with paper and pencils (recommend having a notebook for each student) is suitable with ER learning. To guide their documentation, we provide thinking prompts such as, "What did you learn?", "What questions do you have?", Any ideas for how you would use the new skill in another program? ", "What problems did you solve and how?", etc. This helps students to reflect on their learning practice as well as think about the future application of newly learned knowledge/skills [38]. Students can add artifacts, pictures of codes, drawings, graphics, and recorded robot performances to their reflective writing in the engineering journal.

3.4 Cycling Around

In ER learning environments, learning is not a linear process. Traditional lessons that teachers are familiar with do not support the type of constructionist learning that ER aims to create. Although during a skill mastering phase, where students learn new skills, a lesson with instructions is utilized. However, students will go through some iterative process, which helps them to master the skills. There are various design cycles that can support constructionist ER learning, such as the engineering design cycle [18, 31], Creative Learning Spiral [36] (Fig. 2), and Use-Modify-Create (test-analyze-refine) cycle [40]. With ER learning, following the cycles encourage students continuously to cycle around to improve their project until they are satisfied with the results.

The engineering design cycle follows the following steps to create an iterative process of improving projects:

1. Identify problems or challenges
2. Brainstorm and decide the best possible solution
3. Create or construct a prototype
4. Test and evaluate the prototype
5. Redesign based on the feedback from the test and evaluation
6. Communicate and share the results (and reflect)

During the process, it repeats until the result is satisfactory. Step 6 leads to defining new problems or challenges (back to step 1) and moves naturally onto the following step to improve the result [18].

The Creative Learning Spiral (Fig. 2) promotes students' creativity by going through 5 components: imagine, create, play, share, and reflect [36]. Both cycles help students to improve their projects while sharing and reflecting on their learning. As emphasized with documentation practices, sharing and reflecting on their learning is a key component of constructionist ER learning.

The three-stage progression of Use-Modify-Create (test-analyze-refine) was introduced to engage students in developing computational thinking skills in computational environments [40]. The Use-Modify-Create steps are specifically helpful when new skills are introduced to students. For example, instead of students creating a complex code from scratch, they can first use an existing or sample code to understand what happens, try to modify different components one by one to understand the structure/syntax of the code (personalization of the sample code), then use the knowledge gained through the first two steps, they create a new code. While creating a new code, students go through the iterative test-analyze-refine process. Or students

Fig. 2 Creative learning spiral

use building instruction to create a robot, modify it to personalize the structure, then use the learned new knowledge to create a robot of their own design while going through the test-analyze-refine process. Teachers can use the Use-Modify-Create cycle when using existing linear lessons to create rich constructionist ER learning environments.

4 Conclusion

The paper revisited the chapter published in 2012, reexamined, and updated the pedagogical approaches and strategies that support constructionist ER learning. The paper targets pre- and in-service teachers who are not familiar with the type of learning experiences that ER can create for students, providing an overview of the pedagogical approaches and strategies, to promote their interest in learning more about how to make ER learning effective constructionist learning experiences for students. As it is clear in the paper, both teachers and students need to change the way they understand educational practices in schools and willingly participate in the paradigm shift that it requires to create fun, challenging, and exciting learning experiences. I hope this also benefits educators interested in integrating constructionist ER learning practices in their teacher education or/and teacher professional development programs.

References

1. Eguchi, A.: Bringing robotics in classrooms. In: Khine, M.S. (ed.) Robotics in STEM Education: Redesigning the Learning Experience, pp. 3–31. Springer International Publishing, Switzerland (2017)
2. Papert, S.: Mindstorms: Children, Computers, and Powerful Ideas, 2nd edn. Basic Books, New York, NY (1993)
3. Martin, F., Mikhak, B., Resnick, M., Silverman, B., Berg, R.: To mindstorms and beyond: evolution of a construction kit for magical machines. In: Druin, A., Hendler, J. (eds.) Robots for Kids: Exploring New Technologies for Learning, pp. 9–33. Morgan Kaufmann, San Francisco, CA (2000)
4. Cruz-Martin, A., Fernandez-Madrigal, J.A., Galindo, C., Gonzalez-Jimenez, J., Stockmans-Daou, C.: A LEGO mindstorms NXT approach for teaching at data acquisition, control systems engineering and real-time systems undergraduate courses. Comput. Educ. **59**, 974–988 (2012)
5. Mataric, M.J.: Robotics education for all ages. American Association for Artificial Intelligence Spring Symposium on Accessible, Hands-on AI and Robotics Education (2004)
6. Eguchi, A., Almeida, L.: RoboCupJunior: promoting STEM education with robotics competition. Proc. Robot. Educ. (2013)
7. Eguchi, A.: Educational robotics as a learning tool for promoting rich environments for active learning (REALs). In: Keengwe, J. (Ed.) Handbook of Research on Educational Technology Integration and Active Learning, pp. 19–47. Information Science Reference (IGI Global), Hershey, PA (2015)
8. Ackermann, E.K.: Perspective-taking and object construction: two keys to learning. In: Kafai, Y., Resnick, M. (eds.) Constructionism in Practice: Designing, Thinking, and Learning in a Digital World, pp. 25–37. Lawrence Erlbaum Associates, Mahwah, New Jersey (1996)

9. Ackermann, E.K.: Constructing knowledge and transforming the world. In: Tokoro, M., Steels, L. (eds.) A Learning Zone of One's Own: Sharing Representations and Flow in Collaborative Learning Environments, pp. 15–37. IOS Press, Washington, DC (2004)
10. Martinez, S.L., Stager, G.: Invent To Learn: Making, Tinkering, and Engineering in the Classroom. Constructing Modern Knowledge Press, Torrance, CA (2013)
11. Papert, S.: The Children's Machine: Rethinking School in the Age of the Computer. Basic Books, New York, NY (1993)
12. Duckworth, E.: Critical exploration in the classroom. The New Educator. **1**, 257–272 (2005)
13. Cavicchi, E., Chiu, S., McDonnell, F.: Introductory paper on critical explorations in teaching art, science, and teacher education. New Educ. **5**, 189–204 (2009)
14. Sarfo, K.F.: Learning by design. In: Encyclopedia of the Sciences of Learning. Springer, Boston MA (2012)
15. Hatch, M.: The Maker Movement Manifesto—Rules for Innovation in the New World of Crafters, Hackers, and Tinkerers. McGraw-Hill Education, New York, NY (2014)
16. Halverson, E.R., Sheridan, K.M.: The maker movement in education. Harv. Educ. Rev. **84**, 495–504 (2014)
17. Blikstien, P.: Digital fabrication and 'making in education": the democratization of invention. In: Herrmann, J.W., Buching, C. (Eds.) FabLabs: Of Makers and Inventors. Transcript Publishers, Bielefeld, Germany (2013)
18. Eguchi, A.: Educational robotics theories and practice: tips for how to do it right. In: Barker, B.S., Nugent, G., Grandgenett, N., Adamchuk, V.L. (Eds.) Robotics in K-12 Education: A New Technology for Learning, pp. 1–30. Information Science Reference (IGI Global), Hershey, PA (2012)
19. Eguchi, A., Uribe, L.: Educational robotics meets inquiry-based learning. In: Lennex, L., Nettleton, K.F. (Eds.) Cases on Inquiry Through Technology in Math and Science: Systemic Approaches. Information Science Reference (IGI Global), Hershey, PA (2012)
20. Eguchi, A., Uribe, L.: Integrating educational robotics in elementary curriculum. In: World Conference on E-Learning in Corporate, Government, Healthcare, and Higher Education (ELEARN) (2009)
21. Denner, J., Werner, L., Campe, S., Ortiz, E.: Pair programming: Under what conditions is it advantageous for middle school students? J. Res. Technol. Educ. **46**, 277–296
22. Hanks, B., Fitzgerald, S., McCauley, R., Murphy, L., Zander, C.: Pair programming in education: a literature review. Comput. Sci. Educ. **21**, 135–173 (2011)
23. Salleh, N., Mendes, E., Grundy, J.: Empirical studies of pair programming for CS/SE teaching in higher education: a systemic literature review. IEEE Trans. Softw. Eng. **37**, 509–525 (2010)
24. Lewis, C.M.: Is pair programming more effective than other forms of collaboration for young students? Comput. Sci. Educ. **21**, 105–134 (2011)
25. Eguchi, A.: Educational robotics for undergraduate freshmen. In: Proceedings of the World Conference on Educational Multimedia, Hypermedia and Telecommunications, pp. 1792–1797 (2007)
26. Eguchi, A.: Educational robotics for elementary school classroom. In: Proceedings of the Society for Information Technology and Education (SITE), pp. 2542–2549 (2007)
27. Miller, D.P., Nourbakhsh, I.R., Sigwart, R.: Robots for education. In: Siciliano, B., Khatib, O. (eds.) Springer Handbook of Rootics, pp. 1283–1301. Springer-Verlag, New York, LLC, New York, NY (2008)
28. Rogers, C.: A well-kept secret—classroom management with robotics. In: Bers, M.U. (ed.) Blocks to Robots : Learning with Technology in the Early Childhood Classroom, pp. 48–52. Teachers College Press, New York, NY (2008)
29. Paaskesen, R.B.: Play-based strategies and using robot technologies across the curriculum. Int. J. Play. **9**, 230–254 (2020)
30. Bers, M.U.: Using robotic manipulatives to develop technological fluency in early childhood. In: Saracho, O.N., Spodek, B. (eds.) Contemporary Perspectives on Science and Technology in Early Childhood Education, pp. 105–125. Information Age Publishing Inc., Charlotte, NC (2008)

31. Rogers, C., Portsmore, M.: Bringing engineering to elementary school. J. STEM Educ. **5**, 17–28 (2004)
32. Csíkszentmihályi, M.: Flow: The Psychology of Optimal Experience. HarperCollins Publishers, New York, NY (2008)
33. Prensky, M.: Don't Bother Me Mom—I'm Learning! Paragon House, St. Paul, MN (2006)
34. Yang, Y., Long, Y., Sun, D., Van Aalst, J., Cheng, S.: Fostering students' creativity via educational robotics: an investigation of teachers' pedagogical practices based on teacher interviews. Br. J. Edu. Technol. **51**, 1826–1842 (2020)
35. Alimisis, D.: Teacher training in educational robotics: the ROBOESL project paradigm. Tech. Know. Learn. **24**, 279–290 (2018)
36. Resnick, M.: Ten tips for cultivating creativity. https://mres.medium.com/ten-tips-for-cultivating-creativity-fe79e7ebb83e. Accessed 01 Feb 2023
37. Rinaldi, C.: Making learning visible: children as individual and group learners. In: Zero, P., Children, R. (eds.) Making Learning Visible: Children as Individual and Group Learners, pp. 78–89. Olive Press, Bloomfield, MI (2001)
38. Eguchi, A.: Learner-centered pedagogy with educational robotics. In: Keengwe, J. (Ed.) Handbook of Research on Learner-Centered Pedagogy in Teacher Education and Professional Development, pp. 350–372. IGI Global, Hershey, PA (2016)
39. Weimer, M.: Learner-Centered Teaching: Five Key Changes to Practice, 2nd edn. Josseu-Bass, San Francisco, CA (2013)
40. Lee, I., Martin, F., Denner, J., Coulter, B., Allan, W., Erickson, J., Malyn-Smith, J., Werner, L.: Computational thinking for youth in practice. ACM Inroads. **2**, 32–37 (2011)

Experience Prototyping: Smart Assistant for Autonomous Mobility Concept

Richard Balogh, Michala Lipková, Róbert Hnilica, and Damián Plachý

Abstract The paper describes the subsequent development of two bachelor's degree projects as part of a commissioned research project for the Czech car manufacturer Škoda Auto. The first bachelor's degree project by R. Hnilica from the field of industrial design focused on developing an interaction design concept of a smart personal assistant, suitable for both children and adult passengers in autonomous cars. The follow-up bachelor's degree project of a small robotic assistant by D. Plachý delivered a working prototype demonstrating partial functionality of the design concept. In the paper, we focus more on the design setup and the description of the process than on the technical details of our solution.

Keywords Smart assistant · Experience prototyping · Interactivity · Robotics · Autonomous mobility · Mxlab

---

"Experience prototyping" reffers to a title of the long-term project framework supported by Škoda Auto a.s., contract No. SVS-21-063-STR, SVS-20-017-STR.

R. Balogh (✉) · M. Lipková · R. Hnilica · D. Plachý
Slovak University of Technology, Bratislava, Slovakia
e-mail: richard.balogh@stuba.sk

M. Lipková
e-mail: hello@mxlab.sk

M. Lipková · R. Hnilica
MX lab, Institute of Design FAD STU in Bratislava, Bratislava, Slovakia

R. Balogh · D. Plachý
Institute of Automotive Mechatronics FEI STU in Bratislava, Bratislava, Slovakia

R. Balogh et al. (eds.), *Robotics in Education*, Lecture Notes in Networks and Systems 747, https://doi.org/10.1007/978-3-031-38454-7_9

1 Introduction

1.1 The Industry Cooperation Context

The project Experience prototyping (I. and II.) followed the long-term cooperation of the Slovak University of Technology in Bratislava (STU) with the Czech car manufacturer Škoda Auto. Cooperation with the automotive industry takes place at the recently established MX lab research facility (the acronym 'MX' refers to 'multidisciplinary research' and 'experience design', see also [1]). MX lab was founded and is located at the Faculty of Architecture and Design, but the ambition of all ongoing lab activities is to build an interdisciplinary research group and shared development space for innovative projects in the field of mobility and experience design. The lab focuses on experimentation with emerging design tools (e.g. virtual and mixed reality, eye tracking and attention modelling, generative design, motion capture) with the aim to innovate the design process through the integration of design into technical development [2].

The main research challenge and the umbrella theme of the Experience prototyping project, realised within the time period of two academic years from September 2020 to June 2022, was the development of the next generation of smart car interiors as the key differentiation factor in the automotive industry. The project focused on prototyping the in-car user experience through the tools of both design research and rapid prototyping.

Since a large part of the realisation of the project overlapped with the global pandemic of COVID 19 (especially 2020/09–2022/03), most of the collaboration occurred in a remote setting. On the design development side, the situation accelerated the use of the VR co-creation software Gravity Sketch [3], which allows multi-user presence and collaboration in a shared VR space, as well as the use of visual collaboration tools Mural and Miro, used in the concept development and in the initial research phase of the project. The use of Gravity Sketch co-creation mode allowed the team to use the remote use of lab's physical car interior simulator even in the period of time when students did not have the possibility to access facilities of the university. The simulator enables testing of various car interior packages and architectures in mixed reality, combining virtual reality simulation of CAD designs with physical components from serial production cars. The displayed setup (see Fig. 1) features Škoda Superb Laurin and Klement steering wheel and car seats [4].

1.2 The Brief

Through bachelor's degree projects [5–8], with the support of doctoral research, the project Experience prototyping examined and tested different scenarios of the user experience in an autonomous electric cars of the near future. The aim of the project was to create and implement user testing of concepts of car control interface,

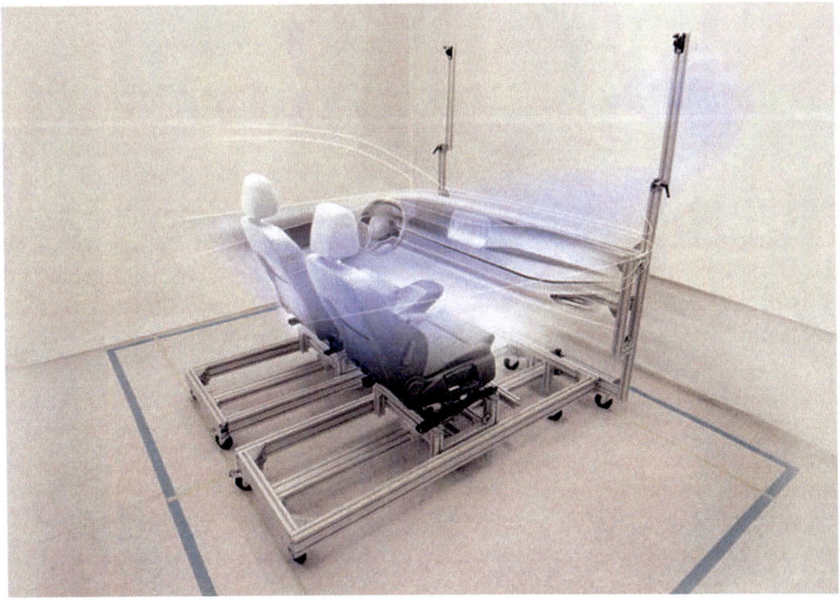

Fig. 1 Adjustable car interior size simulator. Photo: F. Maukš, digital drawing: M. Truben, MX lab 2019

as well as UX and UI concepts of related digital products (mobile applications) through a car interior simulator. Problem areas of the brief were defined as interface personalization, semi-autonomous car interface, digital companion(s), smart home and family infotainment.

The design scenario featured the target group of Generation Alpha customers (born in the time frame of 2010 to 2025 [9]) and a target implementation time of 2035 and beyond. The protagonists of the project scenario were described as living in an urban area, using an electric drive and semi-autonomous vehicles, offering safe space for individual or shared quality time, and becoming an extension of their connected home. The projects developed in the area of interior architecture were supposed to focus on the innovative aesthetics, styling, and layout of the vehicle, with regard to the work or leisure activities that the passenger can perform instead of driving.

"How might we..." questions, which design students at FAD STU use as a regular part of their assignment analysis and problem definition process [10], were defined as the following:

- How might we create interior architecture that offers immersive space for the daily activities of work, play or entertainment?
- How might interior design support long-term use and health safety in shared mobility?

2 Methodology

2.1 Interdisciplinary Project-Based Learning and the Field of Experience Design

Collaboration between two faculties was part of a broader vision of developing competencies for future curriculum in the field of Experience Design. Although multidisciplinary teamwork would have been a great opportunity to develop soft skills of involved students, the project timeline and differences in current study programmes at the involved faculties did not allow such simultaneous co-creation and concept development.

Therefore, the collaboration was planned in advance as a long-term series of two follow-up degree assignments during two academic years 2020/21 (concept development and industrial design) and 2021/22 (prototyping technical development). This way, each of the degree projects kept its autonomy in terms of complexity of the assignment and difficulty of the learning goals, allowing greater freedom to experiment and fail during the process, without negatively influencing the partners.

2.2 Phase No. 1: Design Research and Concept Development

When we talk about the first phase, we use this term for a rather complex, year-long concept development process, which itself consisted of several development stages during two semesters:

- first semester (winter term 2020/21) was solely devoted to problem and target group analysis (qualitative research, development of personas, customer journey), explorative design phase, which resulted in innovative exterior design concept,
- second semester (summer term 2020/21) included detailing of the concept, development of interior design, customer experience concept and design of selected HMI interaction elements (including smart assistant).

The outputs of phase No. 1 included:

- VR simulation of the whole concept (interior and exterior) in Gravity Sketch,
- explanatory presentation and video simulating the customer experience concept, demonstrating the mobility service and its interaction elements.

The results of Phase No. 1 were presented to industry partners, who identified the concept of the smart assistant as the most interesting part of the project, recommending its further development.

2.3 *Phase No. 2: Technical Solution and Rapid Prototyping*

The continuing technical development part of the project was planned for the summer term of the following academic year 2021/22. The process included the following steps:

- analysis of the experience design concept and its translation into technical brief,
- selection and iteration of hardware components,
- prototyping and testing of defined interactions,
- choice of the final technical development strategy,
- evaluation and documentation of the process.

Working in a small research group leads to the requirements for a multidisciplinary approach, various skills are necessary, and at least minimal experience from various fields is a must. Most engineers who learn embedded systems would become experts in technical aspects. As stated in [11], multidisciplinary research also enables the project team to capitalise on a wider range of expertise, drawing from several disciplines on the unique, yet complementary, areas of expertise of the team members. However, students are learning to be experts in their own field, which is a feature that should be combined with embedded systems and robotics to achieve significant results.

The skills needed to design and build very basic robots and embedded systems can be obtained using an already known platform such as an Arduino [12]. Getting started with the programming of robots and embedded systems in general can be eased using a wide variety of Arduino libraries.

The ability to effectively integrate existing and emerging technologies, including CAD/CAM design tools and 3D printing technologies, into the new product development process has the potential to give engineers the ability to produce new high-tech products at an increasing rate [13].

For quick development and rapid prototyping of embedded systems, some common technologies appeared [14]: using powerfull 32-bit ARM based microcontrollers, using solderless breadboard with breakout modules (which makes possible to use otherwise useless SMD and BGA mounted devices) and using object-oriented programming languages (C++) based on the producer provided Hardware Abstraction Layer (HAL) libraries. More and more common is also using a more or less simplified real-time operating systems (RTOS).

The main output of phase No. 2 was an interactive hardware prototype (see Sect. 4).

3 Design Concept "Škoda Alvy"

3.1 Holistic Design Concept of Future Mobility

The aim of the project was to simplify mobility in the urban environment for the four-member family in 2030. Róbert Hnilica's bachelor thesis "Effective Urban Mobility of the Future" [5] presented a future vision of a holistic design concept of a modular, four-piece interior of an autonomous car for four individual passengers. The concept works with two cabin alternatives, for the parent and for the child, while the vehicles are adapted to the users using a mobile device connection. In addition to solving the problem of personal mobility, the concept also introduces the idea of automated services (for example, transporting purchases), which the vehicle performs autonomously and saves users time. Part of the dashboard in the interior of the car is a solution to the basic concept of the user interface, the main carrier of which is an intelligent assistant and a panel with light signaling, informing the user of incoming changes in the direction of travel.

3.2 Adjustable Intelligent Assistant for a Parent and for a Child

The main communication point of each of the two autonomous car cabins was designed in the form of a domestic mini robot that is fixed on the dashboard. The robot has a limited range of motion, one display and it is capable of visual and voice communication. The main motivation for using the concept of smart companion in the car was to overcome the anonymity of autonomous technology and to help build trust and a positive emotional relationship with the mobility service (Fig. 2).

Construction-wise, the assistant consists of a single white cuboid body with rounded edges. In addition to communicating through the animated display, the robot's body can change its colour to emphasise the emotional expression (see Fig. 3).

4 Interactive Prototype

4.1 Experience Design Brief

After finishing and presenting the mobility concept through video's storytelling, the idea of the assistant was identified as a feature to be further developed. Based on feedback from industry partners, the designer has defined the following requirements for the creation of the interactive prototype:

Fig. 2 Visualisations of Škoda Alvy family mobility concept. Design, CAD visualisation, and digital sketches: R. Hnilica, MX lab 2021

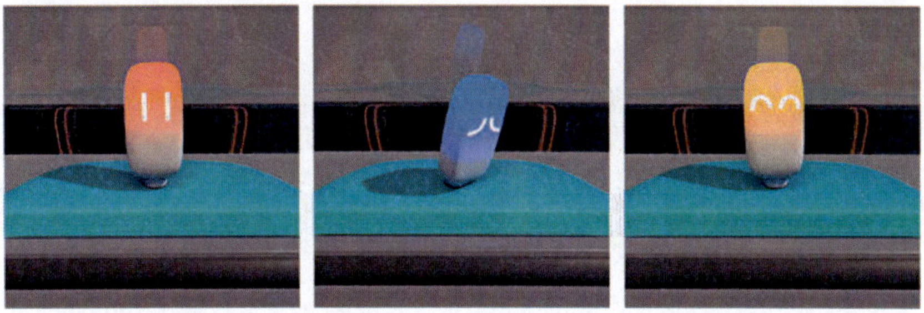

Fig. 3 Stills from video animation of Škoda Alvy concept video. Demonstration of different emotional states of the smart assistant: neutral expression, expression of sadness, happy expression (from left to right). CAD visualisation and design: R. Hnilica, MX lab 2021

- The goal of the physical prototype is to simulate the basic interaction of the assistant with the passenger, which will allow us to collect direct qualitative feedback on assistant behaviour.
- The solution does not have to be implemented in combination with existing voice interfaces (Alexa/Siri), it can present a set of programmed behaviour sequences, determined on the basis of user testing scenarios, executable, for example manually activated by predetermined switches, is also acceptable.
- The aim of the creation of the physical prototype is to verify six interactions of the assistant's behaviour in one scenario sequence: greeting, change of car's direction (in combination with light signalling panel), sad reaction of the assistant to the traffic congestion, sleep mode and waking up, dance and farewell.

- The spectrum of assistant's body movements does not have to be incorporated into the prototype in the full scale, only the necessary movements relevant for the described scenario.

The following elements were defined as necessary to be included in the technical solution of the physical prototype:

- The main body of the assistant can be rotated by 90 °C with interactive elements on a stable base that will be fixable to the car interior simulator.
- Display in the main body of the assistant should work as the main interactive communication element (e.g. abstract display of eyes),
- the illumination of the main body of the assistant should be adjustable, and in order to allow the expression of different emotions,
- assistant's body (or the "base" element) should include a speaker that will allow playing a specific set of audio tracks.

4.2 Translation into the Technical Brief

Based on our previous experience [15] and knowing that the designer's idea is not final, we decided to make as simple as possible that meets the requirements. It is possible to connect a simple TFT display with a cheap Arduino controller and reuse existing libraries to achieve a goal with minimum effort. Connections using wires look a bit messy, but it makes it possible to move them freely and adapt their position in a first shape prototype made on a 3D printer. Playing with some aspects of the design, it was possible to shift the design further, test some interaction problems and solve them quickly (Fig. 4).

Interactivity and communication during the design phase was also a problem. It is quite difficult to convert vague information like "make the movements faster and this colour brighter and more reddish" into exact values for the microcontroller. To make it easier, we control some variables with external potentiometers, which makes it possible to "play" with the speed and colours.

As a result of the first design phase, we have the first prototype more or less specified and accompanied by a list of problems that should be avoided in the final phase.

Using a rapid prototyping technique, we were able to recognise in the very early phase of the project that the proposed design was not ideal. First, the TFT display with resolution only 128×256 pixels was too small and moreover the viewing angle was very narrow, totally changing the colour impression just moving the head a few degrees from the axis. Animations were very simple, using just geometric shapes, but implementation and changes were complicated, so we decided to let experimenting with animations solely on the designer. He was able to create short video sequences with the appropriate sound effects and provide us with the simple videofiles. We changed the simple 8-bit microcontroller with the Raspberry Pi platform which directly enables to play provided videos based on the mechanic switch selector.

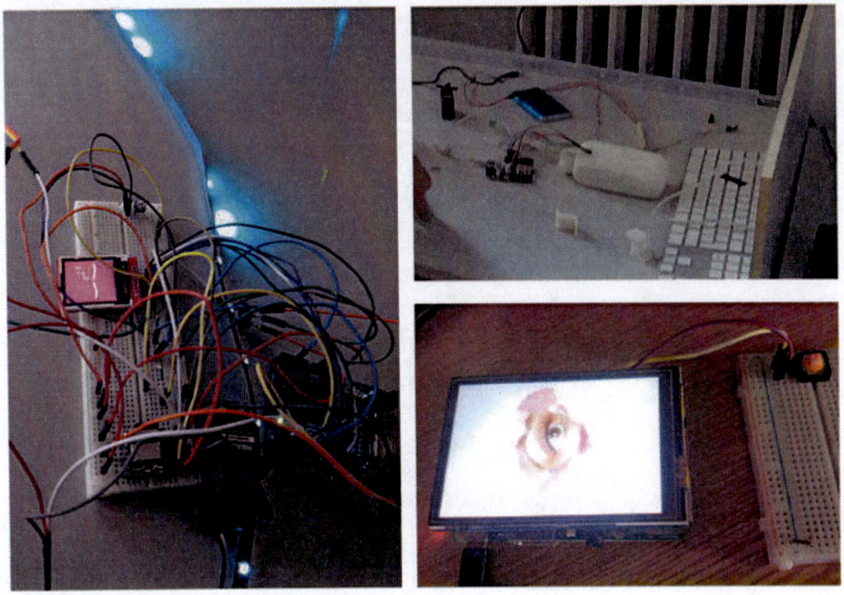

Fig. 4 Long way from a quick prototype to the final model

The remaining colour LED effects both on the desktop and body of the assistant and the simple servomotor for movements were then easily implemented. The only issue remained was the synchronisation of the effects, movements and video, but it was possible to use a synchronisation pulse before each video sequence using the microcontroller's general purpose I/O pins.

The disadvantage of the first solution with the Arduino processor was the inability to synchronise all processes equally. We then decided to use the simplified version of the RTOS operating system [16] which makes task planning much more efficient.

The resulting robotic assistant (see Fig. 5) is capable of presenting various scenarios, which can be easily adjusted or modified according to the designer's requirements. In combination with a simplified dashboard, we can test the acceptance of its behaviour with real customers.

5 Conclusion

The prototyping of assistant interactions was successfully completed, allowing the design concept to move to the testing phase. The prototype contributed to better comprehensibility of the design concept and allowed the audience to experience the interaction in real life. The testing informed the designer about positive and negative aspects of his designs which he could use in this next project from the

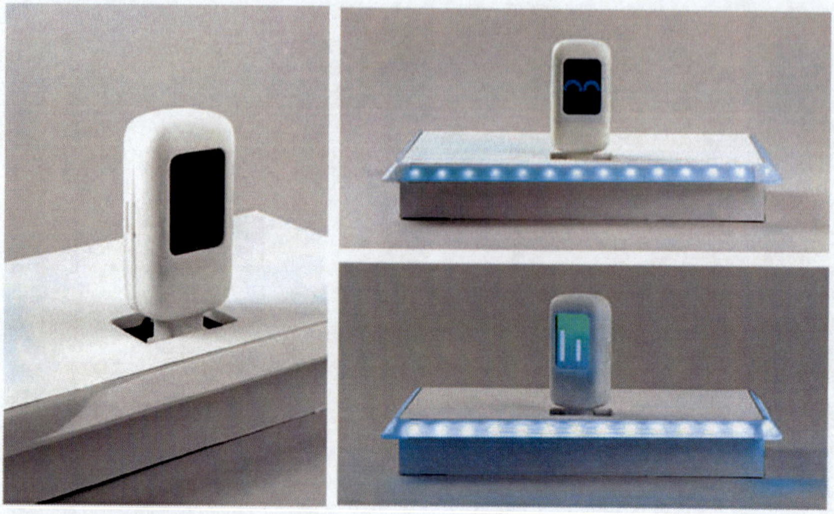

Fig. 5 Interactive prototype of Škoda Alvy smart assistant. Photo: A. Šakový, MX lab 2021

similar field (smart assistant for children's back seats row to help the driving parent). The prototype was therefore useful both in the testing phase and it has informed further development of the design concept.

At the same time, the described case study of two subsequently developed degree projects in two different disciplines (mechatronics and industrial design) will be used as a reference in the future for potential cooperation degree projects developed in parallel. We believe that our experimental setup and experiences may inspire others to reuse it with various robotics and construction-orientated mixed-team projects.

Acknowledgements Publication of this paper was supported from the project *Building of mechatronic laboratory using smart technologies* Nr. KEGA 030STU-4/2021 financed by the Slovak Cultural and Educational Grant Agency KEGA and the project *Support of research activities of STU excellent laboratories* in Bratislava, Nr. NFP313020BXZ1. The presented design project is, at the same time, the result of long-term university support by Škoda Auto a.s. (financed by contracts Nr. SVS-21-063-STR and SVS-20-017-STR). The authors also thank the Fablab Bratislava at the Slovak Centre of Scientific and Technical Information (SCSTI) for its support in rapid prototyping.

References

1. Faculty of Architecture and Design: MX lab. https://www.fad.stuba.sk/mx-lab.html. Accessed 01 June 2022
2. Stokholm, M.D.J.: Design compass. Improving interdisciplinary communication on design. Designmatters **10**, 54–57 (2005). https://www.vbn.aau.dk/ws/portalfiles/portal/3240147/Design_compass.pdf. Accessed 01 June 2022

3. Paredes Fuentes, D.: How we set out to help people bring their ideas to life. https://www.gravitysketch.com/blog/articles/gravity-sketch-genesis/. Accessed 01 June 2022
4. Doci, O., Olah, P., Truben, M.: Digitálne technológie zmiešanej reality v návrhovom procese dizajnu automobilu. In: Architecture Papers of the Faculty of Architecture and Design STU, vol. 25, No. 4, pp. 28–34 (2020). ISSN 2729-7640. https://alfa.stuba.sk/wp-content/uploads/2020/12/04_2020_Doci_Olah_Truben-1.pdf/. Accessed 18 Apr 2023
5. Hnilica, R.: Effective Urban Mobility of the Future. Thesis supervisor: Mgr. art. Michala Lipková, ArtD. Slovak University of Technology in Bratislava. Faculty of Architecture and Design. Registration number FAD-104282-82324 (2021)
6. Matušoviová, S.: Multisensoric experience in car interior. Thesis supervisor: Mgr. art. Michala Lipková, ArtD. Slovak University of Technology in Bratislava. Faculty of Architecture and Design. Registration number FAD-104282-93688 (2021)
7. Plachý, D.: Smart assistant for an autonomous car. Thesis supervisor: Richard Balogh, Slovak University of Technology in Bratislava. Faculty of Electrical Engineering and Information Technology. Registration number FEI-100852-104069 (2022)
8. Fabian,M.: Dashboard with natural user interface. Thesis supervisor: Richard Balogh, Slovak University of Technology in Bratislava. Faculty of Electrical Engineering and Information Technology (2023)
9. Tootell, H., Freeman M., Freeman, A.: Generation alpha at the intersection of technology, play and motivation. In: 2014 47th Hawaii International Conference on System Sciences. IEEE (2014)
10. Turlíková, Z., Pergerová, Z., Otiepková, S.: Methodological guidance for studio design: a way of encouraging the development of design thinking. ALFA-Arch. Pap. Fac. Arch. Des. STU **4**(25), 47–55 (2020). ISSN 2729-7640
11. Cuevas, H.M., Bolstad, C.A., Oberbreckling, R., LaVoie, N., Mitchell, D.K., Fielder, J., Foltz, P.W.: Benefits and challenges of multidisciplinary project teams: "Lessons Learned" for researchers and practitioners. ITEA (International Test and Evaluation Association) J. **33**(1), 58 (2012)
12. Karvinen, Kimmo: Teaching robot rapid prototyping for non-engineers-a minimalist approach. World Trans. Eng. Technol. Educ. **14**(3), 341–346 (2016)
13. Diegel, O., Xu, W.L., Potgieter, J.: A case study of rapid prototype as design in educational engineering projects. Int. J. Eng. Educ. **22**(2), 350 (2006)
14. Hamblen, J.O., van Bekkum, G.M.E.: An embedded systems laboratory to support rapid prototyping of robotics and the internet of things. IEEE Trans. Educ. **56**(1), 121–128 (2013). https://doi.org/10.1109/TE.2012.2227320
15. Balogh, R., Lipková, M., Lukani, V., Ťapajna, P.: Natural notification system for the interior of shared car. IFAC-PapersOnLine **52**(27), 175–179 (2019)
16. Stevens, P.: Using FreeRTOS multitasking in Arduino. https://create.arduino.cc/projecthub/feilipu/using-freertos-multi-tasking-in-arduino-ebc3cc. Accessed 01 June 2022

Educational Robots, Semiotics and Language Development

Dave Catlin and Stephanie Holmquist

Abstract When students interact with robots, they engage in semiotic processes. We ask: what are these, how do they work, and how do students learn from them? We begin with definitions linking pupils, robots and learning. Then we explain basic semiotic ideas introducing the Continental approach started by Ferdinand de Saussure and the American methods of Charles Sanders Peirce. By analysing six successful lessons from this new viewpoint, we find the value of robots beyond coding. We believe the semiotic approach could enrich the sample lessons, help teachers assess students' work and hint at ways to improve robot designs.

Keywords Educational robots · Semiotics · Robot interactions · ERA principles · Roamer

1 Introduction

The Educational Robotic Application Principles (ERA) provide a way to evaluate robots and their applications [1]. In this paper, we examine ERA's Interactive Principle in more depth, starting with its definition, which includes the idea of semiotics. Semiotics, the science of signs, explains how communication works. Our study reviews the communication between students, robots, the environment and how this helps children learn.

Semiotics embraces philosophy, linguistics, anthropology. many different intellectual disciplines. We present a few of its key theories, which we'll use to analyse robot lessons. We've chosen to study activities which took place between 1989 and 2022, in Early Years, Elementary and Middle Schools located in New Zealand, the UK and the USA.

D. Catlin (✉) · S. Holmquist
Holmquist Educational Consultants, Inc, PO Box 3564, Plant City, FL 33563-0010, USA
e-mail: dave@valiant-technology.com

S. Holmquist
e-mail: Stephie@hecedu.com

R. Balogh et al. (eds.), *Robotics in Education*, Lecture Notes in Networks and Systems 747, https://doi.org/10.1007/978-3-031-38454-7_10

Just about anything can act as a sign: the colour red: danger (or good luck if you're Chinese), a person wearing a hard hat (construction worker), bubbling water on a stove (it's hot). However, since language is a major human sign-system, we've chosen lessons aimed at helping students develop their language skills. Where available we've linked short videos of these projects in the citations, which you'll need to view. You can also see a video of the conference presentation, which shows more visual evidence of the essential concepts [2]. The lessons focus on learning English; however, we believe our precepts apply to all language teaching.

In January 1970, Seymour Papert invented the first educational robot; he called it a Turtle, and students controlled it using the LOGO programming language [3]. Papert didn't want to find better ways to teach; he wanted better ways for children to learn [4]. Based on the constructionist ideas of Jean Piaget, he viewed the robot as a tool which allowed students to express themselves and explore the world of ideas: video [5].

Catlin claims some new robots break with Papert's approach [6]. Many of these incorporate smart-robotics; they aim to produce robot teachers. For example: Bodkin quotes Sir Anthony Seldon, Vice Chancellor of the University of Buckingham, "Robots will begin replacing teachers in the classroom within the next ten years in a revolution in one-to-one learning [7]." So we have another objective; how can we add Artificial Intelligence (AI) to the robots and keep Papert's educentric approach? Our method adopts an evolutionary process. We take a successful lesson, review its learning potential and the semiotics involved. Then we ask, "How can AI improve the learning opportunity?" We don't expect this will lead to major advances; instead, we're looking for steady improvement based on good teaching practice.

The lessons used the Classic Roamer and the latest version of the robot: Roamer (R2). Teachers can configure R2's behaviour to support specific lessons and upgrade its firmware and hardware.

2 Semiotic Theory

The ERA Interactive Principle states: "*Students are active learners whose multimodal interactions with educational robots take place* via *a variety of appropriate semiotic systems.*" And Crystal defines semiotics as: "*The study of signs and their use, focusing on the mechanisms and patterns of human communication and on the nature and acquisition of knowledge* [8]".

We can only present a few basic semiotic concepts: enough to ground our proposition. We'll begin with the linguistic approach started by Ferdinand de Saussure (1857–1913), and then the classification methods of Charles Sanders Peirce[1] (1839–1914).

The University of Geneva asked linguist Ferdinand de Saussure to run a course on General Linguistics. His novel method impressed his students so much they published

[1] Peirce pronounced 'Purse'.

his lecture notes after his death [9]. The traditional linguistics approach studied how language changed with time: for example, Shakespearian compared with today's English. Instead, Saussure looked at the structure of language at a given time.

Saussure started with words (written or spoken), which he called a signifier. Saussure considered the signifier a mental model, what he called a Sound Image; Chandler claims the modern view thinks of it as the material part of the sign [10]. By agreement, English speakers take the word dog to mean a canine mammal. The connection between the animal and the word is arbitrary; so different languages have different words for dog: Germans use the word 'hund', and Swahili speakers say 'mbwa'. What happens when you read or hear 'dog' Saussure called the signified. Like coins, which have two inseparable parts (heads and tails), Saussure's sign consists of the conjoined signifier and signified.

Saussure named any verbal or written communication 'parole'; the words used and rules governing their use he called 'langue'. The langue rules cover how you combine words and give parole a structure. For example: "The cat sat on the mat," complies with the preferred subject, verb, and object (SVO) order of English sentences (the order varies with other languages). You can change words, but the structure remains the same. Saussure notes two sorts of change: one that doesn't change the sentence meaning, and one that does. For example: "The moggy sat on the rug" has the same meaning as the original sentence, and "The dog jumped on the sofa." doesn't. Yet, despite these substitutions, the SVO structure is the same. A word's value (meaning) comes from its relationship with other words in a sentence. For example, compare "I like taking my *dog* for a walk early in the morning", with "It was my turn to do the first *dog* watch as we sailed past Gibraltar". The meaning of the sign 'dog' changes because of the context defined by the other signs in the sentence.

Others expanded Saussure's ideas and applied them to more general communications. One of these, Danish linguist Louis Hjelmslev (1899–1965), split the signifier into two parts: its content and expressive nature [11]. The adage: "It's not what you say, it's how you say it", expresses how the speaker's tone of voice changes the meaning. He did the same to the signified: "Argentina beat France in the World Cup Final!" The expressive part of the message will affect Argentinian and French people differently. By introducing this notion of connotation, Hjelmslev showed how sign systems, outside the linguistic sign, influenced the meaning of a parole.

In a series of magazine articles, Frenchman Roland Barthes (1915–1980) examined the messages rooted in our culture, exposing what he called myths [12]. He reviewed subjects as diverse as striptease and soap powder adverts; he uncovered 5 codes:

- **Enigma Code**: a mystery—that makes you want to explore, to find out.
- **Action Code**: suggests something is going to happen.
- **Symbolic Codes**: an image that represents something.
- **Semantic Codes**: the meaning of a word, phrase or image.
- **Cultural Codes**: something that links to a cultural reference.

By adapting the methods of Saussure and Hjelmslev, Barthes provided a holistic way of studying communication; it's not just the text but the total environment that contributes to the establishment of meaning.

Structuralist's take Kant's idea that our minds play a role in organizing our experiences. By inspiring the philosophy of structuralism, Saussure influenced the anthropologist Claude Levi Strauss (1908–2008), to propose that despite their appearance, primitive[2] cultures came from the same mental patterns. His study of myths of various cultures showed while the stories differed at a surface level, they had the same structure. Vladimir Propp (1895–1970) analysed Russian folktales and found similar results. He identified 31 basic structural units (functions) used to outline the story's plot [13]. Propp also discovered seven characters: hero, villain, princess, helper, dispatcher, donor and false hero. Saussure mentioned binary opposites; Propp showed their importance in storytelling. You can't have a hero without a villain, good without evil, happiness without sadness... all essential parts of a fairytale. Resolving the tensions set up between opposites provide the story and its drama. Others extended Propp's ideas beyond Russian folktales when they create films, dramas, video games, television serials...

The pioneering French sociologist, Emile Durkheim (1858–1917) called totemism the most basic form of religion [14]. In totemism, a social group believes they have a spiritual relationship with an animal or a plant which reflect the characteristics they value. Levi Strauss rejected the idea that this was a sign of primitive thinking [15]. The word totemism comes from an American First Nation word referring to totem poles; rejecting Durkheim's religious idea, their descendants explain the totem pole documents important events and people in a tribe's history [16]. Of course, modern societies have their own totems, which we find in heraldry and symbols: England (lion), Scotland (unicorn), Wales and China (dragon), Russia (bear). Ignoring any religious connotations totems stand for a set of ideals, aspirations and values—a sort of community brand image.

So far, we've outlined Continental Semiotics; independently, Peirce's American Semiotics took a different approach: one based on logic [17]. His sign had 3 parts: signifier [Representamen],[3] signified [Interpretant] and object (what the 'sign' stands for). Figure 1 shows how the sign works for a red traffic light; notice how its interpretation depends on the viewer (car driver or pedestrian). Saussure focused on examples of parole, whereas Peirce looked at semiotics as a process [18]. In Peirce's method, the signified can become a signifier. For example, a picture of a beach may make you think of Florida, which starts you thinking of a holiday, and then triggers the thought—Disney World; this is a process known as 'unlimited semiosis'.

Peirce classified signs arriving at 59,049 types, which he thought were too many to remember! Fortunately, semioticians settle on 3 [19]:

---

[2] Levi Strauss emphasises we shouldn't think of the term primitive in a derogatory way **Invalid source specified.**

[3] Because the terminology used by Peirce is archaic, several researchers invented different versions, so on first use, we've done the same and bracketed [Peirce's] language **Invalid source specified.**

Fig. 1 Peirce's triadic sign

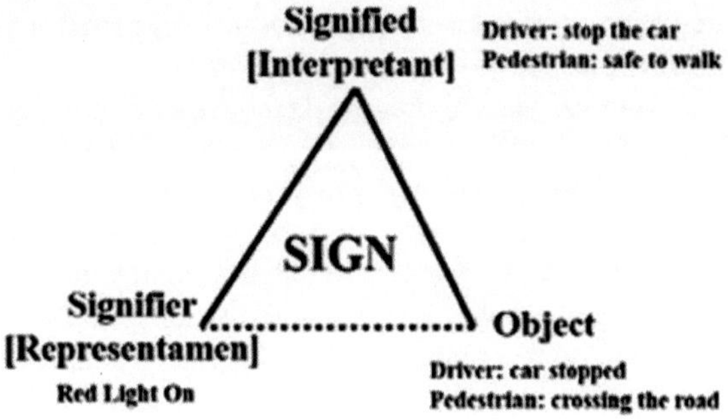

- **Symbol**: arbitrary signifier that people have to learn: language, road signs, numbers, traffic lights, Morse code, semaphore, national flags.
- **Icon**: a signifier that resembles or mimics the signified: a portrait, a cartoon, gestures, onomatopoeic words, sound effects, models, metaphor.
- **Index**: a signifier that's not arbitrary: a footprint, smoke, storm clouds, a clock, a thermometer, medical symptoms, a photograph, the smell of cooking, a baby crying.

3 Lesson Analysis

Between September 1987 and December 1998, the UK ran a national robot research project in 350 schools and 21 districts using Turtle robots programmed by Logo. Most teachers and students had never used a robot before. The following report extracts testify to the improvement of language skills accredited to the use of robots irrespective of the topic focus [20]: "*Many teachers report creative spin-offs associated with the turtle's presence in their classrooms. These might take the form of* ***writing****, drawing and* ***storytelling*** *that involved the turtle and its activities.*" (Durham).

"*Social and* ***language*** *skills develop through working cooperatively in small groups e.g.,* ***talking, thinking listening, discussing and negotiating***". (Northern Ireland).

3.1 Incy Wincy Spider

We'll start by looking at the internationally popular nursery rhyme, Incy Wincy Spider (known in the USA as Itsy-Bitsy Spider). Specialist Early Years teacher Chrissie Dale explains, "*Young children know the words to rhymes but don't think about their meaning.*" The lesson helps them develop their sequencing and comprehension skills. The robot sings and acts out the nursery rhyme in a muddled-up order: video [21]. The children have to teach the robot the correct order. The video shows the student's excitement; the teacher reported a girl suggested a way to solve the problem by showing pictures to the robot.

Semiotics: The children worked out the meaning of the keypad ***icons*** but needed the teacher's help to understand the ***symbol***, **CM** means clear memory. Developing the child's technological literacy requires them to become familiar with such commonly used ***symbols***. By making the robot look like a spider the teacher turned Roamer into an ***icon***. **AI Development**: the technology didn't let the girl try out her idea. A smarter robot would remove this limitation and allow students to invent new ways of 'talking' to their robots.

3.2 The Very Hungry Caterpillar

Report on an after-school project run by Hillsborough County Schools, Tampa, Florida. The idea of a school board member, supported by MacDonald's and YMCA volunteers, the project ran two 3-day workshops with 13–15 boys and girls: Grades 1 to 3. Working in groups of 2–3 students for 3 2 h sessions this STEAM project (Science, Technology, Engineering, Arts and Mathematics) aimed to improve their literacy, mathematics and science skills.

- **Day 1**: The school board member read Eric Carle's story, 'The Very Hungry Caterpillar' to the children [22]. The pupils learnt to program Roamer.
- **Day 2**: Students made their Roamers into caterpillars using maker space skills.
- **Day 3**: The children showed their robot caterpillars to the rest of the group and their parents. They then programmed Roamer to make the caterpillar's journey from one food to the next.

Semiotics: Pictures of the food made ***iconic*** signs. What are the characteristics of a caterpillar (***object***)? "*[the characteristics] serve as the basis upon which the sign can represent the* ***object*** [23]". So when the children make a robot caterpillar (an ***iconic*** sign) they need to work out what makes you think caterpillar. **AI Development**: If the robot knew it landed on the right food, it can pronounce "Yum, yum!"; if it missed, it can shout: "I'm hungry!" This sort of response would make the lesson more engaging (ERA Engagement Principle).

Old MacDonald Had a Farm

In this lesson, children practice their spelling of Constant-Vowel-Constant (CVC) words. The teacher sets up images of different animals and Old MacDonald's farmhouse. The students program their robots to go and feed the animals, but first, they must teach the robot the animal's name by spelling it (C-O-W). When the robot gets to the animal, it says its name (cow); the animal responds by making Roamer play the right sound (moo).

Semiotics: Again, both the robot and child need to recognise the ***iconic*** image of the animals. They need to learn the ***symbolic*** text code (C-O-W), and the animal's

indexical sound effect (moo); these are the 'multi-modal' part of ERA's Interactive principle. **AI Development**: You can change the way Roamer behaves to support specific lessons; you do this by reconfiguring its keypad. AI needs to enable the robot to know its location, check the student's answer and play the right sound effects.

Once Upon a Time

Usually, robot lessons cover a main topic plus one or two other subjects. The ERA Pedagogical Principle identifies 29 characteristics of a robot lesson (challenges, exploration, puzzles and so on) [24]. You can tag Once Upon a Time with 'conceptualising' (students develop a mental model of databases), 'catalyst' (collecting information for a creative writing task) and 'link' (reveals the structure of fairy tales).

The grid (Fig. 2) represents a database; the columns (fields) contain Propp's '***functions and characters***', and each row (records) the details of a fairy tale (not visible to the pupils). The students randomly choose database locations, ensuring they gather information from each row and each column. They then program the robot to collect the data, which they use to write a new story for the robot to act out.

Semiotics: The task will expose students to two semiotic structures: ***storylines***, and perhaps more interestingly—***databases***. We could add 'Experience' to the characteristic tags; we're not trying to teach Propp's ideas: we learn most semiotics unconsciously through cultural exposure: video at 0:54 [5]. **AI Development: now the robot picks up information cards from each location; a smarter robot would automatically collect** this data (***enigma codes***) which it reveals when it reports back to students—making the lesson more exciting.

	Hero	Villain	Prop	Place	Action
Cinderella	Cinderella	Stepmother	Glass Slipper	Kitchen	Marriage
Jack and the Beanstalk	Jack	Giant	Magic Beans	Castle	Escaping
Little Red Riding Hood	Red Riding Hood	Wolf	Food Basket	Cottage	Eating
Snow White	Snow White	Evil Queen	Poison Apple	Forest	Kiss
Aladdin	Aladdin	Evil Magician	Magic Lamp	Cave	Playing a Trick

Fig. 2 Once upon a time database; the grid mat will show row and column numbers. Roamer collects data when stopping on a cell

3.3 Cultural Stories

You feel comfortable if you're talking to someone and they're nodding; it's a sign of agreement—or is it? In some countries, like India and Bulgaria, shaking the head signals agreement: it's unsettling when you first meet it. Nobody specifically taught us to nod or shake—it's the sort of sign you pick up from your culture. Many ancient indigenous peoples feel their culture threatened by modernity's cultural bombardment. **New Zealand**: In the nineties, a Massey University initiative led to Maoris adopting Roamer. After the approval of the Maori Council, they translated the instructions into te reo (Māori language) and created interactive lessons based on several ancient Maori myths. For example, the students created Maori masks for the robots, (complete with their traditional tattoos), programmed the robots to enact the legend of Hinemoa and Tutaneki; they narrated the story while the robots performed [25].

USA: In the 70s, following decades of suppression, American First Nations started reasserting their traditions. Tribes like the Squaxin, based in Puget Sound, Seattle, revived banned traditions like the Potlatch (summer gatherings—festivals of great social and economic importance). They also restarted the great seagoing canoe journeys where the tribes would travel along the North American Pacific coast. Children couldn't join in some of these events, so Sally Brownfield, Squaxin Education Director, set up a summer camp where children could play with the robots. They had challenges like simulating the canoe journey, programming the robot to act out the Squaxin myths and do traditional dances: Video [26].

Sally commented, "*Tribal youth experienced a deeper understanding of their language.*" Initially the Tribal Elders saw the project as curriculum; this transformed, they started to see it as a tool of culture. Sally explained the change began when children asked about traditional dances, blanket-weaving, and the indigenous mathematics and science used to navigate the canoes.

Semiotics: Professor Stuart Hall defines culture as a group of people who share meanings and values [27]. You become a community member through passive participation, but full membership needs an active contribution. The Maori and Squaxin children took the ***myths***, ***symbols***, ***icons*** and ***cultural codes*** of their heritage and re-expressed them through the robot—a modern medium.

We should note that hidden in many of the myths of native cultures lay ancient knowledge and wisdoms acutely relevant to modern life. For example, in the First Nation story, 'The Coyote Stole Fire', you'll find the ***binary opposites***: life with and without energy—a present-day issue worthy of debate, even with young children.

3.4 Robot Movies

In 1989 the children of Southmead Primary school in London made a movie about a circus coming to town using robots instead of puppets. They wrote the script, designed robot characters, created the scenery, programmed the robots and performed the voice-overs. A professional film crew did the filming and editing; many schools can now take over these tasks. The following examples show how projects provide joined-up learning opportunities, helping students see the symbiotic value of subject knowledge. It also showed how careful lesson planning by the teachers played a vital role in achieving this.

Star Wars: Maple Cross Primary school, on the outskirts of Northwest London, within 15 km of 5 film studios, all of whom had made Star Wars movies. Teacher Nick Flint used this local connection to cover the Design Technology (maker space), computing and English Curriculum. He chose Star Wars as a theme, and the students designed and built a 'film set' of Tatooine (the town with the famous scene where the bartender ejects the robots R2-D2 and C3-PO). The children wrote different scenarios (except for a group of girls who decided to write a scene from a soap opera called Eastenders). Later Flint asked the children to write about their experiences in the school blog: video [28].

Peace Pledge: Oxon Hill Middle School Maryland is situated in Prince George's County school district on the fringes of Washington DC. The school district focused on the arts and STEAM. After hearing about the Robotic Performing Arts Project [29], they realized it fitted perfectly into their programme and they embraced the idea of making a robot movie: video [30]. They chose their school motto, 'The Peace Pledge', as a topic: video [31]. Every year the Science-Engineering-Technology Working Group (SETWG) hosts STEM on the Hill in the US Capitol. This gives US Congressmen and staffers a chance to see what STEM projects go on in US schools. They invited Oxon Hill to show their movie: video [32].

Semiotics: The Maple Cross video features a ***binary opposite***: good versus evil and involved the students in discussions and negotiations. The Oxon Hill movie showed some semiotic and creative ideas used by professional film directors: like the ***symbolic*** soaring eagle (America's ***totem***, representing freedom and strength). The school calls their students kings and queens, which they use ***symbolically*** to stand for their ambitions. We also see the ***binary opposites***; the danger imposed by the dragon and the fight to overcome its threat, an ***iconic metaphor*** for the effort they need to put into keeping the pledge. Finally, the video acts out an award ceremony (***indexical sign***) pointing out success and the foundation of the student's future. **AI Development**: There's scope for students to add intelligent objects into the environment, which will provide special effects and interact with the robot.

4 Conclusions

We wanted to examine educational robots and their use from the perspective of semiotics; would this yield new and valuable insights into the design of the robots and their use? We also wanted to ensure we kept to Papert's ideas: children learn by doing and through constructionism. We stress we're not trying to teach children semiotics, although we believe that may take place subconsciously; that is, students will experience how signs work and may have the opportunity to invent them. The robots need to recognise signs, including those made by the students. Many of these will use low-tech art and craft methods typically found in elementary schools, which the robot will need to cope with.

We found 3 types of interactions in the lessons: child-robot, child-environment and robot-environment. When we combine these with the different kinds of signs, we discover multiple learning opportunities. We also realise **a *signifier*** only becomes a ***sign*** when it's ***signified***, that is, only when an interaction takes place, does it unleash the signs learning potential.

Semiotics made us think more about the whole lesson, not just the parts using the robot. We can imagine an extreme case where students build an environment for the Roamer (set up with an exploration behaviour) to explore without programming the robot. In such a task, the students make a tangible computing habitat by considering what signs the robot will interact with. The habitat could include intelligent gadgets like Arduino, which for example, could operate a set of traffic lights to control a 'driverless' robot car. In exercises like this, Barthes' codes will help the pupils' thought processes; for example, how does an ***action code*** change what the robot does? When developing lessons, we need to consider each possible interaction, for example: do the students know the ***symbols***, or do we need to teach them; what ***signs*** will the pupils create; which will they manipulate; what may they learn from the interaction; what will they learn from the design and making of the signs?

In the Incy Wincy Spider task, pupils put a disordered 'story' in order. You'll notice a student programs while the others watch. In the Once Upon a Time lesson, students write a story, which they need to deconstruct into parts the robot can enact. How do they depict each scenario, and what signs do they need to make? In projects like this, the students form a production team: some program the robot, others write dialogue, design robot characters and create scenery. In doing this, they must cooperate and agree on what signs they need and what they mean.

The ERA Equity Principle states robots don't have gender, race or culture; their neutrality allows students to express themselves. The Squaxin and Maori examples show how robots can support different cultures. Applying sign theory can help relieve tensions in multicultural classrooms, like those affected by indigenous-immigrant strife. Carefully planned lessons using a mixture of ***cultural codes*** may help bridge differences, promoting understanding and tolerance.

When we began this project, we didn't suspect semiotics could help with Assessment (ERA Principle). We don't propose this as an extra teacher task, but that sign theory offers a set of heuristic tools that help uncover what learning took place. When we used it to assess the Oxon Hill school's movie, we got a deeper appreciation of the quality of their work.

References

1. Catlin, D., Blamires, M.: The principles of educational robotic applications (ERA): a framework for understanding and developing educational robots and their activities. In: Constructionism 2010, Paris (2010)
2. Catlin, D., Holmquist, S.: Educational Robots, Semiotics and Language Development, Limassol (2023). [Online] Available: https://youtu.be/-GZY0Ep7sMU [Accessed 22 September 2018]
3. Paterson, M.: LOGO Ambulatory Executor—Turtle Robot, 3 August 1969. https://www.goo.gl/VK2DMv. Accessed 1 May 2018
4. Papert, S.: The Children's Machine, pp. 82–105. Basic Books, New York (1993)
5. Papert, S.: Talking Turtle, Part 2 ed.. BBC and the Open University (1983)
6. Catlin, D.: Beyond coding: back to the future with education robots. In: Daniela, L. (Ed.) Smart Learning with Educational Robotics: Using Robots to Scaffold Learning Outcomes, pp. 1–44. Springer Nature Switzerland AG, Cham (2019)
7. Bodkin, H.: Inspirational robots to begin replacing teachers within 10 years. Telegraph Online (2017)
8. Crystal, D.: The Penguin Dictionary of Language, 2nd edn. Penguin (1999)
9. Saussure, F.D.: Course in General Linguistics. Duckworth (1916/1983)
10. Chandler, D.: Semiotics: The Basics, p. 15. Routledge, New York (2007)
11. Hjelmslev, L.: Prolegomena to a Theory of Language. University of Wisconsin Press, Madison (1961)
12. Barthes, R.: Mythologies: Complete Edition. Hill & Wang (2012)
13. Propp, V.: Morphology of the Folktale. University of Texas Press (1968)
14. Durkheim, E.: The Elementary Forms of the Religious Life. Dover Publications (2012)
15. Levi-Strauss, C.: Totemism. Beacon Press (2016)
16. Indigenous Corporate Training: Debunking Misconception about First Nation Totem Poles. https://www.shorturl.at/DIPQ9
17. Atkin: Peirce's Theory of Signs. Fall (2022). https://plato.stanford.edu/archives/fall2022/entries/peirce-semiotics/. Accessed 30 Jan 2022
18. Chandler, D.: Semiotics for Beginners: Chapter 2: Signs. http://visual-memory.co.uk/daniel/Documents/S4B/semiotic.html. Accessed 30 Jan 2022
19. Crow, D.: Visible Signs, p. 33. AVA Publishing (2003)
20. Mills, R., Staines, J., Tabbeerer, R.: Turtling Without Tears. National Council for Educational Technology (1989)
21. Dale, C.: Early Years Incy Wincy (2017)
22. Carle, E.: The Very Hungry Caterpillar. Puffin (1994)
23. Liszka, J.J.: A General Introduction to the Semiotic of Charles Sanders Peirce, pp. 20–21. Indiana University Press, Bloomington and Indianapolis (1996)
24. Catlin, D.: 29 Effective Ways You Can Use Robots in the Classroom: An Explanation of ERA Pedagogical Principle. In: Edurobotics 2016, Athens (2016)
25. Catlin, D., Smith, J.L., Morrison, .: Using educational robots as tools of cultural expression: a report on projects with indigenous communities. In: 3rd International Conference Robots in Education. Prague (2012)
26. Catlin, D., Coode, A.: Squaxin Roamer Cultural Project (2012)
27. Hall, S. (Ed.): Representation: Cultural Representations and Signifying Practices. Sage Publications in association with the Open University (1997)
28. Flint, N.: Maple Crross: A Star Wars Adventure (2013)
29. Catlin, D.: Robotic Performing Arts Project. In: Constructionism 2010, Paris (2010)
30. Williams, K.: Teacher Kenneth Williams Interview (2017)
31. Oxon Hill Teachers. Oxon Hill Peace Pledge 480 (2017)
32. Oxon Hill Students. Oxon Hill Middle School Pledge (2017)

Educational Robotics and Complex Thinking: Instructors Views' on Using Humanoid Robots in Higher Education

María Soledad Ramírez-Montoya, Jose Jaime Baena-Rojas, and Azeneth Patiño

Abstract This paper presents the results of a study that aimed to analyze the perspectives of higher education instructors regarding humanoid robots and the use of educational robotics in higher education for STEM fields. The study employed survey method of data collection to examine the views of 192 instructors on the benefits and challenges of using humanoid robots in higher education, as well as their perceptions of the current state and future trends of educational robotics. Results of the survey reveal instructors' positive attitudes towards the use of humanoid robots in higher education, highlighting the perceived potential for these robots to enhance student engagement, motivation, and learning outcomes. However, the study also identifies challenges such as the high cost of acquisition and maintenance of humanoid robots, as well as the need for ongoing professional development for instructors to effectively integrate these robots into their teaching practices. Overall, this research provides valuable insights into the current state of the field and the future direction of educational robotics in higher education for STEM fields.

Keywords Educational robotics · Higher education · Complex thinking · Humanoid robots · Educational innovation

---

M. S. Ramírez-Montoya · A. Patiño (✉)
Institute for the Future of Education, Tecnologico de Monterrey, 64849 Monterrey, Mexico
e-mail: azeneth.patino@tec.mx

M. S. Ramírez-Montoya
e-mail: solramirez@tec.mx

J. J. Baena-Rojas
Fundación Universitaria CEIPA, 055450 Sabaneta, Colombia

EAE Business School, 08015 Barcelona, Spain

J. J. Baena-Rojas
e-mail: jose.baena@ceipa.edu.co; jbaenarojas@campus.eae.es

R. Balogh et al. (eds.), *Robotics in Education*, Lecture Notes in Networks and Systems 747, https://doi.org/10.1007/978-3-031-38454-7_11

1 Introduction

1.1 Educational Robotics and Humanoid Robots

The use of robots in education has been found to improve student learning. This is why, it has been argued that humanoid robots have the potential to transform the way we approach education by providing new and engaging opportunities for learning [1]. Humanoid robots can serve as interactive teaching aids, providing personalized instruction and feedback to students while also allowing them to experience real world applications of science, technology, engineering, and mathematics (STEM) concepts [2]. Furthermore, the use of humanoid robots in education can help to foster creativity, as students are able to design, program, and control their own robots to perform various tasks. The ability to interact with and control humanoid robots can also help to foster a lifelong interest in STEM fields, inspiring students to pursue careers in these areas and contribute to technological innovation. Overall, the integration of humanoid robots in education holds tremendous potential for enhancing learning experiences and outcomes for students of all ages [3].

In this way, robots in higher education have been clearly introduced as a solution to the lack of motivation and enjoyment faced by students. In the same way, this pedagogical strategy for adopting new technological tools is just a response within current education which may become the social robotics in an essential element of the classroom and of the learning process for future education. This is because robots can contribute to the development of complex thinking [4], and various other cognitive skills related to cutting-edge technologies [5, 6].

Hence, the current paper shows different types of benefits which were perceived by a relevant sample with more than 190 professors of higher education after taking a class based on social robotics. The study was applied in Latin America at the most important university of Bolivia. Instructors from all the fields of knowledge shared their perceptions about this disruptive tool after taking part in a scientific research class mediated with robots. This paper aims to validate from an unusual experience, where certain teachers participated as students in class not only learning a new topic but also, evaluating the role of social robotics and more specifically, humanoid robots in the classroom.

1.2 Instructors' Views on the Use of Educational and Social Robotics

Instructors' views play a crucial role in the adoption and implementation of educational robotics in the classroom. With the increasing interest in the potential benefits of social robots in education, it is important to understand the perspectives of instructors since they are key decision-makers in the integration of these technologies into the learning environment [7]. This section presents the results of some studies that have explored teacher and instructor views on the use of educational robotics in the classroom.

The use of robots in the classroom as a technological resource can become an important tool to facilitate teaching and learning processes. This is why, as time goes by, initiatives to integrate robots in the classroom become more common. This is especially true when evidence shows that even this type of technology can be used with students with learning difficulties [8]. Hence the current need to prepare future teachers to design inclusive digital education. All this, in order to improve not only their digital competencies, but above all to foster their creative ability to design courses using technologies from an inclusive perspective. This situation will allow the introduction of educational and social robotics in their future curricular teaching. This will also make it possible for teachers and their teams (universities and IT personnel) to face all the challenges associated with integrating this type of technology in the classroom. All these, like many others in the past, were at the beginning, at the time of their implementation, unusual and complex, and now they are normal and dynamic for the classrooms. Therefore, it can be said that the interaction of the participants and the continuous feedback from the trainers constitute a source of experiences that will enhance life in the classroom favoring the quality of learning even in higher education [9].

Interest in using robots in various educational contexts is growing; not only because of the needs of current education systems, but also because they are often compatible with a vast number of cutting-edge technologies. In this way, all this represents a great potential to improve the educational experience for both face-to-face and distance learners, as well as to serve as didactic support for teachers. For all this, the literature on social robotics in the classroom tends to grow considering the new practices that are supported by the results of several studies. Most of these studies indicate that university staff and students perceive that robots are useful in the classroom and that this situation is supported by statistical results that show the great potential of these technological resources to improve education [10].

Although in any case, it cannot be forgotten that any technological resource always has limitations around factors as basic as perceived usefulness, perceived ease of use and enjoyment of use. In the first case, it is about the degree to which a person believes that the use of a particular system will improve him or his performance at work. In the second case, it is the degree to which a person believes that using a particular system will relieve him of the effort. In the third and last case, it is about the degree to which a person finds an activity pleasurable when he uses technology. All of the above means that in the event that the technology to be implemented does not bring benefits regarding the previous factors, its adoption will tend to present problems among its potential users [11–13].

Robotics is expected to continue its boom in the coming years, massively surpassing traditional fields of application and spreading to all parts of the planet. Thus, to enable effective international customization of robot designs and facilitate their harmonious introduction into everyday life and educational settings, it is important to study the opinions and attitudes towards robots in different regions of the world. All this because the learning processes are also subject to the cultural imaginary and the idiosyncrasy of each society. Precisely the literature reveals that there are more and more works that reveal the interesting level of acceptance of social robotics in education, especially in the West. However, despite the presumptions that may exist compared to other non-Western cultures where religion, gender, age, education, and worldview may affect the level of acceptance. In any case, the results of studies in these cultures, which include the Middle East, reveal interesting perspectives on the positive cultural acceptance of robots in these other regions leaving aside any prejudice and showing that "Educational and Social Robotics" is practically applicable in any cultural context [14].

In short, the current evidence as well as the new contributions that arise derived from a greater interest of the educational community regarding the implementation of social robotics in class and the advances themselves in this type of technology. They represent a clear and promising scenario where the teacher's work can be improved, which may end up resulting in a higher quality of education; this despite the challenges that imply at a technical and infrastructure level for teachers and universities to implement these resources and tools for classes. Therefore, further research is recommended to understand the contributions of robots to learning in all educational settings and especially in higher education. This is because its acceptance is subject to the impact of factors such as personalization, adaptation, and appearance to achieve true effectiveness for learning in class [15].

All of the above, especially within the framework of complex thinking that, as a construct, seeks to develop higher order skills that allow it to provide high-level solutions to individuals that is involved in any process of formation. In other words, from an educational perspective, different levels of learning are reached that are evidenced in the achievement of comprehensive skills in the student. Competences or abilities that allow them to adapt and make better decisions as a future professional [16, 17].

2 Materials and Method

2.1 Research Design

The survey method implemented in our study was based on Fowler [18] and Fricker Fricker [19] for collecting data on the views of instructors on the use of educational robotics in higher education. A questionnaire was administered to a sample of 192 instructors from different faculties and fields of knowledge. In this way, all of them are nowadays working in a Latin American university. In this case, the biggest public higher education institution, the *Universidad Autónoma Gabriel René Moreno* (UAGRM) located in Bolivia which count with 100 thousand students approximatively.

The survey included questions on the perceived benefits and challenges of using robotics in higher education, the level of comfort and expertise of instructors with the technology, and their views on its impact on student learning outcomes.

It is important to highlight that the design and structure of the survey has focused on analyzing various technological resources used in a pilot class. Thus, in this specific case, the present work or scientific research considered only the third part of the previous questionnaire, which analyzed in depth the possible effects of applying social robotics in higher education classes. The above, according to the perceptions of instructors who act, at the same time, both as students and as professor.

The "Sect. 3—improvement of classes with technological resources and humanoid robots" is comprised of twelve (12) closed questions which have been formulated from a complete theoretical contextualization on the central theme of this research. Similarly, the main objective of this segment of the survey is:

Objective 3: Observe the perceptions of students regarding the inclusion of social robotics to promote teaching within classrooms and academic environments.

Therefore, all collected data was analyzed with descriptive statistics techniques to provide insights into instructors' perspectives on the implementation and integration of educational robotics in higher education (see Fig. 1).

3 Results

3.1 *Attitudes Towards the Use of Humanoid Robots in Higher Education*

The results indicate that most of the participating teachers had a receptive attitude towards the use of humanoid robots in higher education classrooms. This finding is significant since it suggests that the integration of social robotics in education has the potential to be well received from different environments that seek to characterize a class mediated by these technological resources.

Thus, Fig. 2 reveals that 16.15% of all respondents give a maximum score of 5 with respect to how social robotics grants "(a) rigor in the classroom" from this setting. Then, 31.77% give a score of 4 and finally 41.15% give a score of 3. In the case of "(b) classroom entertainment" 25.04% give the maximum rating of 5, 36.98% give a rating of 4 while 29.17% give a rating of 3. For the setting "(c) ease of recall of concepts" 32.81% give a 5, 35.42% a 4 and 19.27% of the other respondents who approve, give a rating of 3. In "(d) possibilities to enhance learning" also 32.81% value with 5, then 39.06% value with 4 and the remaining 18.75% of those who approve the use of robots value this tool with a rating of 3. In the analyzed setting "(e) innovation in teaching processes" 47.92%, almost half of the respondents, score with 5, then 35.42% score with 4 and 9.38% score with 3. In the last case, "(f) information transmission process" 32.81% scored 5 for this item, then 38.02% scored 4 and 18.23% scored 3.

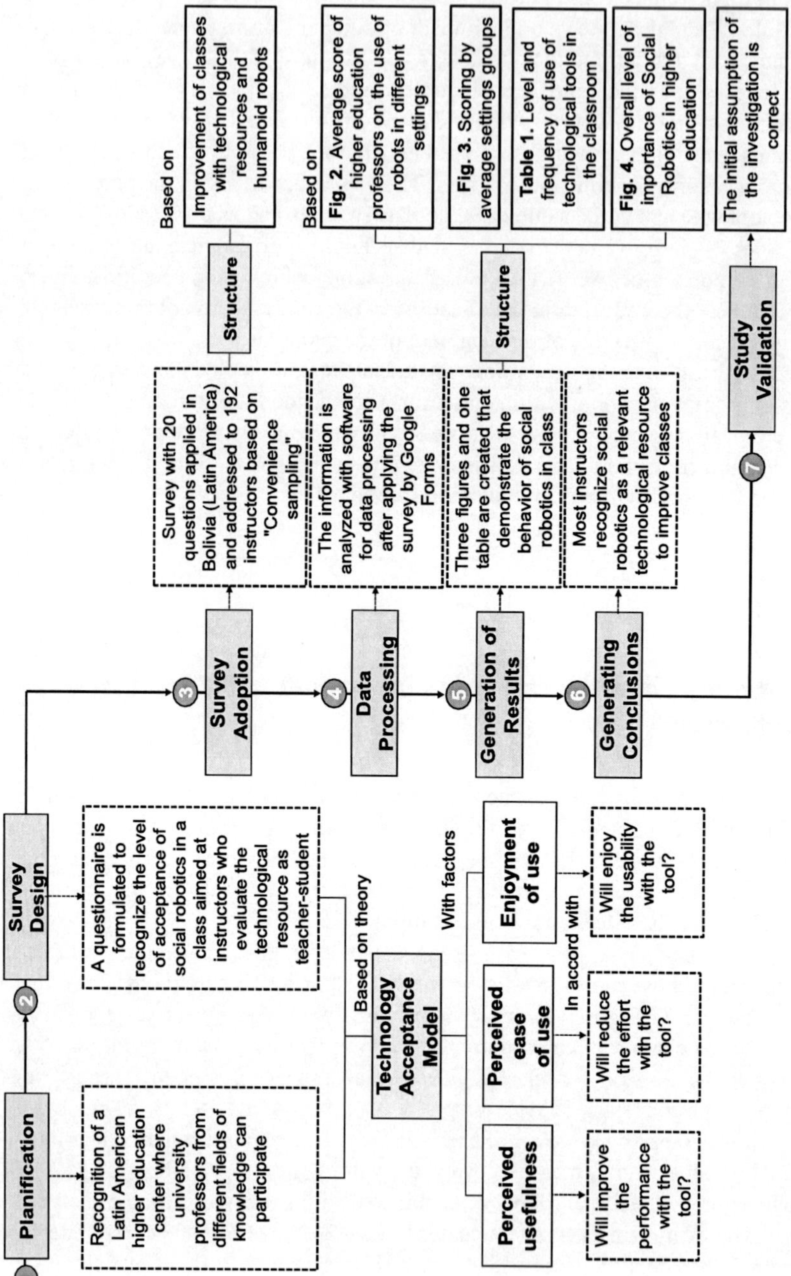

Fig. 1 Representation of the methodology and stages of the entire research proposal

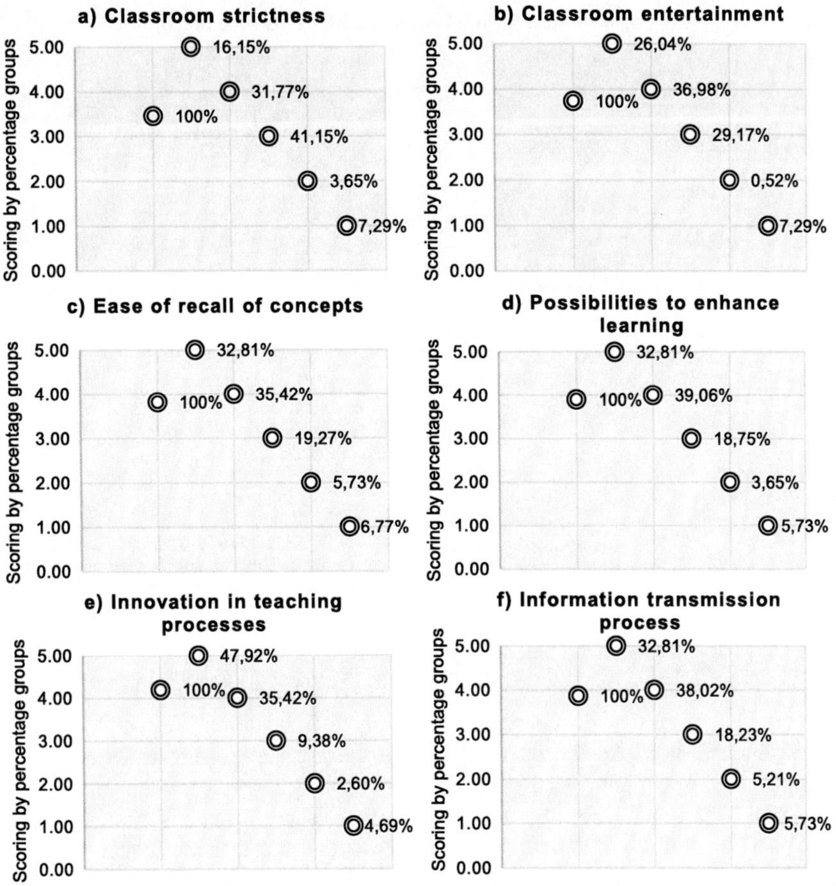

Fig. 2 Average score of higher education professors on the use of robots in different settings

On the other hand, if the average score obtained for each setting is analyzed, it is possible to point out that in all cases the score is high, as shown below.

In this way, Fig. 3 reveals that all the analyzed elements that can enhance teaching are located above 3.4, as can be seen in this part of the results. Thus, the only element that is located above 4 is "(e) innovation in teaching processes" according to the surveyed instructors.

Subsequently, it is important to recognize the level of frequency and use of humanoid robots as a technological resource. This is how, in the following table, these aspects can be recognized from the percentage distribution of respondents according to their field of knowledge.

According to Table 1, it can be said that the majority of respondents come from their professional training, firstly, from the field of Social Sciences with 43.23%, then in second place, from the field of Applied Sciences with 39.58%, in third place,

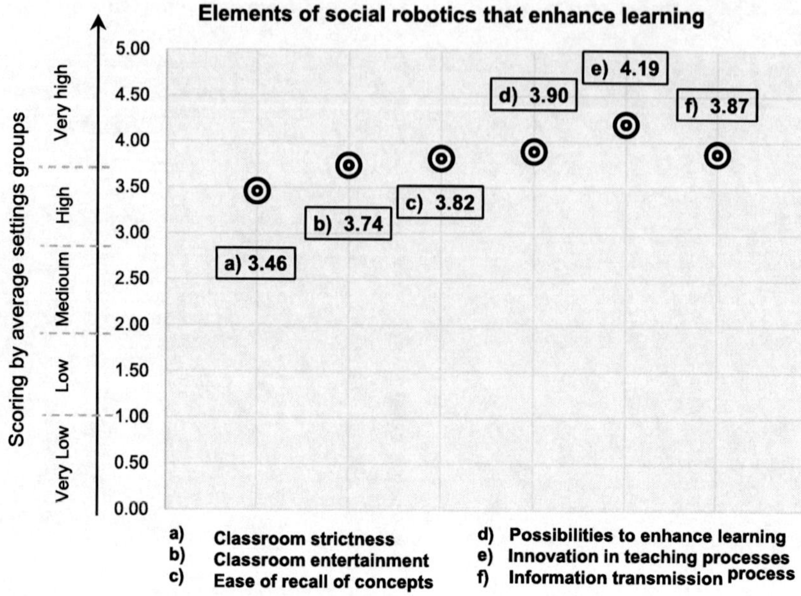

Fig. 3 Scoring by average settings groups

from the field of Basic or Natural Sciences with 13.02% and in fifth and last place in the field of Formal Sciences with 4.17% of the 192 surveyed instructors. Then, among all the respondents and their respective fields of knowledge, all consider, with more than 73% in all cases, that the level of importance of social robotics is very high. Also, in all fields of knowledge they consider, with at least 24%, that the level of importance of this technological resource is high. In fact, in the field of knowledge of Formal Sciences the level of importance is superlative because in this case, all the 8 surveyed instructors value the level of importance of social robotics as very high. Likewise, at the level of frequency of use, each field of study and each level of importance reveals different values. All of this, as can be seen in the last column of Table 1, where many respondents state that they have never used social robotics before, even though they value the tool very highly.

Finally, after all this perception exercise with the surveyed instructors, it can be defined according to their responses whether social robotics can be considered relevant or not, as shown in the following figure.

In summary, according to Fig. 4, after processing the answers of all the respondents, slightly more than three quarters of the Latin American instructors participating in our study consider that humanoid robots or social robotics are, at least, important to make their classes engaging and to develop digital literacy skills. As shown in the results, within the framework of technological tools based on STEM, social robotics can enhance learning in higher education.

Table 1 Level and frequency of use of technological tools in the classroom

Social robotics for the improvement of learning processes								
Percentage and total respondents in the study	Field of knowledge*	Level of importance	Frequency of use**					
			Percentage	Always (%)	Often (%)	Sometimes (%)	Seldom (%)	Never (%)
43,23% 83 respondents	1. Social sciences	Very high	75.90	14.29	17.46	20.63	11.11	36.51
		High	24.10	0.00	20.00	15.00	10.00	55.00
		Medium	0.00	–	–	–	–	–
		Low	0.00	–	–	–	–	–
39.58% 76 respondents	2. Applied sciences	Very high	73.68	5.36	19.64	8.93	12.50	53.57
		High	25.00	5.26	5.26	26.32	5.26	57.89
		Medium	1.32	0.00	0.00	0.00	0.00	100.00
		Low	0.00	–	–	–	–	–
13.02% 25 respondents	4. Basic or natural sciences	Very high	76.00	0.00	10.53	10.53	10.53	68.42
		High	24.00	0.00	16.67	16.67	0.00	66.67
		Medium	0.00	–	–	–	–	–
		Low	0.00	–	–	–	–	–
4.17% 8 respondents	5. Formal sciences	Very high	100	0.00	25.00	50.00	12.50	12.50
		High	0.00	–	–	–	–	–
		Medium	0.00	–	–	–	–	–
		Low	0.00	–	–	–	–	–

*The five fields of knowledge [20] include various professional studies of the respondents grouped in the present categories. Applied Sciences includes architecture, nursing, agricultural engineering, civil engineering, commercial engineering, food engineering, electrical engineering, food engineering, petroleum engineering, forestry engineering, pharmaceutical engineering, industrial engineering, industrial engineering, computer engineering, mechanical engineering, chemical engineering, medicine, and veterinary medicine. "Basic or Natural Sciences", composed of biology. "Social Sciences", consisting of business administration, arts, political science, journalism, accounting, law, economics, education, psychology, and sociology. Finally, "5. Formal Sciences" integrated by computer science

**The frequency of use indicates the percentage for each level of importance for each field of knowledge. It is presented horizontally

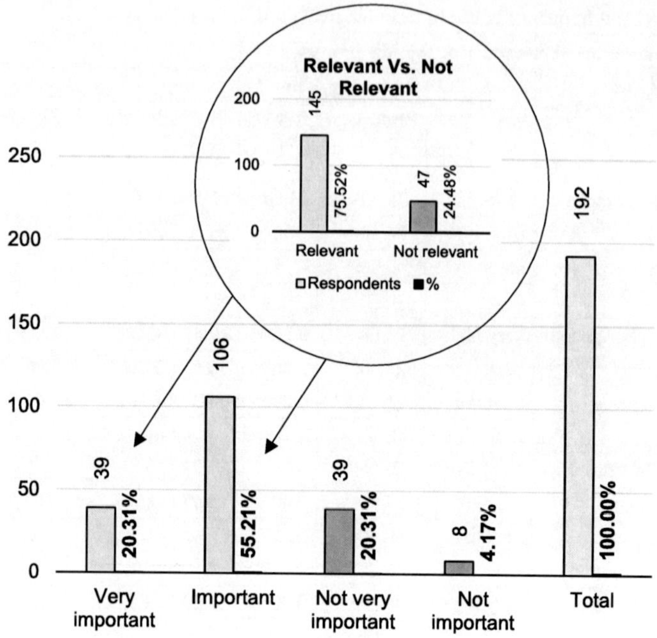

Fig. 4 Overall level of importance of social robotics in higher education

4 Conclusion

In conclusion, this research study provides valuable insights into the perspectives of higher education instructors regarding the use of humanoid robots in higher education in the Latin American region. The results of the survey reveal that instructors have positive attitudes towards the use of humanoid robots, highlighting the potential for these robots to enhance student engagement, motivation, and learning outcomes. This is related to the findings of previous studies [1] as it suggests that the integration of humanoid robots in education can positively impact the students' learning experience. The use of humanoid robots in the classroom has the potential to engage students in unique ways, promoting active learning and encouraging students to take ownership of their education. This, in turn, can increase student motivation, making them more invested in the learning process. The study also identifies challenges such as the high cost of acquisition and maintenance of humanoid robots, as well as the need for ongoing professional development for instructors to effectively integrate these robots into their teaching practices. The high cost of acquisition and maintenance can be a significant barrier for public higher education institutions, while the lack of professional development can result in instructors struggling to effectively integrate the robots into their teaching practices. These challenges must be addressed to ensure the successful implementation and integration of humanoid robots in education.

As is often the case in scientific research, our study is not without limitations. One potential limitation is related to the fact that our instruments captured the perceptions of instructors at a single point in time. Future studies should document the perception of instructors at different points in time when integrating humanoid robots in the classroom. Another limitation could be the fact we analyzed self-reported data, which may not reflect the actual behaviors or experiences of instructors. Further studies should examine the actual use of humanoid robots in classroom interventions in terms of learning outcomes and student engagement. Also, researchers should analyze the strategies deployed by instructors in the Latin American region when integrating humanoid robots in the classroom. However, these findings provide valuable information for educators, administrators, and policy makers to consider as they strive to advance the field of educational robotics in higher education. The results of this study contribute to the ongoing conversation about the role of humanoid robots in education and suggest a promising future for the use of these robots in enhancing student learning and engagement in STEM fields.

Acknowledgements The authors would like to acknowledge the financial and the technical support of the Writing Lab, Institute for the Future of Education, Tecnologico de Monterrey, Mexico, in the production of this work. Also, the authors would like to thank the financial support from Tecnologico de Monterrey through the "Challenge-Based Research Funding Program 2022". Project ID # I003-IFE001-C2-T3-T.

References

1. Chou, H.S., Thong, L.T., Chew, H.S.J., Lau, Y.: Barriers and facilitators of robot-assisted education in higher education: a systematic mixed-studies review. Technol. Knowl. Learn. **1**(1), 1–40 (2023)
2. Guggemos, J., Seufert, S., Sonderegger, S.: Humanoid robots in higher education: evaluating the acceptance of Pepper in the context of an academic writing course using the UTAUT. Br. J. Edu. Technol. **51**(5), 1864–1883 (2020)
3. Bolea-Monte, Y., Grau-Saldes, A., Sanfeliu-Cortés, A.: From research to teaching: integrating social robotics in engineering degrees. Int. J. Comput. Electr. Autom. Control Inf. Eng. **10**(6), 1020–1023 (2016)
4. Noh, J., Lee, J.: Effects of robotics programming on the computational thinking and creativity of elementary school students. Educ. Tech. Res. Dev. **68**(1), 463–484 (2020)
5. Ioannou, Makridou, E.: Exploring the potentials of educational robotics in the development of computational thinking: a summary of current research and practical proposal for future work. Educ. Inf. Technol. **23**(1), 2531–2544 (2018)
6. Daniela, L.: Smart Learning with Educational Robotics: Using Robots to Scaffold Learning Outcomes, vol. 1, pp. 1–41. Springer International Publishing, Berlin (2019)
7. Istenic, I.B., Rosanda, V.: Are pre-service teachers disinclined to utilise embodied humanoid social robots in the classroom? Br. J. Educ. Technol. **52**(6), 2340–2358 (2021)
8. Alcorn, A.M., Ainger, E., Charisi, V., Mantinioti, S., Petrovic, C., Schadenberg, B.R., Tavassoli, T., Pellicano, E.: Alcorn: educators' views on using humanoid robots with autistic learners in special education settings in England. Front. Robot. AI **6**(107), 1–15 (2019)
9. Gratani, F., Giannandrea, L., Renieri, A.: Educational robotics for inclusive design. In: 27th ATEE Spring Conference on Social Justice, Media and Technology, ATEE 2021, Florencia (2021)

10. Leoste, J., Virkus, S., Talisainen, A., Tammemäe, K., Kangur, K., Petriashvili, I.: Higher education personnel's perceptions about telepresence robots. Front. Robot. AI **9**(1), 976836 (2022)
11. Davis, F.: Perceived usefulness, perceived ease of use, and user acceptance of information technology. MIS Q. **13**(3), 319–340 (1989)
12. Davis, F., Bagozzi, R., Warshaw, P.: Extrinsic and intrinsic motivation to use computers in the workplace. J. Appl. Soc. Psychol. **22**(14), 1111–1132 (1992)
13. Venkatesh, V., Bala, H.: Technology acceptance model 3 and a research agenda on interventions. Decis. Sci. **39**(2), 273–315 (2008)
14. Mavridis, N., Katsaiti, M., Naef, S., Falasi, A., Nuaimi, A., Araifi, H., Kitbi, A.: Opinions and attitudes toward humanoid robots in the Middle East. AI & Soc. **27**(4), 517–534 (2012)
15. Papadopoulos, R.L., Miah, S., Weaver, T., Thomas, B., Koulouglioti, C.: A systematic review of the literature regarding socially assistive robots in pre-tertiary education. Comput. Educ. **103924**(1), 1–60 (2020)
16. Ramírez-Montoya, M., Castillo-Martínez, I., Sanabria-Z, J., Miranda, J.: Complex thinking in the framework of education 4.0 and open innovation, a systematic literature review. J. Open Innov. Technol. Market Complex. **8**(4), 1–15 (2022)
17. Baena-Rojas, M.R.-M., Mazo-Cuervo, D., López-Caudana, E.: Traits of complex thinking: a bibliometric review of a disruptive construct in education. J. Intell. **10**(3), 1–17 (2023)
18. Fowler, F., Jr.: Survey Research Methods. Sage publications, Boston (2014)
19. Fricker, R.: Sampling Methods for Web and e-Mail Surveys—The SAGE Handbook of Online Research Methods. SAGE Publications Ltd., London (2008)
20. Baena-Rojas, P.S.-B., López-Caudana, E.: Reflections about complex thought and complex thinking: why these theoretical constructs maters on higher education? Eur. J. Contemp. Educ. **12**(1), 1–22 (2023)

Educational Robotics and Computational Thinking: Framing Essential Knowledge and Skills for Pedagogical Practices

Marietjie Havenga and Sukie van Zyl

Abstract The aim of the study on which this conceptual paper is based, was to provide an updated view and nuanced understanding of educational robotics and computational thinking. A brief overview of knowledge and skills required for educational robotics and computational thinking are provided to inform pedagogical practices. We propose five integrated dimensions of computational thinking, and suggest a problem-driven framework to visualize integrated robotics and computational thinking abilities. Based on the framework, we also suggest pedagogical practices for enhancing computational thinking through educational robotics in the classroom. The proposed framework can be implemented for enhancing computational thinking with robotics as a basis for future research.

Keywords Computational thinking · Digital skills · Educational robotics · Pedagogical practice · Self-directed learning

1 Introduction

Digital skills are important for sustainable development and essential for future demands. UNESCO highlights data literacy, collaboration, communication, the creation of digital content, as well as problem solving and technology as essential aspects of a digital literacy framework [1]. In addition, several scholars view computational thinking (CT) and educational robotics (ER) as crucial in a digital world [1, 2]. CT, populated by Wing [3], is regarded as a "foundational competency" to solve problems innovatively [4]. CT, based on computer science principles, involves abstraction or simplification, problem decomposition, algorithmic thinking ability and pattern recognition [3]. Çakıroğlu and Kiliç [5] also add automation and

M. Havenga (✉) · S. van Zyl
Research Unit Self-Directed Learning, Faculty of Education, North-West University, Potchefstroom, South Africa
e-mail: marietjie.havenga@nwu.ac.za

S. van Zyl
e-mail: sukie.vanzyl@nwu.ac.za

R. Balogh et al. (eds.), *Robotics in Education*, Lecture Notes in Networks and Systems 747, https://doi.org/10.1007/978-3-031-38454-7_12

execution, generalization of solutions, the use of relevant tools as well as classroom practices. The use of robotics in the classroom is important to develop the necessary skills. ER is multidisciplinary, involves collaboration, requires integration of different types of subject content, provides opportunities for innovation, and relates to learning environments, such as maker spaces [6].

Research is still ongoing on how to develop the mentioned skills in the classroom. Active teaching–learning practices and the use of relevant tools, such as unplugged activities, block-based programming and robotics are suggested [5]. Avello et al. [2] consider the writing of code as a developmental process (e.g. formulating ideas, making decisions and evaluating the outcome), which requires high-order thinking abilities, such as creative and critical thinking. Catlin and Woollard [7] believe that there is a "natural symbiotic relationship" between ER and CT which offers exciting opportunities for classroom praxis.

It is however not clear what ER and CT entail [8] and they are interpreted and applied in various ways. Consequently, a more nuanced understanding of ER and CT with regard to essential knowledge and skills is required. In addition, a key aspect is to determine how such research may inform teaching and learning.

In particular, the objectives of this paper were the following:

(1) to provide an updated view and nuanced understanding of ER and CT; and
(2) to inform pedagogical practices for developing ER and CT abilities.

In the following sections, an overview of the literature regarding ER and CT is given. Although we did not follow a formal systematic research review or meta-analysis protocol, the selection of relevant research outputs is outlined as follows:

1. We developed inclusion criteria and used the search phrases or descriptors as well as a combination of Boolean operators, such as "educational robotics", "robotics in education", "computational thinking", "pedagogical practices" and "educational robotics", "pedagogical practices" and "computational thinking", "knowledge" and "skills" and "robotics", "knowledge" and "skills" and "computational thinking".
2. We searched for sources on Google Scholar that would illuminate the research objectives and provide an updated view of ER and CT in this conceptual paper.
3. We checked the usefulness and credibility thereof and excluded resources that were personal opinions or research outputs whose full text was not available.

Initially we focused on outputs with historical value. This was followed by a selection of more recent studies.

2 Knowledge and Skills Involved in Educational Robotics

The concept of ER is defined as "a field of study that aims to improve learning experience of people through the creation, implementation, improvement and validation of pedagogical activities, tools (e.g. guidelines and templates) and technologies, where robots play an active role and pedagogical methods inform each decision" [9].

ER is linked to the theories of constructivism, constructionism and learning-for-use [10]. With constructivism, students construct their own learning, while being actively engaged and interacting with each other [8, 11, 12]. Active learning, based on the epistemological views of constructionism, focuses on learners' involvement in the construction of an artefact and development of a deeper understanding while working on solutions to a problem [13]. In accordance with this view, Papert [14] initiated the idea of active learning where children used programming to move the so-called 'turtle' as an "example of a constructed computational 'object-to-think-with'". With robotics, learners are active participants "in the art of learning by doing" [15]. Proponents of active learning believe that ER provides an environment for creating and implementing activities where students develop the ability to construct their own understanding [9, 10]. Active learning tasks, such as unplugged activities, block-based coding and robot programming, are advantageous, as students cooperate and co-construct new knowledge [5]. The purpose of such activities is not merely to learn about robotics, but to acquire knowledge and skills to be applied in other contexts and to be able to solve complex problems.

Several benefits of including robotics in instructional designs are echoed through research. Robot programming results in improved problem-solving skills, metacognition, spatial ability, scientific inquiry [15] as well as knowledge of computer programming. Additional skills involve the formulation of problems, logical thinking, addressing complexity, interaction with peers, the use of computational tools and the development of CT [4, 16].

ER allows for concrete modelling and visualisation of CT aspects, such as abstraction, which assist with knowledge construction and transfer [17]. The movement of the robot provides immediate feedback and opportunities to reflect and revise solutions [10]. Students "can see their thinking" and identify their mistakes when the robot malfunctions, which enhances their metacognition [18]. Furthermore, students need to be willing to work in collaborative settings where group members trust each other, and where they can realise the possibilities of their actions and imagine what the outcome should be [19]. Creativity is another prominent competency enhanced by robotics. It is considered a crucial skill for the Fourth Industrial Revolution [20] and vital to solve complex problems. The connection between creativity and risk-taking also applies to the context of ER. Students need to experience a possibility to succeed, they need to expect to fail, and they need to experience failed attempts. Robot activities are fun, and students are eager to engage in these activities [21]. Subsequently, increased motivation, engagement and improved performance are reported [8, 15]. In research done by Keane et al. (2016), curiosity, willingness to accept challenges and CT emerged when implementing ER in various subjects [25]. Because of their curiosity about the robots, students were driven to attempt coding of robots and to take risks to solve problems. Moreover, ER encourages students to manage their learning and be self-directed learners [15]. In essence, the incorporation of ER brings a "real-world approach" [15], which results in new interest in the subject being studied and a changed mind-set about traditionally difficult concepts. On the other hand, students may experience problems to work effectively in groups when solving ER problems, or they may even consider ER difficult and challenging, as it requires logical thinking, problem-solving and some programming abilities [22]. Furthermore, teachers may not be trained in ER, they are supposed to follow an overloaded curriculum, and might have negative attitudes towards ER, such as a lack of confidence and uncertainty regarding the implementation of ER [22].

To inform future teaching and learning that involve ER, we provide a summary of knowledge and skills developed by ER in Table 1.

Table 1 reveals that the knowledge and skills developed by ER are comprehensive and comprise skills in the cognitive, interpersonal and intrapersonal domains. The fun element and the curiosity that robotics triggers, pave the way for willingness to engage in complex concepts, such as CT. As indicated by Stewart et al. [18], problem solving is a "significant predictor" of CT. Robotics positively affects variables, such as enjoyment, learning motivation and willingness to engage in groups, all which have been identified as predictors of problem solving [18].

Table 1 Knowledge and skills developed by educational robotics

Knowledge and skills	Description
Cognitive and metacognitive skills	Integrated higher-order thinking, computational thinking, innovation, reasoning, logical thinking, analysis, problem solving, creativity, critical thinking, deeper understanding, mental representation, visualisation, and transfer of knowledge to address challenging problems [2, 4, 6, 10, 13, 15–17, 20, 21, 23–25]. Planning, monitoring, evaluation and reflective thinking [10, 18, 24]
Design, development, innovative thinking	Novel way of thinking, willing to engage in complex problems, experimenting with new ideas, risk-taking, developing original solutions that involve visualisation, modelling, and simulation [6, 15, 17, 21, 25]
Intrapersonal skills	Enjoyment, intrinsic motivation, self-efficacy, self-directed learning, confidence, lifelong learning, satisfaction, willingness to learn, curiosity, take control of own learning, less cultural barriers towards STEM education [15, 23, 25]
Interpersonal skills	Communication, distributed cognition, collaborative problem-solving and creativity, working towards a common goal [4, 6, 10, 21, 25]

3 Knowledge and Skills Involved in Computational Thinking

The concept of CT was first indicated by Papert [14] who suggested the role of computers to be relevant in daily activities [26]. However, Wing [27] populated CT and refers to a mind-set on how to approach and solve problems based on the principles of computer science. CT is described as a fundamental skill and thought process [27] that inspires people to collaborate and engage in developing solutions for real-world problems in integrated disciplines [3].

CT comprises knowledge and skills commonly used in programming as well as creative thinking, reasoning and the application of strategies that can be used in a digitalised society [2]. Distinctive skills associated with CT involve problem representation and problem formulation, separation of concerns, breaking down a complex problem into simpler sub-problems (decomposition), identification of similarities between problems (pattern recognition), development of algorithms and problem abstraction [3]. Additional aspects of CT are confidence in dealing with complexity, persistence, tolerance for ambiguity, dealing with open-ended problems, and the ability to communicate and work with others to achieve a common goal [28].

CT is multifaceted [29] and various frameworks for CT have been proposed. Brennan and Resnick [30] distinguish three dimensions, namely computational concepts (e.g. iteration), practices (e.g. testing) and perspectives (e.g. views regarding programming experiences) as essential aspects in their CT framework. The scope of CT comprises design-based thinking, conceptualisation, automation, analysis, testing and debugging, generalization, mathematical reasoning, modelling and the implementation of solutions [31]. Furthermore, Kotsopoulos et al. [32] structured CT with the aim to involve pedagogical practices, such as unplugged experiences, tinkering activities, construction, and remixing of objects and/or components to be used in other objects. In terms of active pedagogical practices, Grover and Pea [4] aligned the elements of CT with iterative refinement, testing and debugging, incremental development, computational artefacts as well as student cooperation and creativity. CT is also associated with reflective thinking and metacognitive control [33]. Tsai et al. [33] accordingly argue, "CT goes beyond problem-solving contexts" that requires regulation "by higher level metacognitive skills". Hoppe and Werneburg [34] accentuate the role of metacognition in CT by defining "computational metacognition" as "the application of analytic (computational) techniques to learner generated computational artefacts", such as programs. In this regard we propose *meta-computational thinking* to address the need of a higher metacognitive thinking process [33] as crucial strategy to assist in the process of decision-making, recursive reflection, iterative refinement, debugging and testing of the solution. Such thinking is of particular importance when a problem is not solved as anticipated.

Several scholars emphasise the importance of solving problems cooperatively to enhance CT. Korkmaz et al. [35] note that solving challenging problems in cooperative groups can be seen as a crucial skill required for CT. They developed computational thinking scales which proved to be reliable as based on the following five factors, namely creativity, algorithmic thinking, cooperativity, critical thinking and problem solving [35]. Saad [36] also illustrated the development of CT through student-centred and cooperative learning activities. It is therefore not merely joint problem solving that will enhance CT, but also the way in which group learning is structured.

As based on the literature, we suggest five integrated dimensions of CT as shown in Table 2, with essential characteristics and references related to each dimension.

The following quintuple set Venn diagram (see Fig. 1) depicts the integration of the five CT dimensions as summarised in Table 2.

Since there is no agreement yet on a general definition of CT [39], it is important to refer to applications of CT to devise a taxonomy for the dimensions thereof. As a result, we formulated and defined CT as follows, as based on Table 2 and Fig. 1:

> Computational thinking, based on the principles of computer science, is an integrated and systematic mental process and essential competency to solve challenging real-world problems innovatively, as individuals or in cooperative groups, by using active strategies, interactive tools, relevant platforms and integrated spaces to achieve a particular goal that benefits a digital society.

This definition refers to the application of computer science principles where both cognition (in particular higher-order thinking skills, such as problem solving, reasoning, critical, creative and innovative thinking), and metacognitive and reflective

Table 2 Five integrated dimensions of CT as based on the literature

Dimension	Characteristics
(A) Computer science principles (CSP)	Problem decomposition, algorithmic thinking, pattern recognition, abstraction, testing and debugging, and implementing solutions [3, 31, 35–37]
(B) Cognitive and metacognitive skills (CMS)	Multi-levelled thinking, problem formulation and problem solving, analysis, reasoning, creativity, critical thinking, design thinking, innovation, conceptualization, generalization, mathematical reasoning and transfer, reflection, computational metacognition and meta-computational thinking, regulation, planning, monitoring and evaluation, iterative refinement [2, 4, 31, 33, 34, 38, 39]
(C) Real-world problems (RWP)	Developing solutions for real-world problems in integrated disciplines, modelling and implementation of solutions, and automation [3, 4, 31, 35]
(D) Learning spaces, tools and pedagogical practices (LTP)	Unplugged activities, interdisciplinary practices, block-based and robot programming, platforms, tools, tinkering, tools for classroom praxis; Student-centered teaching and learning strategies are essential in facilitating students' understanding of concepts and the application of CT skills; Develop positive attitudes, intrinsic motivation, and enjoyment [5, 26, 32, 40]
(E) Cooperation and social interaction (CSI)	Cooperation, student engagement, collaborative problem solving and co-construction of knowledge are crucial to develop and apply CT when solving complex and challenging problems [35, 36, 40]

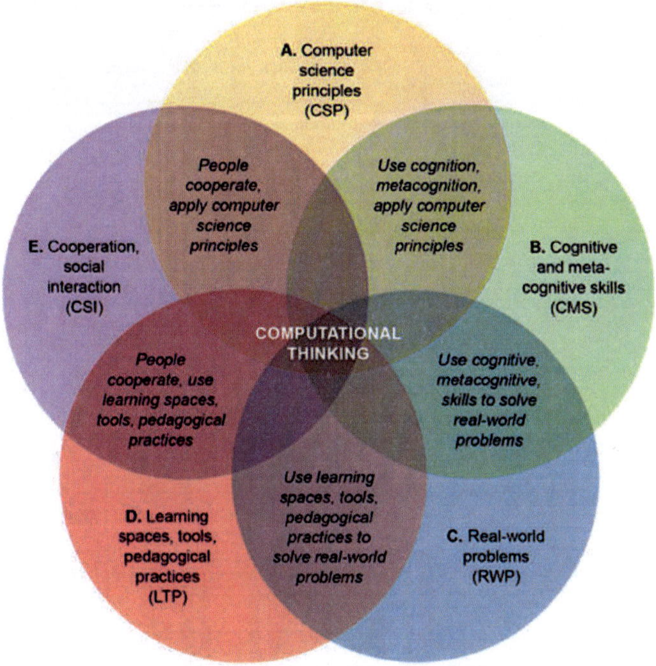

Fig. 1 A quintuple set Venn diagram to depict computational thinking dimensions

thinking (planning, monitoring, evaluation and reflection) are involved. Although CT is not exclusive to individual thought processes, the emphasis is on active co-construction of knowledge in cooperative groups where all members are responsible for specific tasks. In addition, a variety of strategies, interactive physical and/or digital tools, platforms and integrated spaces (e.g. makerspaces) can be used to achieve a particular goal. It is however essential to highlight that these five dimensions (see Fig. 1 and Table 2) are incorporated and integrated to solve challenging real-world problems to benefit a digital society. We are of the opinion that, if these dimensions are not all incorporated, it cannot be seen as CT per se.

4 An Integrated Problem-Driven Framework to Visualize CT and ER Abilities

As based on the previous sections on ER and CT, we suggest the following integrated framework and layered model as depicted in Fig. 2. Our integrated framework applies the five dimensions (see Table 2 and Fig. 1) to emphasise the knowledge, skills and practices involved in ER (Table 1) and CT.

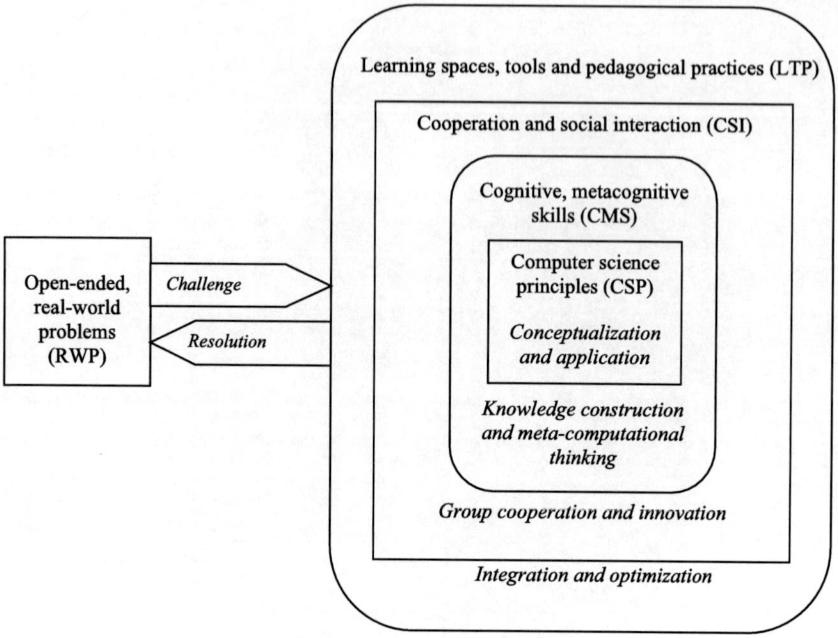

Fig. 2 Integrated framework regarding essential knowledge, skills and practices involved in CT and ER

In this framework, ER can be seen as physical or digital devices as part of learning spaces, tools and pedagogical practices (LTP).

An open-ended or *real-world problem* (RWP) initiates the thinking process in a particular context. At the core of the framework is *computer science principles* (CSP) as a mind-set on how to approach the problem by applying CT.

Cognitive and metacognitive skills (CMS) are involved in knowledge construction and require higher-order thinking, such as reasoning, critical and creative thinking as well as metacognitive control and reflection. Regarding meta-computational thinking, robot execution requires a process of recursive reflection that takes place while monitoring the output, until the problem is solved.

ER is furthermore mainly used in collaborative teaching and learning environments. Group members are challenged to take part in STEM activities, for example, FIRST® LEGO® League competitions. Working in effective groups on challenging, meaningful real-world problems is essential. Cooperation should therefore involve a particular way of group work where members have to take responsibility for activities, learn from each other and be responsible for their final achievement as a group [41]. As a result, group work must be a cooperative effort and should involve *promotive cooperation and social interaction* (CSI) and positive interdependence to maximise members' contribution to the group [35].

Table 3 Some examples of learning spaces, tools, strategies and pedagogical practices for CT and ER

Learning spaces and tools	Strategies and practices
• LEGO® MINDSTORMS® EV3 (robot components, servo motors and sensors—colour, touch, ultrasonic and gyro) and LEGO® MINDSTORMS® Robot Inventor (build and program interactive robots, develop essential skills for STEM) • micro:bit (pocket-sized PC board with light-emitting diode, buttons and sensors) • Makeblock mBot (various robots that can be programmed with block-based code or Arduino C/C++) • Scratch (event-driven and visual programming language). Use block-based or text-based programming language such as Python • NetLogo (integrated programming environment used for simulations) • Tinkercad™ Circuits platform (exploring and learning circuits) • Robot Shield with Arduino (programming environment and learning circuits)	• Represent problems, design and construct models and plan solutions • Apply heuristics for robot movement • Implement debugging and testing strategies • Develop innovative thinking • Apply incremental development and iterative refinement • Provide for active engagement, cooperative decision-making and co-construction of knowledge

The outer layer of the integrated framework in Fig. 2 displays learning spaces, tools and pedagogical practices (LTP) as ways of structuring for meaningful teaching–learning activities. Table 3 shows possible examples of learning spaces and tools that are currently used in robotics education in which the integrated framework can be incorporated to develop CT and ER. It is however important to take into consideration possible differences in the use of ER devices, based on their characteristics, for example the use of micro:bit to introduce students to educational robotics.

The focus should however be on how to implement learning spaces and tools by applying strategies and practices to incorporate the integrated framework (see Fig. 2) to develop the necessary knowledge and skills, and not on which learning space or tool to use.

5 Pedagogical Practices for Educational Robotics and Computational Thinking

Sound pedagogical practices that incorporate the knowledge and skills indicated in Tables 1 and 2 are essential to enhance students' learning of ER and CT. Teachers therefore need to be supported not only in terms of their subject knowledge and robotics skills, but also in terms of being able to design educational environments for optimal learning [8].

We will subsequently give examples of open-ended problems (Table 4) followed by suggestions of pedagogical practices to facilitate the learning of ER and CT (Table 5). For example, where students are working together on ER (See Fig. 3). The main dimensions as described in the integrated framework (Fig. 2) are shown in brackets.

Table 4 Examples of open-ended problems in educational robotics

Examples of open-ended problems	
Example 1: Develop your own game (RWP) and demonstrate sound mastering of the programming concepts involved in micro:bit (LTP). Include coding options such as Loops, Logic, Math and Variables (CSP), (CMS). Submit the code as well as a video of the output. Write a report and reflect on your thinking, challenges and problems you experienced, as well as your responsibilities in the group, based on the game you developed (CSI). Give examples of your thinking, challenges and responsibilities	
Example 2: Formulate a problem (RWP) for the LEGO® EV3 Minstorms robot (LTP) that learners must do in class. Activities must include movement, push/pull actions and a combination of sensors (CSP), (CMS). This must be done in groups (CSI). Assess each other on their active participation. Compile a report, including your problem formulation, photos (start and end position), examples of the programming code and reflection on each robot activity. Also submit a 2–3-minute video of the movement as proof with each activity	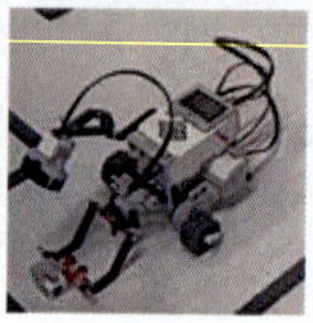

Table 5 Points to consider as part of pedagogical practices for CT and ER

Pedagogical practices
1. Provide meaningful, student-centred learning experiences that incorporate real-world problems within learning contexts. Consider the application of problem-based learning and cooperative learning to drive learning activities
2. Plan class activities that require high-level thinking (e.g. creative thinking and innovation), while taking students' development level into account
3. Tasks should be refreshing, and focus on incremental changes instead of repetition
4. Scaffold students to develop meta-computational thinking and recursive reflection, and give appropriate and explicit feedback. Failure should be seen as an opportunity to achieve success
5. Provide blended learning activities by selecting appropriate technologies and platforms to develop CT and ER (using both physical and digital artefacts and simulation programs)
6. Mediate group cooperation, meaningful student engagement, critical dialogue, ethical and respectful behaviour when solving problems (discuss cooperative learning principles with students)
7. Plan relevant types of assessment (self or peer assessment) as well as assessment tools (e.g. rubrics) with CT principles in mind
8. Support self-efficacy and self-directed learning (a student has to manage his or her own leaning processes, e.g. formulating clear objectives, searching for relevant resources, reflecting on solutions, and curiosity about CT and ER)

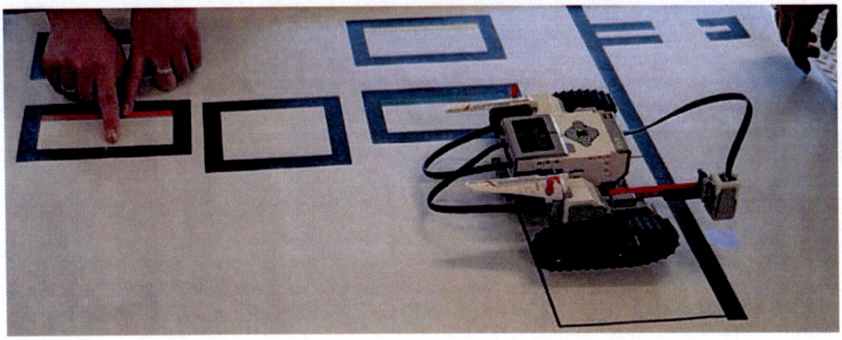

Fig. 3 Students discuss the robot movement

Meaningful learning experiences that incorporate ER to solve real-world problems within learner contexts should be provided (Table 5). Problems should be diversified and expanded to include a variety of real-world scenarios. These should all be done within a collaborative environment through social interaction where students engage responsibly in activities. In addition, teachers should encourage students to take risks and to view failures as opportunities to solve problems by applying recursive reflection.

6 Conclusion

In this paper, we aimed to provide an updated view of ER and CT for informing pedagogical practices and suggest a problem-driven framework for developing ER and CT abilities. The framework and suggested pedagogical practices open the field for further research to develop CT with ER. We conclude that ER could enhance a multitude of knowledge and skills, improve attitudes towards learning of difficult concepts, and therefore provide a powerful scaffold for learning. A limitation involved in this paper is that the proposed framework has not been tested empirically and needs to be implemented in further research.

References

1. Law, N., Woo, D., de la Torre, J., Wong, G.: A global framework of reference on digital literacy skills for indicator 4.4.2. http://uis.unesco.org/sites/default/files/documents/ip51-global-framework-reference-digital-literacy-skills-2018-en.pdf (2018)
2. Avello, R., Lavonen, J., Zapata-Ros, M.: Coding and educational robotics and their relationship with computational and creative thinking. A compressive review. Rev. Educ. Distancia **20**(63), 1–21 (2020). https://doi.org/10.6018/red.413021

3. Wing, J.M.: Computational thinking. Commun. ACM **49**(3), 33–35 (2006). https://doi.org/10.1145/1118178.1118215
4. Grover, S., Pea, R.: Computational thinking: a competency whose time has come. In: Sentence, S., Barendsen, E., Schulte, C. (eds.) Computer Science Education: Perspectives on Teaching and Learning in School, pp. 19–38. Bloomsbury, London (2018). https://doi.org/10.5040/9781350057142.ch-003
5. Çakıroğlu, Ü., Kiliç, S.: Assessing teachers' PCK to teach computational thinking via robotic programming. Interact. Learn. Environ. **31**(2), 818–835 (2020). https://doi.org/10.1080/10494820.2020.1811734
6. Yiannoutsou, N., Nikitopoulou, S., Kynigos, C., Gueorguiev, I., Fernandez, J.A.: Activity plan template: a mediating tool for supporting learning design with robotics. In: Merdan, M., Lepuschitz, W., Koppensteiner, G., Balogh, R. (eds.) Robotics in Education, pp. 3–13. Springer, Cham (2017). https://doi.org/10.1007/978-3-319-42975-5_1
7. Catlin, D., Woollard, J.: Educational robots and computational thinking. In: Proceedings of 4th International Workshop Teaching Robotics, Teaching with Robotics & 5th International Conference Robotics in Education, pp. 144–151. Padova (2014)
8. Zhang, Y., Luo, R., Zhu, Y., Yin, Y.: Educational robots improve K-12 students' computational thinking and STEM attitudes: systematic review. J. Educ. Comput. Res. **59**(7), 1450–1481 (2021). https://doi.org/10.1177/0735633121994070
9. Angel-Fernandez, J.M., Vincze, M.: Towards a definition of educational robotics. In: Proceedings of the Austrian Robotics Workshop 2018, pp. 37–42. Innsbruck University Press, Innsbruck (2018). https://doi.org/10.15203/3187-22-1-08
10. Ioannou, A., Makridou, E.: Exploring the potentials of educational robotics in the development of computational thinking: a summary of current research and practical proposal for future work. Educ. Inf. Technol. **23**(6), 2531–2544 (2018). https://doi.org/10.1007/s10639-018-9729-z
11. Piaget, J..: Piaget's theory. In: Mussen, P. (ed.) Carmichael's Manual of Child Psychology, pp. 703–832. Wiley, New York (1970)
12. Vygotsky, L.S.: Mind and Society. Harvard University Press, London (1978)
13. Kahn, K., Winters, N.: Constructionism and AI: a history and possible futures. Br. J. Edu. Technol. **52**(3), 1130–1142 (2021). https://doi.org/10.1111/bjet.13088
14. Papert, S.: Mindstorms: Children, Computers, and Powerful Ideas, 1st edn. Basic Books, New York (1980)
15. Chahine, I.C., Robinson, N., Mansion, K.: Using robotics and engineering design inquiries to optimize learning for middle level teachers: a case study. J. Math. Educ. **11**(2), 319–332 (2020). https://doi.org/10.22342/jme.11.2.11099.319-332
16. Pollak, M., Ebner, M.: The missing link to computational thinking. Future Internet **11**(12), 263–275 (2019). https://doi.org/10.3390/fi11120263
17. Fadiran, O.A., Van Biljon, J., Schoeman, M.: How can visualisation principles be used to support knowledge transfer in teaching and learning? In: 2018 Conference on Information Communications Technology and Society (ICTAS), pp. 1–6. IEEE, Durban (2018). https://doi.org/10.1109/ictas.2018.8368739
18. Stewart, W.H., Baek, Y., Kwid, G., Taylor, K.: Exploring factors that influence computational thinking skills in elementary students' collaborative robotics. J. Educ. Comput. Res. **59**(6), 1208–1239 (2021). https://doi.org/10.1177/0735633121992479
19. Beghetto, R.A.: Taking beautiful risks in education. Educ. Leadersh. **76**(4), 18–24 (2018)
20. Rahimi, S., Shute, V.J.: First inspire, then instruct to improve students' creativity. Comput. Educ. **174**, 104312 (2021). https://doi.org/10.1016/j.compedu.2021.104312
21. Curto, B., Moreno, V.: Robotics in education. J. Intell. Rob. Syst. **81**(1), 3–4 (2016). https://doi.org/10.1007/s10846-015-0314-z
22. Papadakis, S., Vaiopoulou, J., Sifaki, E., Stamovlasis, D., Kalogiannakis, M., Vassilakis, K.: Factors that hinder in-service teachers from incorporating educational robotics into their daily or future teaching practice. In: Proceedings of the 13th International Conference on Computer Supported Education, vol. 2, pp. 55–63 (2021). https://doi.org/10.5220/0010413900550063

23. Armesto, L., Fuentes-Durá, P., Perry, D.: Low-cost printable robots in education. J. Intell. Rob. Syst. **81**(1), 5–24 (2016). https://doi.org/10.1007/s10846-015-0199-x
24. Jin, Q., Kim, M.: Supporting elementary students' scientific argumentation with argument-focused metacognitive scaffolds (AMS). Int. J. Sci. Educ. **43**(12), 1984–2006 (2021). https://doi.org/10.1080/09500693.2021.1947542
25. Keane, T., Chalmers, C., Williams, M., Boden, M.: The impact of humanoid robots on students' computational thinking. In: Albion, P., Prestridge, S. (eds.) Australian Council for Computers in Education 2016 Conference: Refereed Proceedings, pp. 93–102. The Queensland Society for Information Technology in Education (QSITE), Australia (2016)
26. Soboleva, E.V., Sabirova, E.G., Babieva, N.S., Sergeeva, M.G., Torkunova, J.V.: Formation of computational thinking skills using computer games in teaching mathematics. Eurasia J. Math., Sci. Technol. Educ. **17**(10), em2012 (2021). https://doi.org/10.29333/ejmste/11177
27. Wing, J.M.: Computational thinking benefits society. http://socialissues.cs.toronto.edu/index.html%3Fp=279.html (2014)
28. ISTE-CSTA: Operational definition of computational thinking for K–12 education. https://cdn.iste.org/www-root/Computational_Thinking_Operational_Definition_ISTE.pdf (2011)
29. Allsop, Y.: Assessing computational thinking process using a multiple evaluation approach. Int. J. Child-Comput. Interact. **19**, 30–55 (2019). https://doi.org/10.1016/j.ijcci.2018.10.004
30. Brennan, K., Resnick, M.: New frameworks for studying and assessing the development of computational thinking. In: Proceedings of the 2012 Annual Meeting of the American Educational Research Association, pp. 1–25. Vancouver (2012)
31. Kalelioğlu, F., Gülbahar, Y., Kukul, V.: A framework for computational thinking based on a systematic research review. Balt. J. Mod. Comput. **4**(3), 583–596 (2016)
32. Kotsopoulos, D., Floyd, L., Khan, S., Namukasa, I.K., Somanath, S., Weber, J., Yiu, C.: A pedagogical framework for computational thinking. Digit. Exp. Math. Educ. **3**(2), 154–171 (2017). https://doi.org/10.1007/s40751-017-0031-2
33. Tsai, M.-J., Liang, J.-C., Hsu, C.-Y.: The computational thinking scale for computer literacy education. J. Educ. Comput. Res. **59**(4), 579–602 (2021). https://doi.org/10.1177/0735633120972356
34. Hoppe, H.U., Werneburg, S.: Computational thinking—more than a variant of scientific inquiry! In: Kong, S-C., Abelson, H. (eds.) Computational Thinking Education, pp. 13–30. Springer, Gateway East (2019). https://doi.org/10.1007/978-981-13-6528-7_2
35. Korkmaz, Ö., Çakir, R., Özden, M.Y.: A validity and reliability study of the computational thinking scales (CTS). Comput. Hum. Behav. **72**, 558–569 (2017). https://doi.org/10.1016/j.chb.2017.01.005
36. Saad, A.: Students' computational thinking skill through cooperative learning based on hands-on, inquiry-based, and student-centric learning approaches. Univers. J. Educ. Res. **8**(1), 290–296 (2020). https://doi.org/10.13189/ujer.2020.080135
37. Kale, U., Akcaoglu, M., Cullen, T., Goh, D., Devine, L., Calvert, N., Grise, K.: Computational what? Relating computational thinking to teaching. TechTrends **62**(6), 574–584 (2018). https://doi.org/10.1007/s11528-018-0290-9
38. Topal, A.D., Geçer, A.: Examining the relationship between computational thinking skills and metacognitive thinking skills. In: Kaya, Ö. (ed.) Theory and Research in Social and Administrative Sciences, pp. 65–94. Iksad Publications, Ankara (2020)
39. Troiano, G.M., Snodgrass, S., Argımak, E., Robles, G., Smith, G., Cassidy, M., Tucker-Raymond, E., Puttic, G., Harteveld, C.: Is my game OK Dr. Scratch? Exploring programming and computational thinking development via metrics in student-designed serious games for STEM. In: IDC'19. Proceedings of the 18th ACM International Conference on Interaction Design and Children, pp. 208–219. ACM (2019)
40. Liu, Z., Xia, J.: Enhancing computational thinking in undergraduate engineering courses using model-eliciting activities. Comput. Appl. Eng. Educ. **29**(1), 102–113 (2021). https://doi.org/10.1002/cae.22357
41. Johnson, D.W., Johnson, R.T.: Cooperative learning: the foundation for active learning. In: Brito, S.M. (ed.) Active Learning-Beyond the Future, pp. 59–70. IntechOpen, London (2018)

Pair-Programming with a Telepresence Robot

Janika Leoste, Jaanus Pöial, Kristel Marmor, Kristof Fenyvesi, and Päivi Häkkinen

Abstract Telepresence robots (TPRs) are seen as promising tools for maintaining social presence in distance learning conditions, contributing to student persistence and wellbeing by reducing their feeling of isolation and distress. We examined the challenges that the use of telepresence robots in a pair-programming course presents to the teacher and students. Semi-structured interviews were used to collect data from the teacher and four students about their experience of being mediated via a telepresence robot and having a teacher or student mediated via a telepresence robot. The data were coded and analyzed to map the main challenges. Four general areas of concern were revealed: preconditions for use, justifications for use, robot characteristics, and potential challenges. Using TPRs is justified for students in conditions where their social presence is required (e.g., in discussions, workshops or other educational activities). In this particular case, the use was not recommended for the teacher. TPRs' educational implementation should be planned meticulously to maximize their positive effect and to reduce potential setbacks. In addition, the TPR's features should match as best as possible to the requirements of the educational activities to be carried out in a specific physical and social environment.

Keywords Telepresence robots · Pair-programming · Technology-enhanced learning · Technology implementation · Educational innovation

J. Leoste (✉) · J. Pöial · K. Marmor
Tallinn University of Technology, Tallinn, Estonia
e-mail: janika.leoste@taltech.ee

J. Leoste
School of Educational Sciences, Tallinn University, Tallinn, Estonia

K. Fenyvesi · P. Häkkinen
Finnish Institute for Educational Research, University of Jyväskylä, Jyväskylä, Finland

R. Balogh et al. (eds.), *Robotics in Education*, Lecture Notes in Networks and Systems 747, https://doi.org/10.1007/978-3-031-38454-7_13

1 Introduction

In this paper, we examine a few aspects of the implementation of Telepresence Robots (TPRs) for supporting the pair-programming teaching and learning method. We introduce the background of relevant concepts; describe the research method and related details of our experiment. Finally, we present and discuss our results.

Due to the increasing importance of digital Information and Communication Technologies (ICT) in modern societies, there is a constantly growing need for programmers. Coding has been introduced to the curricula of several countries, and programming courses became generally available in various higher education fields of study [1]. As many students find computer programming challenging, different methods have been implemented to facilitate learning [2]. One of these methods is Pair-Programming (PP), which is especially useful for fostering beginners' programming skills and relevant collaborative teamwork skills [3]. In PP, students work in pairs. One of them is the "driver" and the other one is the "navigator." These roles can be switched during the learning process. The driver writes code, being guided by the navigator. The navigator checks the written code. The programming process is thus based on collaborative discussion, and both students share the same goal. Compared to solo programming, using PP with beginners has several educational benefits [1, 4, 5], for example: developing more effective code, minimizing errors, reducing confirmation bias, verbalizing implicit mental models, supporting knowledge transfer, applying peer pressure, rising self-efficacy, reducing the gender gap, and increasing retention rates. However, despite its benefits, the classroom-based PP method proved vulnerable to the recent COVID-19 pandemic and the subsequent energy crisis. In these years (2019–2022), universities worldwide were forced to implement digital distance teaching and learning methods in order to ensure health safety or to cope with the rising heating costs [6, 7]. Under these conditions, Distributed Pair Programming (DPP) was often implemented [8].

DPP allows students from different geographical locations to develop and write code remotely while maintaining collaboration [5]. According to [9], DPP requires two main categories of tools: (a) screen-sharing applications, including teleconference software; and (b) applications that support collaborative programming. Satrazemi, Stelios & Tsompanoudi [5] suggest in their literature review of studies comparing DPP with PP that in terms of code quality and academic performance, the methods yielded similar results, while students used more time with DPP to accomplish their tasks. However, the studies examined in [5] were not longitudinal (only about 1/3rd of the studies had durations from 4 weeks to 1 semester, the rest were one-time events). This is an important distinction because, especially in the educational context, the wide-scale long-term impact could involve negative aspects. Negative aspects can be similar to the ones related to digital distance learning that were revealed during the COVID-19 forced online education period—such as increased stress, decreased sense of belonging and other consequences of reduced social presence [10].

Social presence can be defined as "the ability to project one's self and establish personal and purposeful relationships" [11], or "the degree to which a person is perceived as a 'real person' in mediated communication" [12]. Thus, social presence allows people to perceive each other as real persons when interacting. In education, social presence affects group learning [13]. For example, as suggested by [14], even naturally developed digital learning environments cannot provide similar levels of social presence compared to physical academic environments, influencing how students develop their learning-related skills and habits. In regular in-person social interaction, students can use their bodily presence to provide social support and create the zone of proximal development where students that are more knowledgeable influence their less capable peers [15]. In online learning settings, the limited non-verbal communication and the different conditions for social interaction can lead to dysfunctions in the zones of proximal development, where students are unable to bond and support each other, causing student distress and isolation [16]. Compared to computer-based distance learning solutions, telepresence robots have been found to provide better social presence [17, 18].

Telepresence robotics is a relatively new technology in the context of education. The concept itself goes back to Minsky's visionary essay from 1980, where the term of telepresence has been coined and the main characteristics of telepresence have been defined [19]. The term's prevalence in the literature started to increase since 2010 [20]. Telepresence robots (TPRs) are reported to provide communication partners with better social presence, leading to more natural and effective patterns of interaction. A TPR's body provides its user with embodied presence, allowing students to be physically present in the zone of proximal development and to participate in discussions or collaborative tasks [21]. Compared to tele-surgery robots with unit prices from $500,000 to $1.5 million, educational TPRs are accessible at a lower price (less than $10,000 for a Double 3 TPR), but, on the other hand, also relatively limited. Most educational robots (e.g., in Fig. 1) consist of a moving base, a body or a (sometimes height-adjustable) neck, and a central unit that has a display, speaker, microphone array and cameras. A TPR typically uses the Internet via a Wi-Fi connection to stream real-time video and audio between the robot and its user's computer. The industrially produced TPRs that are currently used in education cannot be used for direct manipulation of objects or opening doors, nor can they provide their users with the ability to jump or sit down. However, according to literature, even these basic TPRs can support students' and teachers' social presence and facilitate classroom discussion and collaborative learning [17, 18, 22–24].

This paper aims to gain insight into involving TPRs in the PP teaching and learning method. The study is a small-scale pilot of a larger, longitudinal study about TPRs feasibility in higher education, conducted simultaneously at the Tallinn University and the Tallinn University of Technology. In this study, we examine the effects of using TPRs without previous usage planning in a pair-programming course, part of a typical computer science curriculum, mimicking a situation when a student or a teacher is forced to work from a distance.

Fig. 1 The TPRs used in the study (from the left): Ohmni, TEMI and Double 3

To guide our study, we have formulated the following research question: What are the main challenges the teacher and students face while using a telepresence robot for classroom communication in a pair-programming seminar?

2 Method

We conducted our experiment twice in October 2022, in the two seminars of the Algorithms and Data Structures course at Tallinn University of Technology (see also [25]). The overall volume of the subject is 6 ETCS (64 classroom hours). The students had passed 24 academic hours before the experiment. In total, there were 84 students enrolled in the course. These students were already familiar with programming, as they had previously passed an introductory programming course. The TPRs were used during the experiment by four students, all of whom shared their experience with researchers via semi-structured Zoom interviews. While the teacher had previous experience with TPRs, the students had not worked with TPRs before.

In the first experimental seminar, on October 18, the teacher participated in person, two students participated via TPRs (see Fig. 2), and the rest of the students participated in-person (Fig. 3). The students had to solve a task in pairs, show their work to the teacher and receive the teacher's feedback. The work had to be shown on a computer display; the students had to be able to modify their work and to show different views. Both students in a pair had to participate in a discussion with the

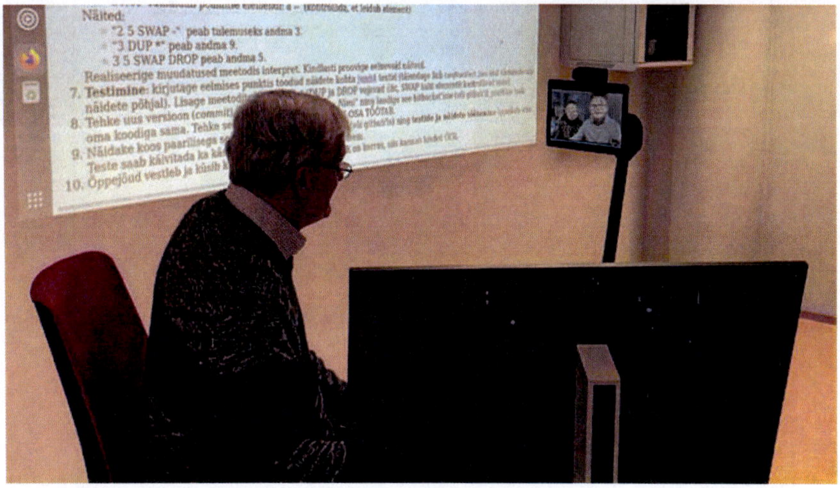

Fig. 2 TPR-mediated students presenting their work to the teacher

teacher. The students in the TPR were asked not to use other means of communication, and were located in a neighboring room. The work assessment criteria were the same for both in-person and TPR-mediated students. As there were many students presenting their work then TPR-mediated students had to compete actively for the teacher's attention. The teacher tried to use his typical in-person teaching methods as much as possible.

During the second experimental seminar on October 18, the teacher participated via a TPR (see Fig. 4) and all students were present in-person. The goal of the seminar and the task to be solved by students was the same as in the October 11 seminar, but the students were different.

2.1 *TPRs Used in the Study*

We used three different types of TPRs in our experiment (Fig. 1). All of these robots consist of a driving base, body and audio-visual module with a display, microphones and speakers for transmitting audio and video from and to the TPR-mediated person. The full specifications of the used robots are available as follows:

- the Ohmni robot—https://ohmnilabs.com/products/customers/faq/#spec;
- the TEMI 3 robot—https://www.robotemi.com/specs/;
- the Double 3 robot—https://www.doublerobotics.com/tech-specs.html.

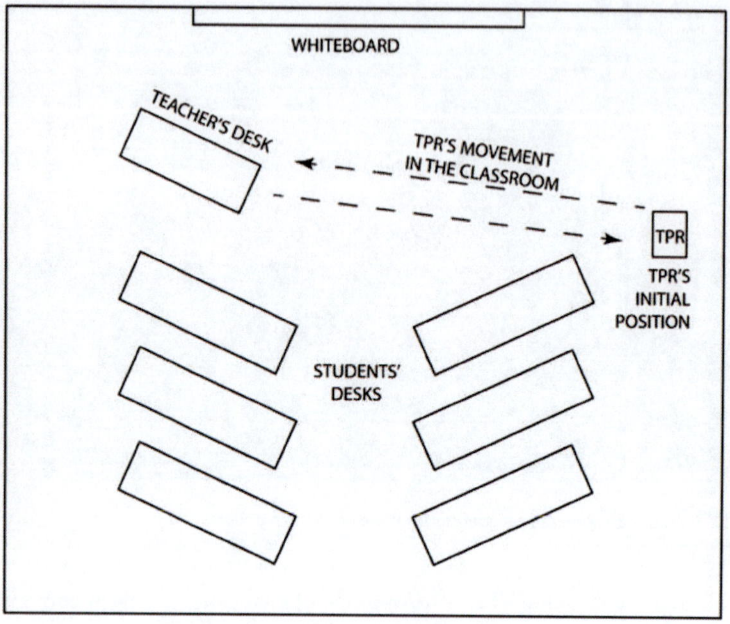

Fig. 3 The classroom's physical setup during the October 11 seminar

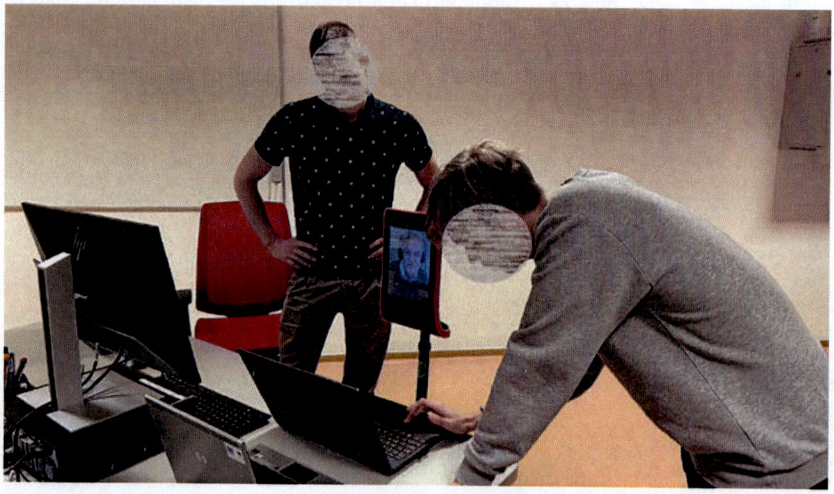

Fig. 4 Students present their work to the TPR-mediated teacher

2.2 *Data Collection and Analysis*

We collected data from the teacher (male) who conducted the course and four students (three female and one male) who attended the course via TPRs during the first experimental seminar. We used the semi-structured interview as the data collection tool. The interviews were conducted via the Zoom video-conferencing software. The time of the interviews was chosen by the interviewees. They also chose the physical location of their end of the Zoom call for the interviews. The interviews were transcribed via Microsoft Word transcription service. The transcriptions were independently analyzed and open-coded by two researchers in order to map the information related to our research question. Coding discrepancies were resolved through discussion between the researchers.

3 Results

Our research question was "What are the main challenges the teacher and students face while using a telepresence robot for a classroom communication in a pair-programming seminar?" The interviews revealed four general areas of concern, for both students and teachers: (a) preconditions for use; (b) justifications for use; (c) robot characteristics; and (d) potential challenges.

In the first area, some **preconditions** should be addressed before starting to use TPRs. First, the use of TPRs should be planned. This involves matching the needs of teaching and learning with the features of available robots, ensuring the availability of required infrastructure resources, especially reliable and stable internet over Wi-Fi, and working out changes in teaching methods. In addition, a technical assistant may have to be present to provide help in case of technical problems.

The second area, **justifications for use**, involves reasoning about the situations where the TPR use offers higher value than the alternatives. At Tallinn University of Technology, all lessons are recorded, allowing students to catch up on the topics discussed. However, both the teacher and students found that participation via a TPR could be more beneficial to the students compared to reviewing the lesson from a remote location. Using TPRs provided students with an increased feeling of social presence in the classroom: they felt that by being able to move around and talk to their classmates via a TPR they could influence the processes and discussion in the classroom. In addition, the students found that, compared to the videoconferencing approach (such as Microsoft Teams), TPRs are better suited for workshops and collaborative work. In students' words, besides being exciting and easy to use, TPRs help maintain eye contact and keep focus on the learning subject, facilitate active participation (ask questions, move in the classroom, show one's work to the teacher—all these activities cannot be conducted when watching a recording of the lesson) and other classroom activities (e.g., seeing the blackboard instead of opening files on the computer). It is interesting to note that according to students, the TPR use would be more justified for students and less so for teachers. In their opinion, remote teaching could be as well conducted with similar efficiency via, for example, the Microsoft Teams software.

The third concern area, **robot characteristics**, focuses on the different features of different robots. Our study used three TPRs: the Double 3, Ohmni and TEMI v3 robots. The feedback indicated that of these TPRs, the TEMI robot was the least useful for classroom use: it is relatively short, making it difficult for the TPR-mediated person to read papers on a table; it moves very slowly and reacts slowly to movement commands; its camera is unsuited for classroom use as it is difficult to read written text due to camera's resolution and its constant seeking of faces. The Double 3 and Ohmni robots were both better suited for our use, although different characteristics distinguished them. The Ohmni robot has three wheels and thus more stable compared to the Double 3 that has self-balancing movement with two wheels. However, the Double 3 robot has an obstacle detection sensor, keeping it from colliding with tables and persons—a feature that the Ohmni robot lacks. It was easier for the physically present persons to read the text and follow the graphical materials on the Ohmni robot display. However, it was easier for the TPR-mediated person to follow classroom events and read the text from a whiteboard using the Double 3 robot.

Both the teacher and students highlighted various **problems** that, in turn, can be divided into several groups of challenges. First, as pointed out by the teacher, the implementation of TPRs would require the teacher to use additional time. Due to the technology being novel for both teachers and students, some time will be used for them to become familiarized with the possibilities and limitations of TPRs. Initially, additional time will be spent on setting up the robots, entering their Wi-Fi settings, adjusting audio levels to avoid acoustic feedback, developing more appropriate classroom use scenarios, preparing materials with large enough visuals that can be seen or demonstrated via a TPR, etc. Later, during the lessons, a need arose for a technical assistant, to deal with ongoing problems: to lift the robot from one area to another, make necessary technical adjustments, etc. Next, the teacher and students brought out several problems, revealed during classroom use. All used robots were sensitive to the quality of the internet connection, losing audio and video quality or moving problematically improperly when the internet connection had problems. In addition, the audio settings had to be adjusted frequently to avoid picking up too much room noise or causing acoustic feedback. On some occasions, it was difficult for the teacher or students to read printed text (both when the text was shown to the TPR or from the TPR). After that, the lack of manipulators, such as hands, limited the teacher's ability to demonstrate learning materials to students (instead, he relied on the help of the students). Compared to being physically present, the limitations of TPR's movement further restricted the teacher's presence: his body language was not visible, it was difficult to keep eye contact with all students, and it was time-consuming to move from one student to another. However, from the point of view of students, using a TPR was not a challenge, because dealing with the subject was significantly more difficult.

4 Discussion

TPRs have been slowly implemented in education in the last few decades. The advancements in technology (high-res cameras, battery life, and widely available high-speed internet) have made TPRs viable in the educational context. TPRs are often suggested as tools to provide students or teachers with alternative physical access to allow social presence from a remote location, using the same teaching and learning methods as in the classroom with only physically present students and teachers [17, 23, 26]. However, due to their physical limitations compared to the human body, special preparations may be needed when implementing TPRs in education. We studied the challenges experienced by a teacher and students who used TPRs in a pair-programming course. Below we will discuss the findings of our study.

First, the use of TPRs is not always justified (e.g., in lessons where the learning content can be shared via a video recording). It seems that TPRs could be better suited for allowing students to participate remotely in practicums or workshops, i.e., in lessons where the nature of classroom activities or existing social norms favor physical presence even if it is rudimentary. When participating in classroom discussions or certain collaborative activities, physical presence could contribute to a person's social wellbeing (similarly to suggestions made by [27]), encouraging TPR-mediated persons to more actively take part in a lesson, allowing their voice to be heard and themselves to be taken more seriously. However, both the teacher and students could not see any value in having a teacher conduct a lesson via a TPR. This opinion could be based on the lack of physical capabilities of the current telepresence robots: it is difficult for a teacher to show something physically without the help from an assistant. Although students have similar restrictions when participating via a TPR, they would have their fellow students acting as assistants. It is possible that this situation could be remedied by either having an assistant in the classroom for the teacher or having more advanced TPRs with manipulators similar to human hands.

Next, technical specifications are decisive when choosing a TPR for classroom use. The robot should be capable to transmit high-quality video and audio in both directions. For this, it needs to have high-resolution zoom-able cameras, a good display and microphones with noise-cancellation feature. The robot also needs to be able to move quickly and safely, while being easy to control. However, it is difficult to recommend one specific type of TPR as in different learning courses different features are prioritized (for example, in our study TPRs with landscape display were more useful for showing programming code). This notion suggests that different TPRs could be needed for different courses, potentially making the implementation of TPRs unreasonable in terms of finances and maintenance.

Then, as using TPRs requires high-quality high-speed internet via Wi-Fi, it seems that implementation of TPRs may result in the need for an organization to make additional investments in infrastructure. In addition, dealing with TPRs causes teachers to spend more time for preparing and conducting lessons and potentially requires changes in teaching methods. For example, in our experiment, the teacher, based on his previous experience with TPRs, had meticulously prepared for the TPR-enhanced seminars and despite his efforts, there were still technical problems and situations that needed his additional attention. These notions imply that implementation of TPRs could require to alterations in teachers' remuneration basis—or, at least, they should be provided with technical assistants. In conclusion, we suggest that the routine use of TPRs in any subjects should be preceded by careful planning that considers all the relevant details [26]. In other words, the implementation of TPRs in education should follow the innovation process model for technology-enhanced learning, as described by [28], where the innovation process is divided into three stages (Awareness, Acceptance, Adoption) in order to address time-critical challenges most efficiently.

4.1 Limitations and Future Developments

We studied the challenges of implementing TPRs in teaching and learning by using them in two seminars of a pair-programming course. As different subjects may present different challenges and requirements to the use of TPRs in different educational situations, then further studies are needed to gain an understanding about the suitability of TPRs in various subjects, including those in the humanitarian field (e.g., law, psychology, arts, etc.). In addition, there is a need to investigate further what kinds of tasks and pedagogical approaches are most relevant to be integrated with TPRs, and how classroom practices should be orchestrated when TPRs are present. For example, learner-centered approaches such as inquiry-, problem-, project- and case-based learning approaches could potentially benefit from social and physical presence enabled by TPRs.

For us, this study was a pilot that confirmed the applicability of TPR testing methods for other subjects at Tallinn University of Technology. At the moment, we have started a study where TPRs are used by ten different higher education teachers in ten different learning contexts in four higher education institutions. That extended study will also address the limitations arising from the small sample size of this pilot study, and should give more information about features that TPRs should have for general use in education.

References

1. Demir, Ö., Seferoglu, S.S.: A comparison of solo and pair programming in terms of flow experience, coding quality, and coding achievement. J. Educ. Comput. Res. **58**(8), 1448–1466 (2021). https://doi.org/10.1177/0735633120949788
2. Li, L., Xu, L., He, Y., He, W., Pribesh, S., Watson, S.M., Major, D.A.: Facilitating online learning via zoom breakout room technology: a case of pair programming involving students with learning disabilities. Commun. Assoc. Inf. Syst. **48** (2021). https://doi.org/10.17705/1CAIS.04812

3. Hawlitschek, A., Berndt, S., Schulz, S.: Empirical research on pair programming in higher education: a literature review. Comput. Sci. Educ. (2022). https://doi.org/10.1080/08993408.2022.2039504
4. Kuttal, S.K., Ong, B., Kwasny, K., Robe, P.: Trade-offs for substituting a human with an agent in a pair programming context: the good, the bad, and the ugly. In: Proceedings of the 2021 CHI Conference on Human Factors in Computing Systems, vol. 243, pp. 1–20 (2021). https://doi.org/10.1145/3411764.3445659
5. Satratzemi, M., Stelios, X., Tsompanoudi, D.: Distributed pair programming in higher education: a systematic literature review. J. Educ. Comput. Res. (2022). https://doi.org/10.1177/07356331221122884
6. Marinoni, G., Van't Land, H., Jensen, T.: The impact of COVID-19 on higher education around the world. IAU Global Survey Report. UNESCO House. Paris, France (2020). https://www.iau-aiu.net/IMG/pdf/iau_covid19_and_he_survey_report_final_may_2020.pdf
7. Upton, B.: European universities cancel classes as energy bills soar. The Higher Education (2022). https://www.timeshighereducation.com/news/european-universities-cancel-classes-energy-bills-soar
8. Simaremare, M.E.S.: Strategies for an effective distributed pair programming. Jurnal Mantik **5**(4), 2531–2535 (2022). http://iocscience.org/ejournal/index.php/mantik/article/view/1963. Accessed 3 Jan 2023
9. Winkler, D., Biffl, S., Kaltenbach, A.: Evaluating tools that support pair programming in a distributed engineering environment. In: Proceedings of the 14th International Conference on Evaluation and Assessment in Software Engineering (EASE'10). BCS Learning & Development Ltd., Swindon, GBR, pp. 54–63 (2010)
10. Leoste, J., Rakic, S., Marcelloni, F., Zuddio, M. F., Marjanovic, U., Oun, T.: E-learning in the times of COVID-19: the main challenges in higher education. In: 19th International Conference on Emerging eLearning Technologies and Applications (ICETA), pp. 225–230 (2021). https://doi.org/10.1109/ICETA54173.2021.9726554
11. Garrison, D.R.: Online community of inquiry review: Social, cognitive, and teaching presence issues. J. Asynchron. Learn. Netw. **11**(1), 61–72 (2007)
12. Gunawardena, C.N.: Social presence theory and implications for interaction collaborative learning in computer conferences. Int. J. Educ. Telecommun. **1**(2/3), 147–166 (1995)
13. Kreijns, K., Xu, K., Weidlich, J.: Social presence: conceptualization and measurement. Educ. Psychol. Rev. **34**, 139–170 (2022). https://doi.org/10.1007/s10648-021-09623-8
14. Messmer, G., Berkling, K.: Overcoming the gap of social presence in online learning communities at university. In: 2021 World Engineering Education Forum/Global Engineering Deans Council (WEEF/GEDC), pp. 563–570 (2021). https://doi.org/10.1109/WEEF/GEDC53299.2021.9657401
15. Vygotsky, L.S.: Mind in Society: The Development of Higher Psychological Processes. Harvard University Press, Cambridge, Massachusetts (1978)
16. Ouellette, T., Ramadan, Y., Muthuraj, M., Pi Noa, D., Yuvaraj, R., Rana, A., Jacobs, R.: The association of stress, collaborative learning, social presence, and social interaction with teaching modality type among medical students. In: ICERI2021 Proceedings, pp. 8121–8126 (2021)
17. Weibel, M., Nielsen, M.K., Topperzer, M.K., Hammer, N.M., Møller, S.W., Schmiegelow, K., Bækgaard Larsen, H.: Back to school with telepresence robot technology: a qualitative pilot study about how telepresence robots help school-aged children and adolescents with cancer to remain socially and academically connected with their school classes during treatment. Nurs. Open **7**, 988–997 (2020)
18. Powell, T., Cohen, J., Patterson, P.: Keeping connected with school: implementing telepresence robots to improve the wellbeing of adolescent cancer patients. Front. Psychol. **12**, 749957 (2021)
19. Minsky, M.: Telepresence. OMNI magazine (1980). https://web.media.mit.edu/~minsky/papers/Telepresence.html
20. Google Ngram Viewer. https://books.google.com/ngrams/graph?content=telepresence+robot&year_start=2000&year_end=2019&corpus=26&smoothing=3&direct_url=t1%3B%2Ctelepresence%20robot%3B%2Cc0. Last Accessed 28 Dec 2022

21. Mennecke, B.E., Triplett, J.L., Hassall, LM., Conde, Z.J.: Embodied social presence theory. In: 43rd Hawaii International Conference on System Sciences. IEEE, Honolulu, USA (2010)
22. Fitter, N.T., Rush, L., Cha, E., Groechel, T.R., Matarić, M.J., Takayama, L.: Closeness is key over long distances: effects of interpersonal closeness on telepresence experience. In: 2020 15th ACM/IEEE International Conference on Human-Robot Interaction (HRI), pp. 499–507 (2020)
23. Gallon, L., Abénia, A., Dubergey, F., Negui, M.: Using a telepresence robot in an educational context. In: 15th International Conference on Frontiers in Education: Computer Science and Computer Engineering (FECS 2019), 2019, Las Végas, United States (2019)
24. Schouten, A.P., Portegies, T.C., Withuis, I., Willemsen, L.M., Mazerant-Dubois, K.: Robomorphism: Examining the effects of telepresence robots on between-student cooperation. Comput. Hum. Behav. **126**, 106980 (2022)
25. Pöial, J.: Challenges of Teaching Programming in StackOverflow Era. In: Auer, M.E., Rüütmann, T. (eds.) Educating Engineers for Future Industrial Revolutions. ICL 2020. Advances in Intelligent Systems and Computing, vol. 1328. Springer, Cham (2020). https://doi.org/10.1007/978-3-030-68198-2_65
26. Lei, M., Clemente, I.M., Liu, H., et al.: The acceptance of telepresence robots in higher education. Int J Soc Robot (2022)
27. Dakof, G.A., Taylor, S.E.: Victims' perceptions of social support: what is helpful from whom? J. Pers. Soc. Psychol. **58**(1), 80–89 (1990)
28. Leoste, J.: Adopting and sustaining technological innovations in teachers' classroom practices—The case of integrating educational robots into math classes. Ph.D. Thesis, Tallinn University, Estonia (2021)

Robots and Children

You're Faulty But I Like You: Children's Perceptions on Faulty Robots

Sílvia Moros and Luke Wood

Abstract This paper presents a study conducted in a United Kingdom primary school with the Maqueen BBC micro:bit robot. The purpose was to explore whether easy-to-fix hardware issues affected the children's perception of the robot or their enjoyment of the session, and whether the children could cope with these failures and/or repair them. As with any piece of technology, robots break down and are in regular need of reparation, but this technical issue could be a disadvantage in a classroom setting, as it might impact the children's enjoyment and confidence in their abilities to carry out the given task; potentially this could deter teachers from using this technology. 128 children participated in this study, aged 7–12 years old (M=9,18; SD=1,061). While children did perceive robots to be faulty less times than the faults were present in the robots, they did consider themselves capable of solving these issues and enjoyed doing so. Their perception of a faulty robot also did not impact significantly in their enjoyment nor in their consideration of the robot as a machine or a friend.

Keywords Children-robot interaction · Education · Failure perception

1 Introduction

Each year, the UK-RAS Network hosts an array of activities related to robotics in what is called the "UK Festival of Robotics" [21], previously "Robotics Week". These events range from talks to live streams to bringing schools to labs or, in the case of our group, bringing robots to a local primary school and delivering a week of different robotics workshops for the students, to provide an overview of the different aspects of the field and inspire them to pursue STEM subjects. To date, our workshops have included sessions covering hardware, programming, interaction scenarios, robotic

S. Moros (✉) · L. Wood
University of Hertfordshire, Hatfield, UK
e-mail: s.moros2@herts.ac.uk

L. Wood
e-mail: l.wood@herts.ac.uk

R. Balogh et al. (eds.), *Robotics in Education*, Lecture Notes in Networks and Systems 747, https://doi.org/10.1007/978-3-031-38454-7_14

toys, 3D printing and more. This collaboration has been established for several years (see, for example, [11, 15, 16]) with mutual satisfaction on both parts.

This paper presents the programming workshop we conducted during the UK Festival of Robotics 2022, along with the experimental data we obtained. The paper is structured as follows: Sect. 2 introduces a brief review of the literature in this topic. Section 3 provides the session plan, along with the methodology we followed. Section 4 introduces the results, while Sect. 5 discusses them in context. Finally, Sect. 6 expands on the limitations and future work.

2 Related Work

In recent years, the use of robots in schools to help the students acquire knowledge and skills has been one of the most researched user-case studies for social robots and Human-Robot Interaction (HRI) outside laboratories. Robots have been deployed in schools to study whether they could provide companionship and support [4], along with the more "traditional" roles of learning STEM-related topics [1–3, 22], enhancing creativity and motivation [1], and even helping with learning of a second language [6] or improving skills like collaboration and teamwork [18]. Apart from helping the children, it has been discussed that the use of robots also helps the teachers to introduce these topics by making them more confident in teaching certain topics, like the aforementioned computational thinking [5]. However, teachers in primary schools seem to have specific concerns about the use of robots in a classroom setting, linked to accessibility and fairness of use of this technology, but also its robustness in student's hands [8, 19, 20] and sometimes in the perceived lack of skills from the teachers themselves [7].

Similarly, block programming languages such as Scratch have increased their presence in classrooms due to their easiness of use and relative low requirements. Scratch is a language [12] developed by the Lifelong Kindergarten MIT Media Lab that consists in coloured blocks that interlock with each other in specific ways; this avoids compilation errors, as the code doesn't run if the blocks are not attached. Aside from their usefulness to teach programming as such [10, 14], Scratch has helped to foster improvement in areas such as problem-solving, creativity and critical thinking skills [13] or some areas of computational thinking [9, 17].

Of course these two aspects are sometimes interlinked, as some educational robots allow students to program them; if so, this programming tends to be conducted via a block programming language. This happens in robots like the LEGO Mindstorms Education v3, the Maqueen BBC micro:bit robot and others. In this case, the potential for educational robots to fail while in use by students might bring undesired consequences, such as loss of children's perceived capability of solving issues or enjoyment of the activity altogether. We aim to study if these consequences actually occur in the classroom.

3 Experimental Design

The session presented in this paper was primarily designed to provide the children some hands-on experience of programming a small robot with its own native system, which is a block programming language named MakeCode. This language is similar to Scratch [12], which is the language the students use during their regular lessons.

Integrated into this activity, we designed an experiment that would allow us to establish whether the children perceived faults within the robots and, if so, whether this impacted their perception and particularly enjoyment of the activity.

3.1 Session

Design Each session was designed to program most features of the robot, using its different functionalities and incrementally building on the previous programs throughout the lesson. The children worked in pairs and had one robot and a computer per pair. The activity lasted for an hour and 20 min, and comprised of 4 parts:

1. The first part was an introduction to the session, the robot and its sensors, and connecting the BBC micro:bit Maqueen robot with the computer and its native programming language, MakeCode.
2. Following that initial introduction, the children were shown how to program the LED board to show messages and make the robot "sing" or emit sounds. The researcher guided the students using a screen and the children were instructed to follow the instructions presented. Upon completion, the children were given 5 min to alter the program to personalize the message and sound.
3. The children were then shown how to make the robot move, again following the instructions from the researcher. Again, once this had been shown they were given 10 min to make it move by using the motors and controlling its speed.
4. To conclude the session, the children were shown how to use the ultrasound sensor to avoid colliding with obstacles, being it other robots, furniture or their own hand; they were shown how to make the robot stop when the sensor detected an obstacle in range. As this part was more complicated to program, the children were instructed to just follow instructions without customisation.

If the time permitted, the children were provided with a demonstration of the line-following feature of the robot, but only as a demonstration of other functionalities that we did not have time to cover.

Robot The robot used was the BBC micro:bit board with a Maqueen type body. This robot had different sensors and hardware (see Fig. 1), including:

- An LED board
- An ultrasonic sensor
- 3 light sensors used for line-following (not used in the session)
- Speaker

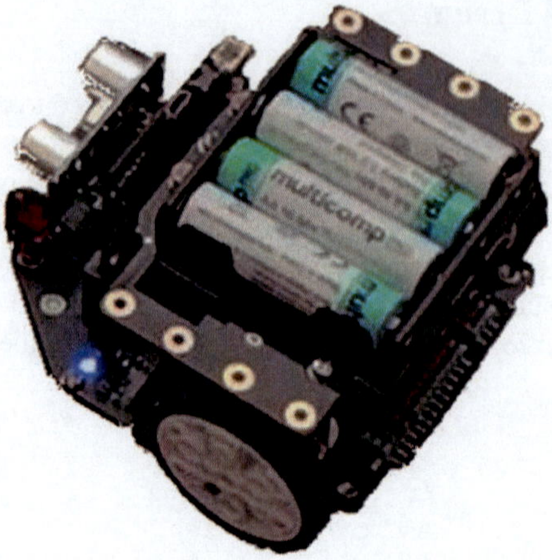

Fig. 1 The Maqueen robot

- Pins to connect other sensors/peripherals (not used in the session)

3.2 *Participants*

To present these results we obtained the parent's or tutor's consent to use the children's data. This study was approved by the Ethics Committee of the University of Hertfordshire under number aSPECS SF UH 05014(1).

We had 128 responses from children, 70 male and 58 female, with a range of ages from 7 to 12 years old (M=9,18 and SD=1,061).

3.3 *Conditions*

There were four conditions in this experiment; three of the conditions contained a "failure" of some condition that the children would become aware of and have to fix during the session. The conditions were as follows:

- **Condition 1 (C1)**: *No failure*. The robot was in perfect condition, without any introduced issues.
- **Condition 2 (C2)**: *Sound off*. The sound is controlled by a toggle in the micro:bit board, and the toggle was moved to the "off" position, so no sound was coming out. This was not immediately obvious without instruction of exploration of the device.

Table 1 Distribution of conditions in the experiment

Condition	Number of answers	Percentage (%)
No Failure (C1)	35	27,3
Sound off (C2)	32	25,0
Pin sensor moved (C3)	31	24,2
Battery slightly out (C4)	30	23,4

- **Condition 3 (C3)**: *Pin sensor moved.* The ultrasound sensor has 4 pins (see Fig. 1); one of them was out of place, so the sensor would not be able to detect anything and the robot would not stop when close to an obstacle.
- **Condition 4 (C4)**: *Battery slightly out.* One of the four batteries in the case was taken slightly out of the case so it was not making contact, therefore the robot would not work.

With the 128 questionnaires, the distribution of conditions was as follows in Table 1.

3.4 *Questionnaire*

The questionnaire consisted of 10 questions, which were related to the failures of the robot and other questions related to their enjoyment of the session and the children's perception of the robot. The questions were:

1. Did your robot have any fault?
2. If yes, did you manage to repair it?
3. Did you enjoy looking for faults on this robot?
4. If your robot had a fault, did you enjoy repairing it?
5. If your robot had a fault, was it easy to repair it?
6. Was it fun playing with this robot?
7. Did you find this robot lovable?
8. Do you think this robot is more like an animal or a machine?
9. Do you think this robot could be a real playmate?
10. Do you thinks this robot could be more like a toy, a friend or a machine?

Questions 3, 4, 5, 6, 7 and 9 were answered using a Likert scale of 5 points, with possible answers being "Definitely No", "No", "Maybe", "Yes" or "Definitely Yes". Questions 4 and 5 had an extra box to tick if the robot didn't have any faults. Question 1 was a Yes/No type of question, while Question 2 added the option to tick the box if the robot didn't have any fault. Question 8 had the options "Animal" and "Machine", and Question 10 added the option "Toy" to the two options mentioned before.

Table 2 Average answer for the Likert-scale questions

	Q3	Q4	Q5	Q6	Q7	Q9
Condition 1	4,03	3,91	3,26	4,77	4,23	4,09
Condition 2	3,61	3,96	2,88	4,83	3,88	3,56
Condition 3	4,27	4,16	3,33	4,90	4,23	4,06
Condition 4	3,90	3,76	3,00	4,60	3,93	3,55

Table 3 Cross-tabular between Condition x Q1

Q1: Did your robot have any faults?

		N/A	No	Yes
Condition	No failure	1	17	17
	Sound OFF	2	10	20
	Sensor pin moved	0	11	20
	Battery slightly out	0	7	23

4 Results

As a starting point, we calculated the average of answers per condition for all the questions that were answered with a Likert scale, with 1 being "Definitely No" and 5 being "Definitely Yes". The "Not Faulty" tickbox was not counted towards this average. The results are all aggregated in Table 2.

4.1 Failures

Question 1 For question 1, "*Did your robot have any faults?*", 35,2% of the children answered "No", while 62,5% answered "Yes" and the rest left it blank.

As we were interested in knowing if the faulty robot had an impact on the children's perception, we performed a Chi-Square Test between the conditions and the answer to Q1 with a $p=0.188$, which shows that these two variables are independent from each other. Nevertheless, as it can be seen in Table 3, when the robot does not have any fault (Condition 1), the children are as likely to think that it had a fault as to think it was non-faulty; however, when a fault is introduced in the robot, the children thought it was faulty about twice as much as they thought it was non-faulty, sometimes even more.

Question 2 For question 2, "*If the robot had a fault, did you manage to repair it?*", 10,9% of the children answered "No", while 62,5% answered "Yes" and the other 22,7% said it did not have any faults.

Table 4 Cross-tabular between Condition x Q3

Q3: Did you enjoy looking for faults on this robot?

		Definitely No	No	Maybe	Yes	Definitely Yes
Condition	No failure	5	0	1	7	15
	Sound OFF	0	3	7	9	5
	Sensor pin moved	1	0	6	6	13
	Battery slightly out	2	0	8	7	10

Table 5 Cross-tabular between Condition x Q4

Q4: If your robot had a fault, did you enjoy repairing it?

		Definitely no	No	Maybe	Yes	Definitely yes	No failures
Condition	No Failure	3	0	2	6	6	11
	Sound OFF	0	4	2	4	10	5
	Sensor pin moved	1	2	3	5	11	5
	Battery slightly out	4	0	1	13	5	4

Questions 3 and 4 Questions 3 and 4 asked if the children enjoyed looking for faults in the robot and if they enjoyed repairing it.

As for looking for faults, over 2/3 of the children said they enjoyed looking for these faults (42,2% said "Definitely Yes" and 26,6% said "Yes"); 18,8% said "Maybe", 4,7% said "No" and 6,3% said "Definitely No".

Coming into repairing it, 24,2% claimed the robot didn't have any fault to repair, while 32% said they "Definitely Yes" enjoyed repairing it. 25,8% said "Yes", 6,3% said "Maybe", 5,5% said "No" and 6,3% said "Definitely No". Overall, their enjoyment was slightly less, but close to 4 ("Yes") in all the conditions, except in Condition 3, when it surpassed 4.

When running the Chi-Square tests between the condition and their enjoyment looking for faults and repairing them, the results were statistically significant: p=0,014 for Q3 and p=0,018 for Q4. The cross-tables are found below, in Tables 4 and 5.

Question 5 Lastly, question 5 asks if the faulty robot was easy to repair. In this question answers were more mixed, as 8,6% said "Definitely No", 14,8% said "No", 20,3% said "Maybe", 20,3% said "Yes" and 10,2% said "Definitely Yes". Lastly, 25,8% of the children said there wasn't any fault to repair.

4.2 Fun

Question 6 For question 6, "*Was it fun playing with this robot?*" 83,6% of the children answered "Definitely Yes", while 10,9% answered "Yes", 3,9% answered "Maybe" and only 1 person answered "Definitely No" (0,8%). 1 other children left this question blank.

Again, we were interested in knowing if being in any specific condition lead to a variation of enjoyment, or if the perception that the children had of the faultiness of the robot had an impact on their fun. We conducted Chi-Square Tests on both these assumptions but we found that none of these factors were significant ($p=0.193$ and $p=0.861$ respectively).

4.3 Others

Other questions asked whether the robot was seen as "lovable", whether the children think it could be a real playmate or whether it was more like an animal, a toy, a friend or a machine.

Question 7 For question 7, "*Did you find this robot lovable?*" 44,5% of the children answered "Definitely Yes", while 28,1% answered "Yes", 17,2% answered "Maybe", 7,6% said "No" and only 1,6% of the children answered "Definitely No" (0,8%). 1 other children left this question blank. The Chi-Squared test conducted shows that the different failures don't have a significant impact in the children's answers ($p=0{,}37$).

Question 8 This question asked the children to decide whether the robot was more like an animal or a machine. 72,7% of the children thought it was like a "Machine", and only 24,2% thought it was like an "Animal". 4 other children left this question blank.
We also run some Chi-Squared tests to see whether the different faults present in the conditions impacted their answers ($p=0{,}781$) and whether their perception of the robot being faulty had an effect on how they viewed the robot ($p=0{,}824$). None of these are significant.

Question 9 This question asked the children to decide whether the robot could be a real playmate of theirs. 43% of the children answered "Definitely Yes", while 19,5% answered "Yes", 21,9% answered "Maybe", 6,3% said "No" and 8,6% of the children answered "Definitely No". 1 other children left this question blank. A Chi-Squared test was conducted to determine whether the different failures affected the children's view, but it was not significant ($p=0{,}343$).

Question 10 This question, similarly to question 8, asked the children to decide whether the robot was more like a toy, a machine or a friend. Although the instructions said to circle one, some children thought this was not representative enough of their opinions (see Sect. 6 for a more detailed explanation). However, for those children that circled only one answer, 46,1% said it was more like a "Friend", where 25% said it was a "Machine" and 21,1% said it was a "Toy". The rest either left this question blank or answered circling more than one option.

As in question 8, we run the Chi-Squared tests to see whether the actual failures or the perception of them had an impact on the answers, but they were not significant either ($p=0,832$ and $p=0,976$ respectively).

5 Discussion

Children perceived the robot to be less faulty than what they actually were, because only around a quarter of the robots used were non-faulty robots (27,3%), but children perceived them to be non-faulty 35,2% of the time. However, there wasn't any significant difference between the conditions and the perception of failure, which suggests two things: on the one side, when a robot wasn't faulty children were probably ascribing other things to failures that were not the ones we introduced. On the other side, children might have assumed that the robot wasn't defective when there was a failure in place.

Despite the children thinking the robot was faulty, they claimed they managed to repair it in more than 60% of the cases. It is worth noting that, since the activity was primarily devised to engage them with programming the robot, there were staff on hand (including their own school teachers) to help the children repair these "faults" so they could continue with the task. Despite that, the children didn't think that repairing the faults was particularly easy, with only a 30,5% answering "Yes" or "Definitively Yes" to that question. The lowest score for this answer was in Condition 2; although not significant, the average answer was lower than in the other conditions. Children in this condition also statistically gave less enthusiastic answers to the question of whether they enjoyed looking for faults in the robot, compared to the other conditions. This might have been because Condition 2, where the sound toggle was turned into the OFF position, was probably somewhat easy to notice if it didn't work, but may not be particularly easy to see: the toggle is located in the board, where the children might not think to look. Is it interesting to note too that 25,8% of the children answered their robot didn't have any failure, contrasting with the 35,2% that answered that their robot didn't have any fault in Question 1. However, although it may not have been easy to repair, it did seem that repairing them was enjoyable for most of the children, as more than 55% of children said they did enjoy repairing the robot, with differences within condition, as children in Condition 4 (battery slightly out) were less enthusiastic than their peers in other faulty conditions.

Taking the experience of the in-class session into the discussion, a possible explanation for these results and the variation of the percentage of non-faulty robots in different conditions is that some children might have mistaken software or coding mistakes for "failures". Since software errors are quite easy to "fix" by following the guiding screen, but they are also very easy to perceive, as the robot didn't work properly, this might have muddled the understanding of what a "failure" is; we also didn't provide any explanation, as we didn't want the children to be primed to think there were faults before the experiment. This could be an explanation that would help to clarify why about half of the children in the non-faulty condition still thought that their robot was faulty.

Children also had a lot of fun playing with the robot, with only 1 person giving a fully negative answer to that question while more than 90% of the children said they did had fun. This fun wasn't impacted by the actual "faultiness" of the robot, so all the different conditions performed about the same in terms of fun; in fact, their average score in all cases was 4,6 or higher. Furthermore, the children's perception of the robot's faultiness also didn't affect the fun that they had.

Another point of interest is the (possible) attachment between the children and the robot. Almost all children identified the robot as a machine when asked specifically between animal and machine (73%), and this did not depend on whether they perceived the robot to be faulty or whether the robot actually was faulty. Perhaps surprisingly given the previous statement, more than 50% said it was lovable and it could be a real playmate (62,5%). However, if the question was between machine, toy or friend, almost half of them decided it was more like a "Friend", not a machine; again this wasn't related to the perception of the faultiness nor to the actual failures present in the robot.

6 Conclusions and Future Work

6.1 Conclusions

Children expressed enjoyment of the session and fun working with the robot, even if their perception was that the robot was faulty to some degree. Their enjoyment of the activity was unrelated to their perception of the robot's failures, so it seems a pretty robust variable that doesn't get affected by a one-off failure. On top of that, most children expressed that they enjoyed repairing the robot, which might have impacted on the activity as a whole too.

6.2 Limitations and Future Work

The limitations on this work include the session being a one-off session; it is possible that some of the findings (like the enjoyment of the session) would change if the robot kept being faulty over a span of time. The novelty effect could also come in play, with children being very interested and engaged with a novel piece of equipment (the robot) and then gradually losing interest with time, as it has been discussed in previous literature.

As in the classroom it also seemed like some of the children were unable to differentiate between a failure and a software error, it would have been interesting to ask them specifically what did they think the failure was, to be able to distinguish whether they were aware of this differentiation or not and their specific insights on what is a "failure".

In the class, some children expressed difficulties in understanding what was the meaning behind the use of the expression "lovable", as well as difficulties in answering to Q8 and Q10 due to them being similar. Some of them also argued that it could be a "friend machine" or "toy machine", so they could not choose one before the other. We need to better specify our questions so they come across unambiguously.

For future work, it would be interesting to continue in the direction of children having to explain (or select from a list) what they think is a failure, as this would let us prepare and correct it in other scenarios, should we want to. Also, it would also be helpful to know whether this would impact their willingness to keep working with the robot on a long-term basis, as this was not explored in our session.

References

1. Anwar, S., Bascou, N.A., Menekse, M., Kardgar, A.: A systematic review of studies on educational robotics. J. Pre-Coll.E Eng. Educ. Res. (J-PEER) **9**(2), 2 (2019)
2. Baxter, P., Ashurst, E., Read, R., Kennedy, J., Belpaeme, T.: Robot education peers in a situated primary school study: personalisation promotes child learning. PloS One **12**(5), e0178126 (2017)
3. Benitti, F.B.V.: Exploring the educational potential of robotics in schools: a systematic review. Comput. Educ. **58**(3), 978–988 (2012)
4. Broadbent, E., Feerst, D.A., Lee, S.H., Robinson, H., Albo-Canals, J., Ahn, H.S., MacDonald, B.A.: How could companion robots be useful in rural schools? Int. J. Soc. Robot. **10**, 295–307 (2018)
5. Chalmers, C.: Robotics and computational thinking in primary school. Int. J. Child-Comput. Interact. **17**, 93–100 (2018)
6. Chang, C.W., Lee, J.H., Chao, P.Y., Wang, C.Y., Chen, G.D.: Exploring the possibility of using humanoid robots as instructional tools for teaching a second language in primary school. J. Educ. Technol. Soc. **13**(2), 13–24 (2010)
7. Chevalier, M., Riedo, F., Mondada, F.: How do teachers perceive educational robots in formal education? A study based on the Thymio robot. IEEE Robot. Autom. Mag. **1070**(9932/16), 1–8 (2016)
8. van Ewijk, G., Smakman, M., Konijn, E.A.: Teachers' perspectives on social robots in education: an exploratory case study. In: Proceedings of the Interaction Design and Children Conference, pp. 273–280 (2020)
9. Fagerlund, J., Häkkinen, P., Vesisenaho, M., Viiri, J.: Computational thinking in programming with scratch in primary schools: a systematic review. Comput. Appl. Eng. Educ. **29**(1), 12–28 (2021)
10. Good, J.: Learners at the wheel: Novice programming environments come of age. Int. J. People-Oriented Program. (IJPOP) **1**(1), 1–24 (2011)
11. Lakatos, G., Wood, L.J., Zaraki, A., Robins, B., Dautenhahn, K., Amirabdollahian, F.: Effects of previous exposure on children's perception of a humanoid robot. In: Social Robotics: 11th

International Conference, ICSR 2019, Madrid, Spain, November 26–29, 2019, Proceedings 11, pp. 14–23. Springer (2019)
12. Maloney, J., Resnick, M., Rusk, N., Silverman, B., Eastmond, E.: The scratch programming language and environment. ACM Trans. Comput. Educ. (TOCE) **10**(4), 1–15 (2010)
13. Moraiti, I., Fotoglou, A., Drigas, A.: Coding with block programming languages in educational robotics and mobiles, improve problem solving, creativity & critical thinking skills. Int. J. Interact. Mob. Technol. **16**(20) (2022)
14. Moreno-León, J., Robles, G.: Code to learn with scratch? a systematic literature review. In: 2016 IEEE Global Engineering Education Conference (EDUCON), pp. 150–156. IEEE (2016)
15. Moros, S., Wood, L., Robins, B., Dautenhahn, K., Castro-González, Á.: Programming a humanoid robot with the scratch language. In: Robotics in Education: Current Research and Innovations, vol. 10, pp. 222–233. Springer (2020)
16. Rossi, A., Moros, S., Dautenhahn, K., Koay, K.L., Walters, M.L.: Getting to know kaspar: effects of people's awareness of a robot's capabilities on their trust in the robot. In: 2019 28th IEEE International Conference on Robot and Human Interactive Communication (RO-MAN), pp. 1–6. IEEE (2019)
17. Sáez-López, J.M., Román-González, M., Vázquez-Cano, E.: Visual programming languages integrated across the curriculum in elementary school: a two year case study using "scratch" in five schools. Comput. Educ. **97**, 129–141 (2016)
18. Scaradozzi, D., Sorbi, L., Pedale, A., Valzano, M., Vergine, C.: Teaching robotics at the primary school: an innovative approach. Procedia - Soc. Behav. Sci. **174**, 3838–3846 (2015)
19. Serholt, S., Barendregt, W., Leite, I., Hastie, H., Jones, A., Paiva, A., Vasalou, A., Castellano, G.: Teachers' views on the use of empathic robotic tutors in the classroom. In: The 23rd IEEE International Symposium on Robot and Human Interactive Communication, pp. 955–960. IEEE (2014)
20. Serholt, S., Barendregt, W., Vasalou, A., Alves-Oliveira, P., Jones, A., Petisca, S., Paiva, A.: The case of classroom robots: teachers' deliberations on the ethical tensions. Ai Soc. **32**, 613–631 (2017)
21. UK-RAS Network: Uk festival of robotics - uk-ras network (2023). https://www.ukras.org.uk/robotics-week/. Accessed 29 Jan 2023
22. Yesharim, M.F., Ben-Ari, M.: Teaching computer science concepts through robotics to elementary school children. Int. J. Comput. Sci. Educ. Sch. **2**(3) (2018)

Effects of Introducing a Learning Robot on the Metacognitive Knowledge of Students Aged 8–11

Marie Martin, Morgane Chevalier, Stéphanie Burton, Guillaume Bonvin, Maud Besançon, and Thomas Deneux

Abstract Artificial Intelligence (AI) and Machine Learning education are entering the classrooms, and yet, the link between their introduction and the development of metacognitive components in students still needs to be addressed. We conducted an experiment with 138 elementary school students (aged 8–11) and tested how the manipulation of a learning robot affected their understanding of the basics of AI, as well as their metacognitive knowledge such as growth mindset, status of error, learning by trial-and-error, and persistence. Results show a positive shift both in students' AI knowledge and learning beliefs, and thus demonstrate the value of teaching the basics of how AI works to develop solid metacognitive knowledge that promotes learning. Future works should measure a lasting effect on students' learning behavior and focus on teacher training in new AI activities.

Supported by the FIRE Ph.D. program funded by the Bettencourt Schueller foundation and the EURIP graduate program (ANR-17-EURE-0012).

M. Martin (✉) · T. Deneux
Université Paris-Saclay, CNRS, Institut de Neurosciences Paris-Saclay, Paris, France
e-mail: marie.martin.2@cnrs.fr; marie.martin@cri-paris.org

T. Deneux
e-mail: thomas.deneux@cnrs.fr

M. Martin
Université Paris Cité, Paris, France

M. Chevalier · S. Burton · G. Bonvin
Haute École Pédagogique du canton de Vaud (HEP-VD), Lausanne, Switzerland
e-mail: morgane.chevalier@hepl.ch

S. Burton
e-mail: stephanie.burton@hepl.ch

G. Bonvin
e-mail: guillaume.bonvin@hepl.ch

M. Besançon
Rennes 2 University, LP3C, Rennes, Rennes, France
e-mail: maud.besancon@univ-rennes2.fr

R. Balogh et al. (eds.), *Robotics in Education*, Lecture Notes in Networks and Systems 747, https://doi.org/10.1007/978-3-031-38454-7_15

Keywords Robotics · Artificial intelligence education · Machine learning · Learning robot · Metacognitive knowledge · Elementary school

1 Introduction

Artificial Intelligence (AI) is invading students' daily lives, both at home and at school (e.g., with Amazon's Alexa AI, OpenAI's Chat GPT, etc.). While the added value of AI is compelling in many fields, its arrival raises ethical questions, such as human autonomy, data collection, algorithmic bias, and deepfakes [4]. This is why a part of the scientific community has recently been interested in the question of teaching AI in elementary school. The current research aims at understanding the nature of the concepts that can be taught to the elementary age group [7, 9, 11]. However, no studies so far have taken advantage of the similarities between the learning strategies implemented into AI processes and those present in human cognitive processes to create new links between AI education and the development of metacognitive components in elementary school students, another major issue for education [8].

The present study addresses this lack by examining the relationship between the introduction of a learning robot and the development of metacognitive knowledge in students aged 8–11.

2 Background

Previous studies have shown that students, as early as elementary school, can understand the basic concepts of AI, and especially of Machine Learning (ML) [3, 10]. However, although AI systems are inspired by theories of human cognition [13], the link between the introduction of AI in the classroom and the development of metacognitive components in students needs to be clarified.

2.1 *Metacognitive Knowledge*

Metacognition, that is, the awareness, understanding, and control of one's own cognitive processes [5], is considered essential to student success [8]. Metacognitive knowledge (MK), one of the components of metacognition, includes knowledge that a person has about their own and others' cognitive processes. It involves knowledge of one's own abilities, tasks, learning strategies (declarative knowledge), how to use these strategies (procedural knowledge), and when and why to use them (conditional knowledge) [5, 8]. To implement and monitor effective learning strategies, students, therefore, need to have this meta-knowledge in advance [8] and feel able to improve their performance. Our study thus focuses on the following MK:

- **Growth Mindset** consists in the belief that, despite initial ability, gender, or social background, everyone's intelligence can be developed; it has been shown to strongly influence one's motivation and behavioral strategies [2].
- **Error Status**: Even though studies have shown the benefit of making mistakes during learning when these are followed by feedback [12], for example, to promote exploration strategies, many students still perceive error as a fault and prefer to avoid it.
- **Learning by trial-and-error** is a strategy that students use when faced with a new learning situation and that requires a reflexive process based on experimentation. This metacognitive process promotes understanding insofar as the student uses planning, error monitoring, and the regulation of his or her actions to achieve the goal [6].
- **Persistence** is another key to academic success because it requires the student to keep working on a task even if it is difficult. It involves the student overcoming its discouragement [8]. It can be measured by the time or number of repetition a student stays engaged in an activity [1, 2].

The MK described above is an important component of academic success. However, several studies have shown that elementary students have only a partial grasp of it and may have erroneous beliefs [8] that can negatively impact their learning process. For this reason, students should have the opportunity to develop this knowledge through explicit teaching beginning in elementary school. Educational robotics seems promising in this respect.

2.2 Learning Robots for Teaching AI and MK: The Case of AlphAI

Reinforcement learning (RL) which is a part of ML, is inspired by human cognition [13], that's why we could establish links between some MK and RL functions, such as for instance trial-and-error learning. However, while educational robotics is considered as a lever for developing metacognition in elementary school students [4, 13], it seems that the very link between AI learning and the development of metacognitive components has not yet been addressed.

To date, the STEM literature has provided evidence that (1) the youngest students can understand the RL basics [13], (2) the use of learning robots, which are tangible objects, enables students to manipulate complex concepts in real-world situations, making them visible and, thus, better understood. Consequently, educational robotics seems an effective tool for introducing AI and its sub-fields as early as elementary school [4, 13].

In this regard, the learning robot AlphAI (developed at the CNRS, France, and commercialized by the Learning Robots company, France) and its activity panel offer

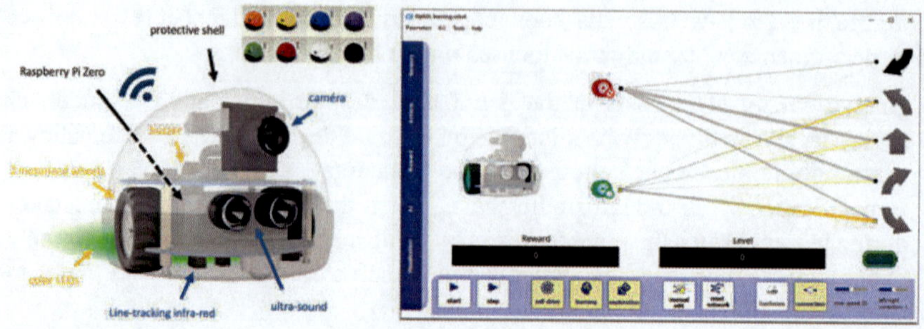

Fig. 1 AlphAI: a learning robot and its graphical interface

real learning opportunities to link MK and AI basics. The AlphAI kit (Fig. 1) consists of a learning robot and a graphical interface running on a computer from which one can manipulate the AI driving the robot, but also see in real-time algorithmic details such as neural connections, predictions, and the consequences of robots actions (reward system in RL). Thus, it makes ML explicit for the students, thanks to the tangible aspect of the robot (the student is then engaged in the "trial") and, thanks to its interface (the student can access to the error monitoring). In particular in the context of RL, students can observe the role of the reward system (returning rewards or penalties following actions) and the need to consider future rewards rather than only the immediate reward (persistence), the step-by-step changes of connections inside the neural network (requiring repetition for efficient learning), they can enable or disable exploration (interfering with the trial-and-error strategy).

The research question asked in this study is thus: What are the effects of introducing a learning robot to 8–11 year old students on their following MK: growth mindset, status of error, learning by trial-and-error and persistence?

3 Methods

3.1 Experimental Protocol

To answer the research question, we set up an experimental protocol with two equivalent groups of participants (see below "Participants"). After both groups passed a pre-test (T0, see Fig. 2), only the experimental group was subjected to pedagogical intervention teaching AI with the educational robot and split into 3 sessions. Following this intervention, the two groups (experimental and control) took a second test (T1) which was repeated 10 d later (T2) in order to ensure the permanence of the results obtained in T1. Right after this third test (T2 + 1 day), the control group was subjected to the pedagogical intervention in the same way as the experimental group (a few weeks earlier) and completed a fourth test (T3) together with the exper-

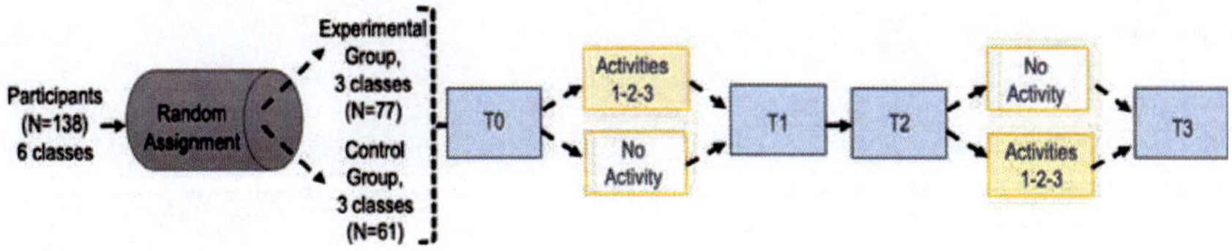

Fig. 2 Experimental protocol

imental group in order to compare all the participants afterwards. At the end of these post-tests, a semi-structured interview was offered to 6 students from each class (N = 36).

The 3 sessions of the pedagogical intervention (yellow boxes in Fig. 2), of a total duration of 6 h, were carried out as:

1. Introduction [1,5 h]: Teacher introduces concepts and questions the class about learning, intelligence, artificial intelligence. She presents the learning robot and a demo of RL.
2. Manipulation [2 h]: Observation of these concepts during an AI activity where students manipulate AlphAI learning robots and experiment 4 levels of autonomy (more details below in "AI activity").
3. Discussion [1,5 h]: Comparison of the behavior of the learning robot with the functioning of the human being, through a philosophical debate.

3.2 Participants

The sample consisted of 138 students in 6 classes of 3 different grades (Table 1) in an elementary school of the Paris area (France). As research on metacognition has shown that students are better able to make explicit knowledge of their own mental processes from the age of 8 [8], we selected elementary school grades relative to the associated age groups (8–11 years). These refer to the French school system (3rd grade, 4th grade and 5th grade).

Due to Covid restrictions we were not able to mix classes to homogenize the experimental and control groups in terms of level and gender, and rather assigned full classes to either the experimental or control groups. These groups appeared quite balanced though in terms of gender, but we have to keep in mind that the samples are not fully independent since all students from one class may share for example a history of teaching that other classes do not have and that can be beneficial with respects to our tests. Note also that the control grade 3 class could not attend the T1 test. The students' participation in this study was approved by their parents, teachers and principal. The recommendations provided by the General Data Protection Regulation (GDPR) in terms of collecting, analyzing, and disseminating sample data were rigorously followed.

3.3 AI Activity

The AI activity was an adaptation of the "Introduction to AI" activity that can be found on the Learning Robots website[1]. We proposed a robot to groups of 2 or 3 students. Students manipulated the robot inside a large wooden arena, and experimented 4 increasing levels of autonomy:

1. Remote control [15 min]: Students played remote controlling the robots; the robots had no autonomy.
2. Programming [20 min]: Students drew connections inside a mini artificial neural network and by doing so "coded" the program"if robot is blocked, turn backward, otherwise go forward". There was no AI yet at this stage.
3. Supervised ("Imitation") learning [40 min]: During a first "training" phase, students remotely controlled their robots inside a racing track and their AIs learned which action to choose as function of robots' camera image. In the second "use" phase, the trained AIs controlled the robots and, after some performance checking, a race was held between the autonomous robots.
4. Reinforcement ("Trial-and-error") learning [30 min]: Students observed the learning robot learning and the real-time impact of that learning on the neural network: The AI controlled the robot and received high bonus scores when it moved straight, but penalties when it hit a wall. Within about 10 min it was able to take actions that gave more bonus and less penalties (e.g. turning before walls). Students were able to speed-up learning by sometimes taking control of their robot and thereby "showing" the AI which actions to explore.

After each of the 4 manipulations the teacher discussed with the class and explicitly noted how much the robot was autonomous, which new features were demonstrated, and what was still missing as compared to human intelligence.

Table 1 Distribution of participants by grade level and age, gender, intervention group.

Class level and age of students	Total number of students	Percentage of male students (%)	Experimental group	Control group
grade 3 (8–9 y.o.)	49	49	26	23
grade 4 (9–10 y.o.)	47	51	23	24
grade 5 (10–11 y.o.)	42	57	28	14
Total (8–11 y.o.)	138	52	77	61

[1] See more details in activity G1 https://learningrobots.ai/resources/?lang=en.

3.4 Data Collection and Analysis

The questionnaire used in pre (T0) and post-test (T1, T2, T3) is identical. It is composed of 4 sections, respectively about (i) robotics, (ii) AI knowledge, (iii) students' interest in robotics and AI, and (iv) MK. This study focuses only on sections (ii) and (iv) (see individual questions in Fig. 5). To remove certain comprehension biases linked to reading difficulties, we read each item aloud as the students answered the paper questionnaire.

In the AI section students were given the choice between 3 responses (yes, no, I don't know). They were awarded 1 point per correct answer, 0 point per wrong answer, and 0.5 point for choosing "I don't know" (which is a more correct answer than choosing the wrong answer). In the MK section, a 4-point Likert scale was used, then a score between 0 and 1 was calculated as follows: for example if the correct answer was "I completely agree" the score was 0 for "I completely disagree", 0.33 for "I rather disagree", 0.67 for "I rather agree" and 1 for "I completely agree". Students' global scores in AI and metacognitive knowledge were obtained by averaging over questions.

We used two different statistical tests in our statistical analysis. When comparing scores between the experimental and control groups (Fig. 3) we compared the means over these groups using a bootstrapping method as follows. We randomly shuffled 200,000 times the membership of data points to these two groups, and each time computed the difference of the two means. The p-value of the actual difference between experimental and control was obtained as the fraction of random shuffles where the obtained difference was superior or equal. When comparing the scores of the same classes (Fig. 4c) or individuals (Fig. 5) on different days, we performed a sign test (Matlab function signtest). In both cases such nonparametric tests were preferred to tests making assumptions on the distribution shape, because as shown in Fig. 3 these distributions are far from being Gaussian. Since the control grade 3 class did not attend the T1 test, results in Fig. 3a, b have less control points for T1 but remain valid; and in Fig. 4a, b no average is calculated for the control classes at T1.

4 Results

4.1 Increase in Artificial Intelligence and Metacognitive Knowledge Following the Activity

We first compared the scores of students after the activity was proposed to the experimental group, but not to the control group (Fig. 3a). We found statistically significant increases in scores of experimental group students following the activity (T1) and 10 d later (T2) as compared to those of students in control groups. We also controlled that differences between the two groups were small before the activity (T0), and

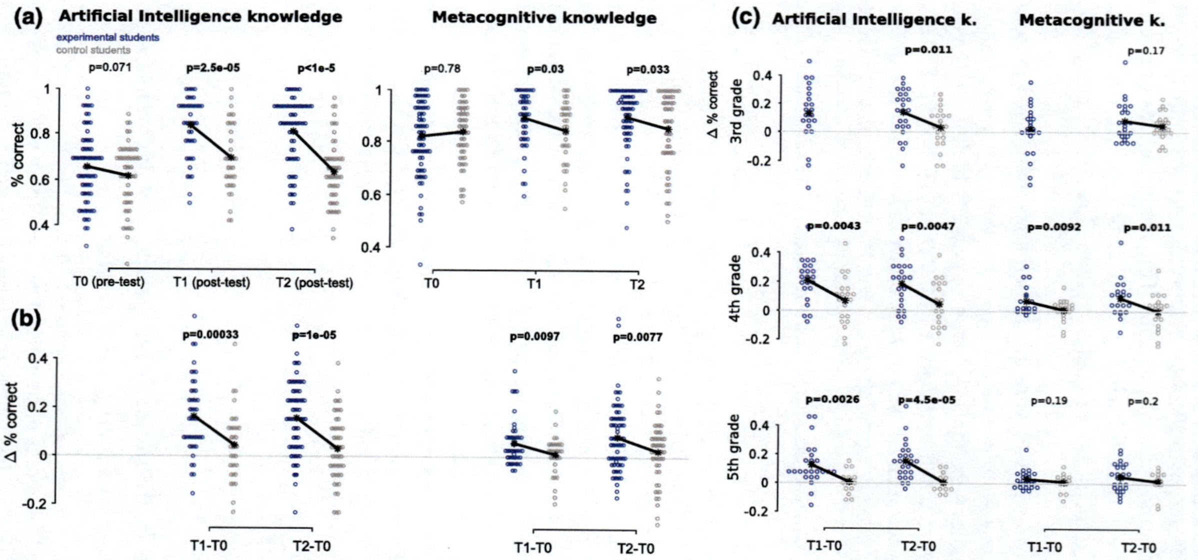

Fig. 3 Increase in AI and metacognitive knowledge for individual students between before and after the activity. **a** Average percentage of correct responses to the AI and MK tests conducted before the activity (T0), one week (T1) and two weeks (T2) later. Students belonging to the experimental group (blue points) followed the activity between T0 and T1, but not those belonging to the control group (gray points). Here and thereafter, p-values of the experimental versus control groups were obtained using a right-sided bootstrap method on the mean and are indicated in bold font when the difference is significant (i.e. $p < 0.05$). **b** Evolution of each student's score between T0 and T1, and between T0 and T2. **c** Evolution of the students' scores, displayed separately for the three levels

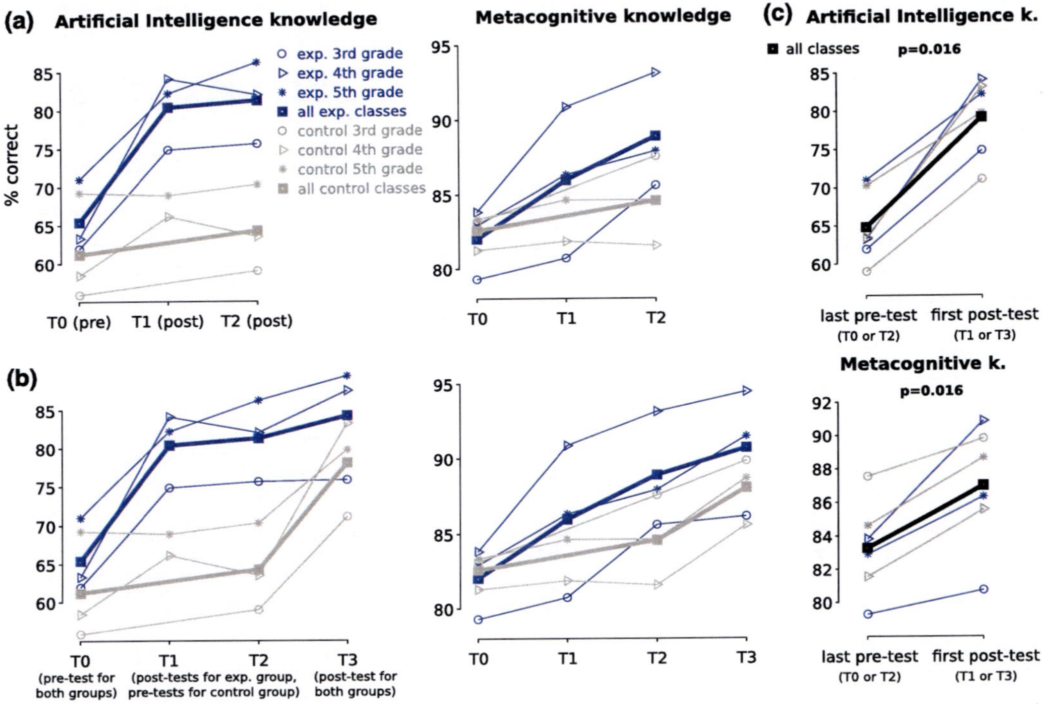

Fig. 4 Increase in AI and metacognitive knowledge averaged over classes. **a** Scores averaged over questions and over students for each individual class, and for the whole experimental and control groups. **b** The activity was also proposed to the control group and the tests were conducted a third time on all students (T3) a week after T2. An increase is observed between T2 and T3 for the control classes that is comparable to the increase between T0 and T1 for the experimental classes. **c** These increases are aligned and the average of the test scores over all classes immediately before and after the activity is displayed as well. The activity has a positive impact on the scores over all classes, which appears to be statistically significant (right-sided sign test)

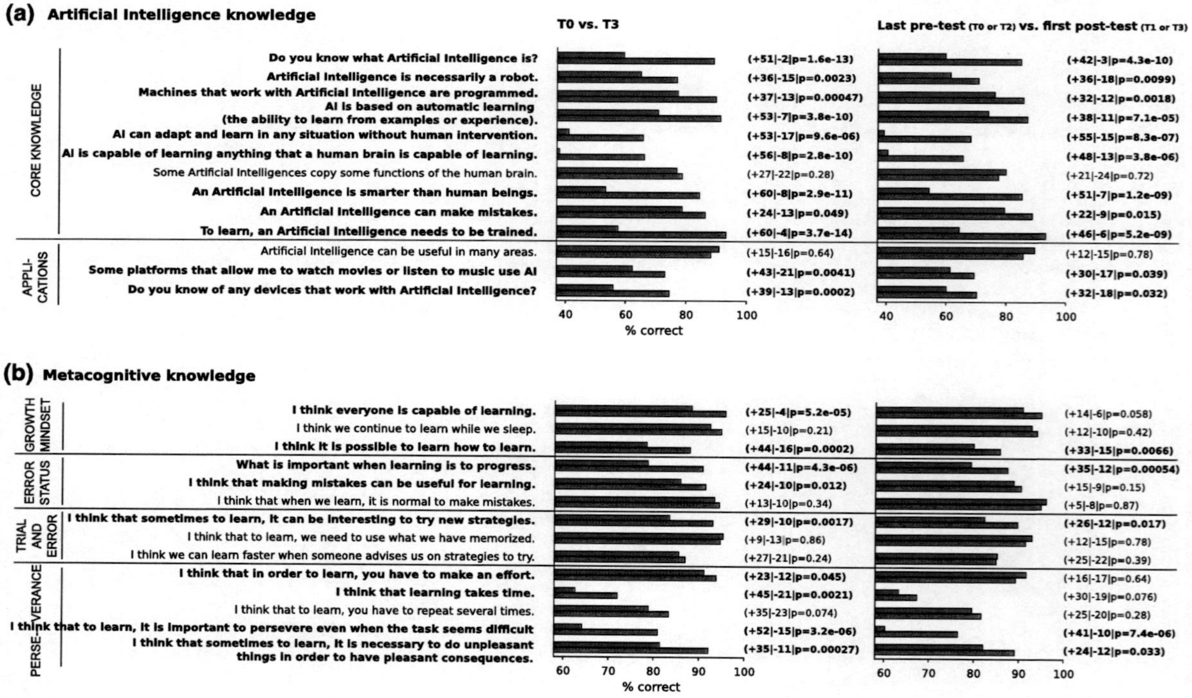

Fig. 5 Increase in correctness of the answers to individual questions related to AI (**a**) or metacognitive knowledge (**b**). For each question are indicated: the scores to the tests T0 and T3 (center) and immediately before and after the activity (right; defined as in Fig. 4), averaged over all students; the number of students who increased or decreased their score between the two tests (for example +51 increases and -2 decreases for the first AI question between T0 and T3); and the statistical significance of the improvement (right-sided sign test; bold font if $p < 0.05$)

not statistically significant (which is important since the classes could not be mixed between experimental and control groups).

To limit the effect of heterogeneity within groups, we subtracted for every student' score at T1 and T2 his or her score at T0 (Fig. 3b). First, we observed that on average students increased their score, in the experimental group as expected, but also in the control group. The increase in AI knowledge by control students was not found to be significant (sign test, $p = 0.11$ between T0 and T1, $p = 0.19$ between T0 and T2). But interestingly, their increase in MK was found to be significant ($p = 0.01$ between T0 and T1, $p = 0.009$ between T0 and T2), which indicates that the sole exposure to our questionnaire led the students to reflect on the metacognitive questions and answer them better a couple of days later. In any case the comparison between experimental and control groups showed that increase in both AI and MK were very significantly stronger for the students who attended the activity ($p < 0.01$ in all cases).

We then scrutinized how these increases were dependent on age groups and repeated our analysis separately for the three class grades (Fig. 3c). The 3 experimental classes progressed more than the 3 control classes, both for AI and MK, and with strong statistical significance in the case of AI. In the case of MK, statistical significance was obtained only in the case of grade 4, i.e. students of median age. We did not observe a trend as a function of age; therefore, despite the established knowledge that metacognition develops with age [8], observed differences between classes were probably due to other unknown variables (e.g. class learning history).

To better appreciate variability between classes, we plotted average results over classes as a function of time (i.e. test number, Fig. 4a). We observe that despite varying initial scores at T0, clear improvements in AI and MK occurred for the experimental group as compared to the control group. To compare and reinforce these results, the same activities were proposed to the control group two weeks later (Fig. 4b). They led to score increases between T2 and T3 noticeably steeper than the mild increases between T0 and T2. Altogether, despite their heterogeneities, the 6 classes showed score increases in both AI and MK between the tests immediately before and after the activity (Fig. 4c), which again is a significant result (sign test, $p = 0.016$).

4.2 Responses to Individual AI Knowledge Questions

We then investigated the students' scores at the level of individual questions, to understand more qualitatively the effects of the activity (Fig. 5). In terms of AI knowledge, the students improved their scores significantly on all but two questions, regardless of whether the comparison was between the very first (T0) and last (T3) tests, or between the two tests immediately before and after the activity. This proves that they understood AI learning's basics very well.

The two questions without notable result were not directly addressed by the activity: AI applications ("AI can be useful in many areas": +15 students increased their scores between T0 and T3, while 16 others decreased it) and AI inspiration ("Some Artificial Intelligences copy some functions of the human brain.": $+27| - 22$).

Among the answers that interest us to make the link with MK, students have well understood that an AI learns through training (from examples $+53| - 7|p = 3.8$e-10 or through trial-and-error $+38| - 11|p = 7.1$e-05) and by repetition ($+60| - 4|p = 3.7$e-14 and $46| - 6|p = 5.2$e-09), and that it can therefore make mistakes ($+24| - 13|p = 0,049$ and $+22| - 9|p = 0,0154$).

4.3 Responses to Individual MK Questions

In terms of metacognitive knowledge, students, on average, already had positive beliefs at T0 (+50% correct answers), yet their scores after our intervention increased for 13 of the 14 questions between T0 and T3, with a significant impact for 9 of them. Note that students in the experimental group benefited from the activities earlier than the control group (between T0 and T1 for the experimental group and between T2 and T3 for the control group), which allowed them to improve their scores not only at T1 but progressively until T3 (Fig. 4b). Comparing only between the tests immediately before and after the activity, we measure significant improvements for 5 of the 14 questions. To track changes in students' meta-knowledge, we assigned their responses to the following 4 categories described in the Background section.

Growth Mindset. Students' scores were already high at T0 and continued to progress, with significant improvements regarding the possibility of learning to learn (T0/T3: +44|−16|p=0.0002) and the idea that everyone is capable of learning ($+25| - 4|p = 5.2$e-05). The question "I think we continue to learn when we sleep", though addressed during the activity (the robot itself continues to learn once stopped) did not show improvement.

Error status. Here also students significantly improved on 2 out of 3 questions ("what is important when learning is to progress": $+44| - 16|p = 4.3$e-06; "mistakes can be useful": $+24| - 10|p = 0.012$. The third question ("it is normal to make mistakes") already had a very good score already at T0 and therefore was less keen to further improve.

Learning by trial-and-error. Students were receptive to the main message of the "trial-and-error" strategy ("I think that sometimes, it can be interesting to try new strategies": $+29| - 10|p = 0.0017$) but were less sensitive to details provided during the activity such as memorization of the attempted strategies and guidance by a master of the learner's trials (during the RL part of the activity, the students could guide the robot exploration and thereby accelerate their learning). In future experiments these aspects will deserve being more stressed by the teacher.

Persistence. Persistence is the topic for which students had initially the lowest scores and where they improved the most (T0/T3 comparison: improvement on 5/5 questions, statistically significant on 4/5). Watching the robot strive for some time and eventually succeed was certainly a key element for improving the students' beliefs.

5 Discussion

5.1 Children's Understanding of Machine Learning After Participating in Educational Activities About AI

Similar to recent work showing that elementary school students understand the AI basics such as supervised machine learning [3, 10], our results show that as early as age 8, students understand that (1) supervised learning depends on the example data provided during the training phase, (2) that the AI can make errors that are due to the quality and accuracy of the provided example data, and (3) that an AI is not more intelligent than a human. Moreover, in line with previous hypotheses [13], our study also contributes to proving that students as young as 8 years old are capable of understanding also more complex operating principles like RL. We hypothesize that our device contributed strongly to this development thanks to a "mise en abyme" or a mirroring effect. Indeed, the AlphAI kit allowed students to "tinker" with AI [9], placing manipulation at the center of the activity. Moreover the pedagogy of the AI activity (Sect. 3.3) follows a structured scaffolding. Indeed, the first two phases of the AI manipulation (remote control and edition of a mini neural network) not only showcase robot usages without an AI but also prepares for the two subsequent appropriation phases: (i) remote control accustoms students to the interface; (ii) network edition lets them play the role of the AI without knowing it yet, as they try different connections to optimize the reward. Playing the role of the AI is also part of the MK pedagogy as it places the AI as a mirror of their learning.

This objectification of ML modalities [4, 13], coupled with discussions to challenge the concept of intelligence, allowed for an examination of the similarities but also differences that exist between artificial intelligence and human intelligence, and thus contributed to a clear understanding of AI and its limitations [7, 9, 11]. This underscores the importance of AI education starting in elementary school: enlightening students' everyday uses, but also giving them the keys to make informed decisions later on and to develop AI consistent with shared values according to the 2021–2025 UNESCO plan.[2]

[2] BEIJING CONSENSUS on artificial intelligence and education https://unesdoc.unesco.org/ark:/48223/pf0000368303.

5.2 *Changes in Metacognitive Knowledge After Participating in AI Educational Activities*

Consistent with the literature, our study shows that students' MKs can develop or improve through explicit instruction [8]. Thus, through guided and commented observation, we made these examples tangible [4], allowing students to identify strategies used by the robot in situation. In other words, by allowing students to see AlphAI learn "in real life" through trial-and-error, and repetition, we embodied the positive effect of these strategies [6, 12] on the robot's self-regulated learning, and through a mirror effect, on students' strategies. Similarly, the real-time visualization of the feedback about incremental changes in neural connections made the concept of the growth mindset concrete [2]. These concepts were then discussed (and thus made explicit) in the philosophical debate in light of their own learning experience, thereby promoting their deep and long-term integration (Fig. 4b: continuous and significant increase in the experimental group's scores until post-test 3).

Our study amply demonstrates the value of teaching the basics of how AI works both for its own sake and for learning, thus enabling the development of a robust MK to support learning.

6 Conclusion, Limitations, and Perspectives

The purpose of the present study was to evaluate the impact of AI education on the development of MK in elementary school students. To do so, we developed and implemented an experimental protocol with 138 students aged 8–11 years. We relied on the statistical analysis of the responses to the questionnaire, comparing the results of our two groups (experimental and control) at T0 and T1, and then on those of the control group after they had also benefited from the educational activities (T2 versus T3).

Given our results, we state the benefits of carrying out at least 6 h of activities with the AlphAI device with students aged 8 to 11, in terms of understanding AI and the relevance of this teaching for the development of specific meta-knowledge such as growth mindset, the status of error, the benefit of experimental trial-and-error and the importance of persistence. Although the introduction of learning robots into schools and the curriculum is beneficial to students, more time (>6 h) needs to be devoted to such activities so that learning strategies go beyond the students' beliefs. Moreover, the students' relatively high score in the pre-test points to limit the fact that we did not mix the classes in our sample. A new experiment in another school will allow us to compare and consolidate these initial results. In addition, new activities involving the AlphAI kit will be developed and tested in order to deepen both the concepts of AI functioning and to act on the behaviour of students during activities. In the perspective of a longitudinal and translational study applied to the needs of the field, future studies will also have to follow students over several years and integrate teacher training.

References

1. Bernacki, M.L., Nokes-Malach, T.J., Aleven, V.: Fine-grained assessment of motivation over long periods of learning with an intelligent tutoring system: methodology, advantages, and preliminary results. In: International Handbook of Metacognition and Learning Technologies, pp. 629–644. Springer (2013)
2. Burleson, W.: Affective learning companions and the adoption of metacognitive strategies. In: International Handbook of Metacognition and Learning Technologies, pp. 645–657. Springer (2013)
3. Druga, S., Ko, A.J.: How do children's perceptions of machine intelligence change when training and coding smart programs? In: Interaction Design and Children, pp. 49–61 (2021)
4. Eguchi, A.: Ai-powered educational robotics as a learning tool to promote artificial intelligence and computer science education. In: International Conference on Robotics in Education (RiE), pp. 279–287. Springer (2022)
5. Flavell, J.H.: Metacognition and cognitive monitoring: a new area of cognitive-developmental inquiry. Am. Psychol. **34**(10), 906 (1979)
6. Freinet, C.: Méthode naturelle de lecture. FeniXX (1961)
7. Long, D., Magerko, B.: What is AI literacy? Competencies and design considerations. In: Proceedings of the 2020 CHI Conference on Human Factors in Computing Systems, pp. 1–16 (2020)
8. Simons, C., Metzger, S.R., Sonnenschein, S.: Children's metacognitive knowledge of five key learning processes. Transl. Issues Psychol. Sci. **6**(1), 32 (2020)
9. Touretzky, D., Gardner-McCune, C., Martin, F., Seehorn, D.: Envisioning AI for k-12: what should every child know about AI? In: Proceedings of the AAAI Conference on Artificial Intelligence, vol. 33, pp. 9795–9799 (2019)
10. Williams, R., Park, H.W., Oh, L., Breazeal, C.: Popbots: designing an artificial intelligence curriculum for early childhood education. In: Proceedings of the AAAI Conference on Artificial Intelligence, vol. 33, pp. 9729–9736 (2019)
11. Wong, G.K., Ma, X., Dillenbourg, P., Huan, J.: Broadening artificial intelligence education in k-12: where to start? ACM Inroads **11**(1), 20–29 (2020)
12. Yang, C., Potts, R., Shanks, D.R.: Metacognitive unawareness of the errorful generation benefit and its effects on self-regulated learning. J. Exp. Psychol.: Learn. Mem. Cogn. **43**(7), 1073 (2017)
13. Zhang, Z., Willner-Giwerc, S., Sinapov, J., Cross, J., Rogers, C.: An interactive robot platform for introducing reinforcement learning to k-12 students. In: International Conference on Robotics in Education (RiE), pp. 288–301. Springer (2022)

Non-verbal Sound Detection by Humanoid Educational Robots in Game-Based Learning. Multiple-Talker Tracking in a Buzzer Quiz Game with the Pepper Robot

Ilona Buchem, Rezaul Tutul, André Jakob, and Niels Pinkwart

Abstract While most research and development in social human–robot interaction, including interactions of learners with educational robots, has focused on verbal communication through speech (synthesis and recognition), research on non-verbal signals has just started to gain attention. Non-verbal sounds such as clapping, laughter, and mechanical sounds are often overlooked elements in design and research in social human–robot interaction. This paper presents the design of an educational buzzer quiz game for groups of learners, which is facilitated by the humanoid educational robot Pepper, detecting non-verbal sounds as part of the game-play. The paper describes the scenario and a prototype of a robot-supported buzzer quiz game in classroom settings, outlines the challenges in the real-time detection of a mixture of sounds, and presents the approach to real-time auditory and visual multiple-talker tracking, which has been applied in the buzzer quiz game with the Pepper robot.

Keywords Non-verbal sound detection · Game-based learning (GBL) · Humanoid robots · Educational robot · Pepper robot · Buzzer games · Technology-enhanced learning (TEL) · Multiple-talker tracking

I. Buchem (✉) · R. Tutul · A. Jakob
Berlin University of Applied Sciences, Luxemburger Str. 10, 13353 Berlin, Germany
e-mail: buchem@bht-berlin.de

R. Tutul
e-mail: rezaul.tutul@bht-berlin.de

A. Jakob
e-mail: andre.jakob@bht-berlin.de

N. Pinkwart
Humboldt University, Unter den Linden 6, 10099 Berlin, Germany
e-mail: pinkwart@hu-berlin.de

R. Balogh et al. (eds.), *Robotics in Education*, Lecture Notes in Networks and Systems 747, https://doi.org/10.1007/978-3-031-38454-7_16

1 Introduction

While most research and development in social HRI, including interactions of learners with educational robots, has focused on diverse aspects of verbal communication through speech (synthesis and recognition), research on non-verbal signals has just started to gain attention [1]. Non-verbal sounds such as clapping, laughter and non-verbal feedback such as "hmm-mm" and other non-verbal vocalizations (NVVs) [2], are important but often overlooked elements in design and research on technology-enhanced learning and human–robot interaction (HRI). In the context of social HRI, non-verbal signals comprise both non-verbal sounds produced by robots, e.g. motor sounds, music, non-linguistic audio signals such as beeps, squeaks, and clicks, and non-lexical sounds produced by humans, e.g. interjections such as "psst", "oops", "aha", and paralinguistic aspects of speech such as intonation patterns used to communicate affect and intent [1]. The review of semantic-free utterances (SFU), i.e. auditory interaction elements without semantic content, in the context of social HRI, has shown that SFUs play an important role in enhancing the expression of emotions and intentions [3]. Semantic-free utterances can be divided into four categories, i.e. gibberish speech (GS), non-linguistic utterances (NLUs), musical utterances (MU), and paralinguistic utterances (PU) [3]. SFUs play an important part in creating a robot's personality, which affects the learning experience [4].

While traditional speech recognition methods usually discard non-verbal vocalizations (NVVs) as non-speech "garbage" sounds, which have to be filtered out, the current researcher has explored the importance of NVVs in conversational interaction, for example in regards to their communicative and affective meaning. [2]. NVVs can be modeled for the development of dialogue systems for HRI [2]. From the perspective of robots in education, some types of semantic-free utterances (SFUs) and non-verbal vocalizations (NVVs), such as laughing, weeping, cheering, crying, or screaming [2], seem to be especially relevant in the design of multimodal interaction of learners with robots in naturalistic learning settings. The recognition and synthesis of affect sounds such as laughter, which is an important expressive behavioral cue for positive emotions [5], is an important yet overlooked aspect in the design of learning enhanced by educational robots. This may be resulting from the complexity of non-verbal communication, such as the use of laughter, the need for extensive datasets of non-verbal sounds, such as laughter episodes, and the lack of comprehensive methods for the detection and generation of non-verbal sounds. The study by [6] may serve as an example of the complexity of non-verbal communication on the human level. This exploratory study on the frequency of laughter as an interactional signal in cooperative and competitive game-based settings found that the frequency of human laughter depended on the familiarity with the task, i.e. there was a significant decrease in laughter frequency while playing a second round of a game [6]. The study by [7] may serve as an example of the complexity of non-verbal sound detection on the technical level. This study focused on Automated Laughter Detection (ALD) and investigated the problem of automatic identification and extraction of human laughter as a variable-length event from audio files in noisy environments [7].

To overcome the mismatch between clean training data in controlled environments and messy test data from naturalistic settings, the researchers used segment-based metrics, with segments defined as individual frames lasting 23 ms, and implemented a ResNet-based model for laughter detection as a robust machine learning method to detect human laughter in noisy environments [7].

Machine-produced non-verbal signals, which may be easier to detect, play an equally important role in social HRI. For example, studies focusing on the effects of robot-generated sounds, have found that the motor sounds of a robot can influence human impressions and expectations of a robot [8]. Depending on the context, the mechanic sounds of a robot can be perceived as disturbing. The study by [8] showed that in the context of social interaction with the robot, a silent robotic arm was rated as more human and more aesthetically pleasing compared to a robot arm producing sounds [8]. Also, the study by [9] explored aural impressions of the servo motors of a robot and found that sounds produced by servos may significantly interfere with interactions. In the context of education, other types of non-verbal sounds are used in game-based settings that could be potentially detected by educational robots and included in the gameplay. For example, buzzer quiz contests make use of the detection of buzzer sounds as an element in competitive game settings. The technical challenge is recognising which person or team presses the buzzer first [10].

This paper presents the design of an educational buzzer quiz game for groups of learners, which is facilitated by the humanoid educational robot Pepper, detecting non-verbal sounds as part of the game-play. The paper describes the scenario and a prototype of a robot-supported buzzer quiz game in classroom settings, outlines the challenges in the real-time detection of a mixture of sounds, and presents the approach to real-time auditory and visual multiple-talker tracking, which has been applied in the buzzer quiz game with the Pepper robot.

2 Quiz Design

Quizzes have been widely used in education as an interactive, game-based method of teaching and learning, which utilize diverse question-answer formats [11]. Typically, an educational quiz includes a set of questions and a choice of answers. Compared to trivia quizzes used for entertainment, educational quizzes are designed and applied to support and motivate learning, review learning material, retrieve knowledge, as well as retain and assess what has been learned. The key benefits of applying quizzes in education have been associated with repeated exposure to the material and the retrieval practice, which lead to improved long-term memory. Other techniques frequently applied by learners, such as restudying the material, highlighting texts, and rereading notes, have been considered less effective because they do not allow students to monitor learning and test understanding [11–13].

Educational quizzes may employ a range of different question types such as multiple choice, true/false, multiple responses, matching, sorting and ordering, sequencing, filling in the blanks, text response, and numeric response [11, 12]. There are different variations of how quizzes can be applied in educational settings, for example as paper-based, online or mobile remote or in-class activities. Furthermore, educational quizzes can be applied in at least three ways, i.e. (a) as a diagnostic method before the beginning of a learning period, e.g. to test the existing knowledge and understanding before the beginning of a course or lesson, (b) as a formative method during the learning period, e.g. as weekly reviews throughout the semester in relation to the material covered in classes, and/or (c) as a summative method at the end of a learning period, e.g. at the end of a semester to test what has been learned.

Quizzes can be designed as a learning and/or assessment activity for single learners or as a group activity [14]. Group quizzes can be used to promote interaction between students, enhance teamwork and healthy competition, support collaborative learning, and reduce anxiety associated with exams and individual quizzes [12, 14]. Group quizzes can encourage students to work together, discuss and reach a consensus, especially if they (a) allow time for a group discussion before attempting a solution, (b) aim to reduce anxiety associated with individual quizzes, e.g. by allowing time to review, (c) provide an appropriate level of difficulty [12, 14]. Moreover, groups can be arranged for added value, e.g. by bringing students with different interests or competencies together [12]. Another important aspect of educational quizzes is providing appropriate feedback to increase the beneficial effects for learning and motivation [13]. Research has shown that retrieval practice with feedback is beneficial for all age groups, and repeated retrieval practice leads to higher exam grades and higher final course grades compared to students without repeated retrieval practice [13].

The buzzer quiz game with the Pepper robot is designed as a group activity with diverse feedback types from the robot. The following sections describe the design of the buzzer quiz game with the Pepper robot along three main aspects, i.e. application context, didactic design, and technologies applied in the quiz game.

2.1 Application Context

The application context for the buzzer quiz game with the Pepper robot was defined to be a classroom or a laboratory room at the university campus. The buzzer quiz game with Pepper is designed for students in higher education but can be easily extended to other target groups and outside of higher education, for example in schools or activity centers. The Pepper robot plays a role of a quiz master and facilitates the gameplay. The buzzer quiz game with Pepper differs from other quiz designs and technologies. For example, classroom response systems (CRS) afford web-based responses with instant feedback without the need for an embodied facilitator. The buzzer quiz game with Pepper is designed as a general template that can be accommodated for different contents and question types, depending on the field of study and/or selected knowledge domain. Two groups of students play the quiz game and the task of the robot is to identify which buzzer sound is heard first. It is also not necessary to localize the sounds, since these are uniquely assigned to both player groups (Fig. 1).

Fig. 1 Set up of the "Buzzer Quiz Game with Pepper"

2.2 *Didactic Design*

The buzzer quiz game with Pepper is designed as a short group-based, in-class learning activity of 10–15 min, which can be planned at the beginning, in the middle, or at the end of a class. The quiz game can be used as a diagnostic (before the begging of a course or a course unit), formative (periodically, e.g. every week), and/or summative method (as a concluding event, e.g. at the end of the semester) for self-assessment and practice in small groups of students. The buzzer quiz game with Pepper can be used to support learners in reviewing the material covered in previous classes, in self-assessment of acquired knowledge, and/or in engaging in group-based activities to promote active learning in a seminar or lecture. In the envisaged scenario, the Pepper robot stands in front of a smart board, on which the content of the quiz is mirrored from Pepper's tablet. The mirroring of Pepper's tablet is important for visually cueing students participating actively in the quiz game and observing the game.

The didactic design of the quiz is rooted in the gamification approach [15] and informed by the Octalysis gamification framework [16]. The didactic design aims to make reviewing and assessing of what has been learned an enjoyable and motivating social learning experience in class. The design of the buzzer quiz game with Pepper follows a popular multiplayer approach of a "quiz party", in which groups of players challenge each other. The core element of the quiz design is the possibility to answer questions in a group-based competition. This allows to engage students in competing against each other, which may positively affect both motivation and learning experience. The social aspect of playing in groups of students is seen as a motivating factor and at the same time a way to reduce anxiety associated with individual assessments.

The buzzer quiz game is based on game rounds, each having a time limit for answering the questions. Setting time limits aims to enhance motivation and enjoyment through the thrill of competition. Setting time limits is planned to be combined with the visualization of a count-down timer, which is an effective gamification element addressing the motivational core drive "scarcity" according to the Octalysis gamification framework [16]. Furthermore, the buzzer quiz game is planned to include point scores per group and a leaderboard displayed for students on the tablet and mirrored on the smart board. Both points and leaderboards address the motivational core drive of "accomplishment" according to the Octalysis framework [16]. In addition, point scores are planned to show both the group score and the cross-game score for all questions. The score describes the increase in correct answers per additional question attempt. The cross-game score of individual questions will be used for quality assurance, i. e. questions with very high or very low scores will be revised [17].

The quiz game starts with a very simple question to guarantee "beginners luck" or a "lucky strike" which is another motivating gamification element related to the motivational core drive "epic meaning" in the Octalysis framework [16]. The difficulty level of quiz questions is progressively increased and the entire process of game-play is divided into small sections, each with a clearly defined learning objective. These design decisions are again informed by the Octalysis framework [16], which recommends combining explicit goals with dynamic feedback for a higher level of effort and motivation, both addressing the core drive "empowerment" [16]. The feedback is provided in different formats both on Pepper's tablet app (visual and textual feedback) and by the Pepper robot in the form of verbal and non-verbal signals such as music and gestures. Dividing quiz questions into short thematic sections aims to create a clear structure, improve retrieval of knowledge and focus attention on the key topics in a given course and/or lesson. Furthermore, we plan to add illustrations in both questions and answers to support different learning modalities, which may be especially helpful for testing more complex learning materials [17].

2.3 *Applied Technologies*

The buzzer quiz game with Pepper makes use of several technologies. First of all, the humanoid robot Pepper is used as a quiz master facilitating the game-play in class and as an instrument for the detection and recognition of non-verbal sounds. In the first version of the buzzer quiz game, the Pepper robot is designed to detect and recognize only buzzer sounds. However, it is foreseen to extend the non-verbal sound detection to human-made non-verbal vocalizations (NVVs) such as vegetative utterances (e.g. coughing) and affect bursts (e.g. laughter). In these cases, it might be necessary to identify the direction from where the recognized sound was emitted, because these sounds are very similar for all players and would be very time-consuming to be trained on the robot in advance. For direction estimation, more microphones of the robot than only Pepper's microphones have to be used and their signals have to be evaluated together. In addition to the Pepper robot, the game uses custom-made game buzzers, which are mechanical buttons used as devices for players to signal that they want to attempt an answer to a quiz question. Because the sounds are clearly distinguishable and the buzzers are assigned uniquely, it is not necessary to detect the direction of sound emission. The game buzzers produce different sounds which are detected by the robot in order to facilitate the gameplay in the manner described below.

3 Technical Implementation

The buzzer quiz game is implemented using the humanoid robot Pepper, two game buzzer buttons that produce two different non-verbal sounds, and a convolutional neural network (CNN). The non-verbal sounds of the game buzzers were recorded with the Pepper robot in a quiet room to prepare the dataset. In the next step, a classification algorithm for detecting the button that gets pressed first was developed. The application, which is still under development, will use a text-to-speech (TTS) service to ask participants questions, record non-verbal sounds when participants press buttons, and use the classification algorithm to detect the button pressed first to take participants' responses to the question. Based on participants' responses, the Pepper robot will display the point scores and happy-sad animations, including gestures and sounds. After a few rounds, the final scores of the participants will be displayed. During the buzzer quiz, the Pepper robot will remain in one position, turning around its own axis to address the two player groups positioned on the right and on the left.

3.1 The Pepper Robot

Pepper is a humanoid robot introduced by Aldebaran Robotics in 2014. It has four microphones in its head, a three-dimensional depth camera in its eyes, and a 10.1-inch tablet in its chest with an Android operating system to facilitate human–robot interaction. Pepper version 1.9 is used in the implementation of the buzzer quiz game. This version only supports Android applications to control the Pepper robot. There is no additional camera or microphone for the Pepper tablet, as it is connected to the camera and microphone in the robot's head. Built-in obstacle detection prevents collisions and damage to Pepper when it is moving. The Pepper plugin in Android Studio provides many pre-made happy, sad, and welcoming gestures that can be used to simulate robot activity depending on the use case before deploying on the real robot.

3.2 Audio Classification with CNN

The classification of sounds has recently attracted the interest of the research community. In contrast to speech recognition, which focuses on the segmentation and recognition of speech signals, sound event classification is mainly concerned with the classification of ambient sounds into one of several known classes [18]. Potential applications are many, including acoustic monitoring [19], classification of animal sounds, classification of environmental sounds [20], and machine hearing [21]. In the classification of sound events, traditional speech recognition methods were first used in many approaches such as Hidden Markov Model (HMM) based on Mel-Frequency Cepstral Coefficients (MFCC). Later, many other features have been proposed to complement MFCC features, e.g. MPEG-7 audio features [22] and spectro-temporal signature [23]. For instance, time–frequency representations such as the spectrogram generated by Short-Time Fourier Transform (STFT) of sound events provide rich visual information that allows a person to visually identify the sound class. As a result, many recent approaches have treated the spectrogram as a texture image and used image processing techniques on the spectrogram to classify sound events [21, 24].

Two types of sounds from the game buzzers were recorded corresponding to the names of the sounds provided by the producer, i.e. "laser" and "charge". The sounds were recorded as one-channel signals with a sample rate of 16 kHz and a resolution of 16 bits. The initial and final silence in both recorded sounds was trimmed. The trimmed charge signal was overlaid with the untrimmed laser signal to create a new signal called "chargeFirst", which has a delay of random milliseconds in the range of 0.5 s between the "charge" and "laser" signals. Similarly, "laserFirst" signals were produced and overlaid with an untrimmed "charge" signal onto the trimmed "laser" signal. The noise signals from Pepper's servo motors were recorded, and generated further noise signal samples by varying the amplitude of the recorded signals within a range of 0.05. This additional class is called "silence". Altogether five classes were prepared for the classification algorithm, with each class consisting of 400 samples of sounds. Examples of sound samples and their spectrograms are shown in Table 1.

The spectrograms were computed via the Short-Time Fourier Transform (STFT) using a frame length of 1024 samples and a hop size of 128 samples. In order to train the CNN model faster, the spectrogram images were first resized and the resulting smaller images were used as inputs to the CNN. The model was trained over 25 iterations and used three convolutional layers with the 2nd and 3rd convolutional layers followed by the pooling layer and a dropout of 25%. We also used a normalization layer and a Rectified Linear Unit (ReLU) activation function. The trained model can predict unseen data of these five classes with very high accuracy above 99.7.

Table 1 Sound samples and spectrogram

Classes	Untrimmed	Trimmed	Spectrogram (trimmed)
laser	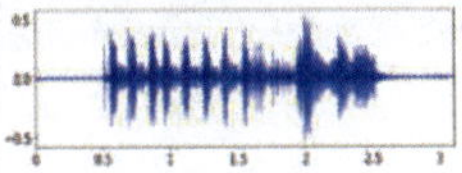	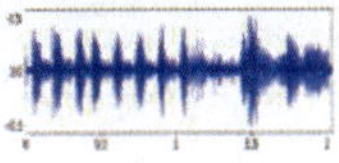	
charge	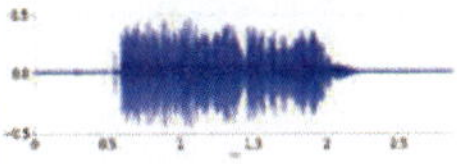	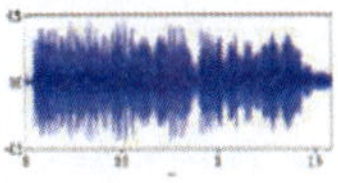	
laserFirst	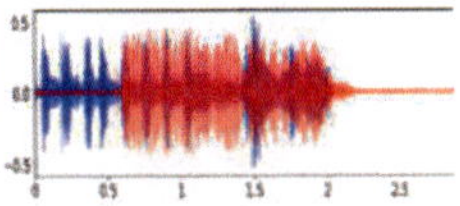	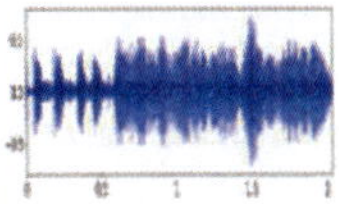	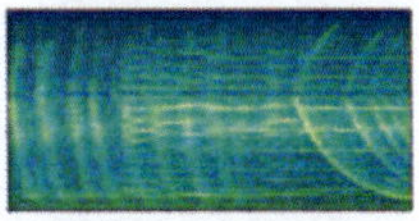
chargeFirst	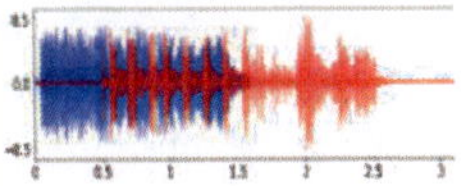	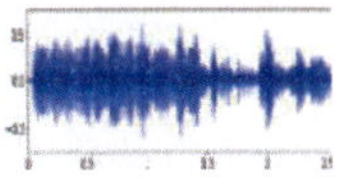	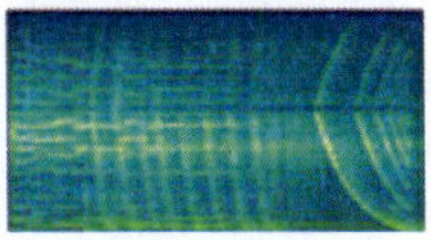
silence		No trim	

3.3 Audio Classification with CNN

The buzzer quiz game with Pepper has been developed using Android Studio, Kotlin, OkHttp communication protocol, and Python. In the application, the Pepper robot asks quiz questions to the game participants, offers a time slot to press the buttons once the groups of players know the answer, recognizes the buzzer sound pressed first using the CNN model described above, takes the response of the recognized group, produces a happy-sad animation based on the responses of the recognized group and offers a point to the respective group, and continues for 5 iterations to display the result of the groups. The algorithm under development on is shown in Fig. 2.

The server-client-based application is still under development. The Pepper robot is a client that records sounds when participants press the buzzer game buttons and sends the recorded data to the server via OkHttp protocol to get the result of the first button pressed. Once the server receives the recorded signal from the Pepper robot, it computes the corresponding spectrogram image, passes it through the CNN, and returns the recognition result. Based on the server response, Pepper takes the question response from the participants and performs a happy, sad animation on a correct or

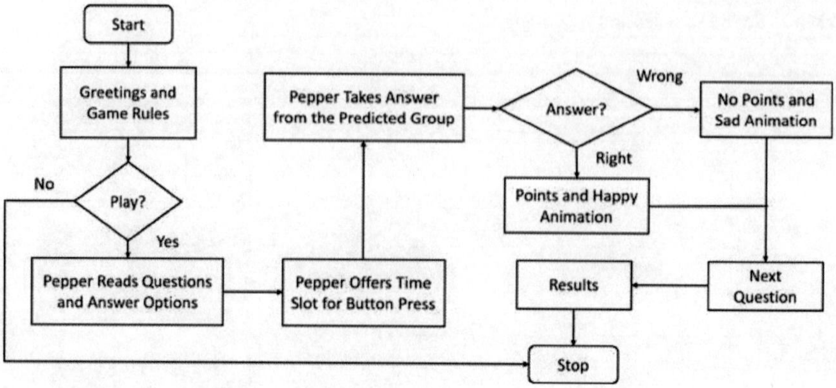

Fig. 2 Flowchart of the application "Buzzer Quiz Game with Pepper"

wrong answer as described in [25]. During the development process, a number of challenges had been faced such as the audio signal recorded by the Pepper robot may contain many human voices and other disturbances at the beginning and end of the audio data which was not included in the trained data, generation of datasets for the "laserFirst" and "chargeFirst" classes with a perfect delay between the "laser" and "charge" datasets, selection of the range of amplitude variations for the "silence" class based on the recorded noise of the Pepper servo motors, selection of the appropriate frame size and hop length for our data set to generate the spectrograms.

4 Conclusions

This paper described the design of the prototype of the quiz buzzer game facilitated by the humanoid, educational robot Pepper in its role as a quiz master. The quiz buzzer game with Pepper is designed as a group-based, in-class activity and the design is informed by the Octalyis gamification framework as an approach to designing engaging and motivating learning experiences. The technical development has focused on the detection and recognition of non-verbal sounds, starting with the buzzer sounds used in the game to signal that a group of players wants to attempt an answer to a quiz question. The recognition of buzzer sounds was tested and the results show that the Pepper robot can detect which buzzer sound was initiated first, even when the buzzer sounds are overlaid. Since testing has been done so far without additional sounds, further tests in classroom settings will be necessary. Further development will focus on the detection of other non-verbal sounds, such as laughter, in order to detect effect and enhance engagement. Once the development of the application is completed, the quiz buzzer game with Pepper will be tested with university students and students' opinions and experiences will be captured using questionnaires with items related to the learning experience. Furthermore, the experience from the development of the buzzer quiz game will be used to design further educational scenarios, in which educational robots will be used to detect a range of diverse non-verbal signals.

References

1. Pelikan, H., Robinson, F.A., Keevallik, L., Velonaki, M., Broth, M., Bown, O.: Sound in human-robot interaction. In: HRI '21 Companion: Companion of the 2021 ACM/IEEE International Conference on Human-Robot Interaction, pp. 706–708 (2021)
2. Trouvain, J., Truong, K.P.: Comparing non-verbal vocalisations in conversational speech corpora. In: Devillers, L., Schuller, B., Batliner, A., Rosso, P., Douglas-Cowie, E., Cowie, R., Pelachaud, C. (eds.) Proceedings of the 4th International Workshop on Corpora for Research on Emotion Sentiment and Social Signals (ES3 2012), pp. 36–39. European Language Resources Association (ELRA) (2012)
3. Yilmazyildiz, S., Read, R., Belpaeme, T., Verhelst, W.: Review of semantic-free utterances in social human-robot interaction. Int. J. Hum.-Comput. Interact. **32**, 63–85 (2016)
4. Aylett, M.P., Vazquez-Alvarez, Y., Butkute, S.: Creating robot personality: effects of mixing speech and semantic free utterances. In: Companion of the 2020 ACM/IEEE International Conference on Human-Robot Interaction (2020)
5. Lascio, E.D., Gashi, S., Santini, S.: Laughter recognition using non-invasive wearable devices. In: Proceedings of the 13th EAI International Conference on Pervasive Computing Technologies for Healthcare (2019)
6. Rychlowska, M., McKeown, G., Sneddon, I., Curran, W.: Laughter during cooperative and competitive games. In: SMILA (2022)
7. Gillick, J., Deng, W.H., Ryokai, K., Bamman, D.: Robust laughter detection in noisy environments. In: Interspeech (2021)
8. Tennent, H., Moore, D.J., Jung, M.F., Ju, W.: Good vibrations: how consequential sounds affect perception of robotic arms. In: 2017 26th IEEE International Symposium on Robot and Human Interactive Communication (ROMAN), pp. 928–935. IEEE, Lisbon (2017)
9. Moore, D.J., Tennent, H., Martelaro, N., Ju, W.: Making noise intentional: a study of servo sound perception. In: 2017 12th ACM/IEEE International Conference on Human-Robot Interaction (HRI), pp. 12–21 (2017)
10. Khan, M.M., Tasneem, N., Marzan, Y.: 'Fastest finger first—educational quiz buzzer' using Arduino and seven segment display for easier detection of participants. In: 2021 IEEE 11th Annual Computing and Communication Workshop and Conference (CCWC), pp. 1093–1098 (2021)
11. Dengri, C., Gill, A., Chopra, J., Dengri, C., Koritala, T., Khedr, A., Korsapati, A.R., Adhikari, R., Jain, S., Zec, S., Chand, M., Kashyap, R., Pattan, V., Khan, S.A., Jain, N.K.: A review of the quiz, as a new dimension in medical education. Cureus **13** (2021)
12. Andrés, B., Sanchis, R., Poler, R.: Quiz game applications to review the concepts learnt in class: an application at the University context (2015)
13. Kenney, K., Bailey, H.: Low-stakes quizzes improve learning and reduce overconfidence in college students. J. Sch. Teach. Learn. **21**(2), 79–92 (2021)
14. Yokomoto, C.F., Ware, R.: Variations of the group quiz that promote collaborative learning. In: Proceedings Frontiers in Education 1997 27th Annual Conference. Teaching and Learning in an Era of Change, vol. 1, pp. 552–557 (1997)
15. Deterding, S., Khaled, R., Nacke, L.E., Dixon, D.: Gamification: toward a definition. In: Proceedings of CHI 2011. Workshop on Gamification, Vancouver, BC, 7–12 May 2011, pp. 12–15

16. Chou, Y.: Actionable Gamification: Beyond Points, Badges, and Leaderboards. Leanpub (2016)
17. Pawelka, F., Wollmann, T., Stöber, J., Lam, T.: Successful learning through gamified E-learning (Erfolgreiches Lernen durch gamifiziertes E-Learning). In: Lecture Notes in Informatics (LNI), Proceedings - Series of the Gesellschaft fur Informatik (GI), pp. 2353–2364 (2014)
18. Dennis, J., Tran, H., Li, H.: Spectrogram image feature for sound event classification in mismatched conditions. IEEE Signal Process. Lett. **18**(2), 130–133 (2011)
19. Gerosa, L., Valenzise, G., Tagliasacchi, M., Antonacci, F., Sarti, A.: Scream and gunshot detection in noisy environments. In: Proceeding of the European Signal Process Conference, pp. 1216–1220 (2007)
20. Chu, S., Narayanan, S., Kuo, C.: Environmental sound recognition with time–frequency audio features. IEEE Trans. Audio Speech Lang. Process. **17**(6), 1142–1158 (2009)
21. Lyon, R.: Machine hearing: an emerging field [exploratory DSP]. IEEE Signal Process. Mag. **27**(5), 131–139 (2010)
22. Muhammad, G., Alghathbar, K.: Environment recognition from audio using MPEG-7 features. In: Proceedings of the International Conference on Embedded and Multimedia Computing (EM-Com), pp. 1–6 (2009)
23. Tran, H., Li, H.: Sound event recognition with probabilistic distance SVMs. IEEE Trans. Audio Speech Lang. Process. **19**(6), 1556–1568 (2011)
24. Costa, Y.M., Oliveira, L., Koerich, A.L., Gouyon, F., Martins, J.: Music genre classification using LBP textural features. Signal Process. **92**(11), 2723–2737 (2012)
25. Buchem, I., Mc Elroy, A., Tutul, R.: Designing and programming game-based learning with humanoid robots: a case study of the multimodal "Make or Do" English grammar game with the Pepper robot. In: ICERI2022 Proceedings, 15th Annual International Conference of Education, Research and Innovation (2022)

Promoting Executive Function Through Computational Thinking and Robot: Two Studies for Preschool Children and Hospitalized Children

Shrieber Betty

Abstract This article presents two case studies that examine the impact of using computational thinking and robots on executive functions. The first study assesses the impact of a COLBY mouse robot on working memory, while the second examines the effects of combining computational thinking and a Lego robot on executive functions such as planning and emotional regulation. The results suggest that the use of robots for preschool children enhances their ability to sequence actions for coding the robot and improves auditory working memory to a greater extent than phonological working memory. The findings of the study with hospitalized children indicate that a high-technology learning environment fosters engagement, challenge, and motivation to learn. The use of the WEDO2 application in building Lego robot models had a positive impact on the students' executive functions, including emotional regulation and self-control.

Keywords Executive function · Working memory · Computational thinking · Robotic program

1 Introduction

1.1 Executive Functions

Executive functions refer to a collection of innate skills that enable to regulate our behavior in various situations. These functions encompass three interrelated core abilities: Inhibitory Control, Working Memory, and Cognitive Flexibility [10, 12]. These core skills, enable us to resist impulses, maintain focus, solve problems, adjust to changing demands, and view things from different perspectives. They are crucial for success in all areas of life and are more predictive of success than IQ or socioeconomic status. Cognitive Flexibility: This ability is based on working memory and

S. Betty (✉)
Kibbutzim College of Education, Technology and Arts, 6250769 Tel Aviv, Israel
e-mail: betty.shrieber@smkb.ac.il

R. Balogh et al. (eds.), *Robotics in Education*, Lecture Notes in Networks and Systems 747, https://doi.org/10.1007/978-3-031-38454-7_17

inhibition. It allows one to think creatively, to consider alternative solutions to situations and social interactions. It enables one to adapt to unforeseen events or circumstances, even if this means deviating from original plans [11]. Inhibitory control is critical in various contexts that require conscious consideration, before speaking, while playing, and in order to maintain focus and avoid engaging in harmful or destructive acts [12].

Working Memory: This is one of the most crucial executive functions, enabling us to absorb and temporarily store a small amount of information, which can be used for planning, reasoning, and problem-solving [9, 16]. Working memory plays a central role in performing a sequence of actions and is a significant factor in daily functioning and learning processes [1–3].

The information stored in working memory is updated in real-time during the performance of routine tasks and activities. As a result, working memory has a significant impact on higher-order cognitive functions, including abstract thinking, learning, problem-solving, investigating relationships between information, breaking down and recombining information, and understanding the cause-and-effect relationships between events over time [12].

Computational Thinking and Robots

Computational thinking is seen as a key competence for the twenty-first century. It has become increasingly important, as an educational approach, particularly in the teaching of science and subjects like mathematics, engineering, and technology [6]. This approach emphasizes the use of algorithmic operations to create sophisticated solutions and effectively solve problems through the application of computer science concepts, with or without technology [19]. The integration of computational thinking into education has been shown to improve problem-solving skills in other aspects of life [23]. It is built on a range of cognitive abilities, including verbal thinking, visuospatial abilities, and numerical abilities. Principles of this approach are commonly taught through coding and educational robot programming, such as Scratch video programming and educational robot programming [8]. Research findings show that students who studied computer programming outperformed those who did not. This was manifested in programming skills and other cognitive skills, such as creative thinking, mathematical skills, and metacognition [20]. It also indicates that implementation of computational thinking and educational robot programming leads to an increase in students' motivation levels, as well as their active participation in lessons. Additionally, this approach enhances students' ability to sustain attention and retain interest [13].

Using computational thinking along with robot programming might enhance some executive functions during the training. The utilization of visual programming, such as Scratch, requires the student to engage in a series of sequential tasks that stimulate their capability for prospective memory, planning, problem-solving, and decision-making. Furthermore, the programming and planning process often involves encounters with programming failures, which cause him frustration and thereby allow him to cope and regulate his emotional reactions attention and retain interest [13].

Studies have demonstrated that the use of computational thinking in programming and operating educational robots [5] can enhance the executive functions, such as inhibition, planning ability, and working memory, among students [14, 21]. Additionally, it has been shown to positively impact the spatial abilities of the population [14, 21].

This paper presents two empirical studies designed to apply computational thinking and robotics to improve executive functions, including working memory, in both preschool and hospitalized children's populations.

2 Two Case Studies

2.1 First Study: Promoting the Working Memory of Kindergarten Children Through Mouse Robot and Its Effect on Auditory and Phonological Skills

The early childhood years are critical in the development of cognitive abilities such as mathematics and spatial ability [4, 24]. Therefore, the development of such abilities is important among kindergarten children. Spatial ability in kindergarten children predicts achievements in reading, mathematics, science, and technology in elementary school, and therefore, constitutes an important set of skills for entering school [7, 22]. Findings from a study that tested the effect of a robot intervention on spatial awareness indicated that children who participated in the intervention program improved significantly in spatial representation scores and acquired knowledge of directions (i.e., left and right) [17].

Method. Qualitative action research. Two, five-year old boys with developmental delay study in preschool.

The research tools. Behavior Rating Inventory of Executive Function (BRIEF) test [15], and participant observation.

Technology. COLBY mouse robot. Colby Robot helps students develop thinking competence and problem-solving skills in a fun and engaging way. The robot provides audio cues, turns on its light and has two speeds. Also, on the back of the robot are colored arrow buttons so that the student can code it. Attached to the robot are activity cards that include colored arrows for coding [18].

Intervention. The Intervention phases followed by two training sessions, along 6 sessions. Each session lasts about 30 min. The intervention itself divided into two types of working memory: auditory and phonologies.

Enhancing auditory working memory through a series of auditory commands. The kindergarten teacher gave spatial concept instructions orally to the children. The children controlled the robot based on cards placed on the table. The teacher sequentially increased the number of commands according to the children's success.

Developing phonological awareness by using the robot. The children selected a picture from a collection and divided it into letter sounds. They needed to remember the number of sounds in the word and analyze the sounds into a word. In the visual channel, the students identified letters placed on the table and matched them to the corresponding sound. The objective was to improve phonological awareness through analysis and synthesis. When the children successfully programmed the robot to move between two sounds in a word, the difficulty level increased to three sounds.

Findings

1. Enhancing auditory working memory. Both Students learned to code the robot by designing the tabs. After mastering the coding concept, we put magnets on the table and selecting a puppet as the target object familiar to the children. The children asked to plan the puppet's path on the table using arrow tabs and direct the robot to reach the endpoint. The children successfully coded and navigated the mouse robot along the track, leading to increased excitement and initiative to build and code their own tracks. During the intervention, a decrease in reliance on assistance and an increase in children's independence and spatial orientation using the robot observed. The two children were able to plan the sequences of the puppet and mouse robot on the constructed tracks. Y. successfully remembered seven actions for coding the robot, while D. remembered five with the support from the visually presented tabs.
2. Developing phonological awareness by using the robot. During the initial meeting with me, partial support was necessary while accommodating to the student's needs. Subsequent sessions gradually reduced this support, adjusting to the student's progress and success in assembling the robot's track. Observing their participation showed that the two students were capable of dividing and analyzing words appropriately for their age. At times, they faced difficulties coding the robot and became frustrated. However, they took control and attempted to fix the coding, resulting in emotional experience of frustration and eventual success. Student Y was able to combine sounds in synthesizing words, found joy in the task, and effectively planned. Student A required my assistance as a researcher in internalizing the coding of the robot.

2.2 Second Study: Development and Review of Program Designed to Promote Executive Functions Through Robot Model Building for Hospitalized Children

This initiative included the development and review of a unique teaching program that focused on executive function (EF) and computational thinking through robotics classes conducted for children continuing their education during their hospitalization. This teaching unit incorporates innovative technological learning environments by using tablets and the WeDo 2.0 app to plan and build motorized Lego models.

Methods. The hospitals make it possible to provide an educational solution to students aged 3–21 who are hospitalized due to physical or mental pain. The teaching strategies and curricula for hospitalized students is individually adapted teaching.

The intervention program. Four students aged 9–12, participated in the program. The program plan was implemented during eight lessons. It includes four motorized Lego robotics models, each model is divided into two sets: the first set refers to the assembly of the model, while the second set refers to the programming of the model.

The program assessment. The assessment of the activity was performed by the students' self-reports assessment along with an evaluation chart of the teacher. This chart included several key components: (a) the activity; (b) The executive function that this activity aimed to promote (c) an appraisal of the students' ability to engage with the activity (d) the teacher suggestions for further development, whether an additional practice was required, and any potential interventions that could be implemented during future meetings. In addition, the activities were documented, videos of student activities were taken during the computations and construction of the models, and conversations were held with students to ascertain their feelings and experiences.

Teaching hospitalized students is challenging especially because most of them are non-permanent students. They are hospitalized for various reasons and knowing that some of them may be released from the hospital after a short period of time (perhaps even in the middle of the robot train).

Findings

The students received comprehensive information about robotics, chose a model, watched a tailored video, and experienced building and programming the model. Using the WEDO2 app and building robots as instructed through the app, encouraged students and inspired them with motivation and passion for learning. Students who are diagnosed with learning disabilities and have difficulty reading, bypassed this obstacle, since completing instructions through the application could be done using pictures.

In terms of executive functions, it was possible to notice a real change in 'self-control' among students who made mistakes in the initial stages of building the Lego models. When it was evident that as they progressed through the stages, they made sure to do a self-check to make sure they assembled correctly before the next stage. In addition, the use of the tablet helped the students who had mobility disability. It was challenging to place the tablet, so that their disability did not affect their performance.

Another important finding that came up as part of the experience concerns the significant improvement in executive function—'emotional regulation', among students who demonstrated difficulty and frustration following mistakes in building the model. The difference in the students' response stood out, especially during later stages of the construction process. Despite encountering mistakes, these students displayed increased control over their emotions and persisted in completing the task, resulting in a finished product. It can be concluded that constructing robotics models helps to develop executive function and emotional regulation in students.

In summary, the robotics classes were far more significant than the simple act of Lego construction. To successfully build the model, students were required to display digital literacy, learning competencies, and practice executive functions. Executive functions are exercised at almost every step of the program, from student's choice of the model (decision making), watching videos on model assembly during planning (attentional control), constructing the Lego blocks (sequencing), providing commands and specifications for building (planning), and emotional regulation (managing frustrations), and more.

3 The Studies Limitations

These two studies are rather a description of two experiences in using robots for promoting executive functions. Due to the small number of participants and no control group the results can however not be generalized and used as evidence for the usefulness of the intervention.

References

1. Baddeley, A.: The episodic buffer: a new component of working memory? Trends Cogn. Sci. **4**(11), 417–423 (2000)
2. Baddeley, A.: Working memory: theories, models, and controversies. Annu. Rev. Psychol. **2012**(63), 1–29 (2012)
3. Baddeley, A., Hitch, G., Allen, R.: A multicomponent model of working memory. In: Working Memory, pp. 10–43 (2020)
4. Bierman, K.L., Torres, M.: Promoting the development of executive functions through early education and prevention programs. In: Griffin, J.A., McCardle, P., Freund, L.S. (eds.) Executive Function in Preschool-Age Children: Integrating Measurement, Neurodevelopment, and Translational Research, pp. 299–326. American Psychological Association (2016)
5. Bottino, R., Chioccariello, A.: Computational thinking: videogames, educational robotics, and other powerful ideas to think with. In: KEYCIT: Key Competencies in Informatics and ICT, vol. 7, p. 301 (2015)
6. Castro, E., Di Lieto, M., Pecini, C., Inguaggiato, E., Cecchi, F., Dario, P., Cioni, G., Sgandurra, G.: Educational robotics and empowerment of executive cognitive processes: from typical development to special educational needs. Form@re-Open J. Form. Rete **19**(1), 60–77 (2019)
7. Cheng, Y., Mix, K.S.: Spatial training improves children's mathematics ability. J. Cogn. Dev. **15**(1), 2–11 (2014)
8. Chevalier, M., Giang, C., El-Hamamsy, L., Bonnet, E., Papaspyros, V., Pellet, J.P., ... Mondada, F.: The role of feedback and guidance as intervention methods to foster computational thinking in educational robotics learning activities for primary school. Comput. Educ. **180**, 104431 (2022)
9. Cowan, N.: Working memory underpins cognitive development, learning, and education. Educ. Psychol. Rev. **26**(2), 197–223 (2014)
10. Diamond, A.: Executive functions. Annu. Rev. Psychol. **64**(1), 135–168 (2013)
11. Diamond, A.: Executive functions: insights into ways to help more children thrive. Zero Three **35**(2), 9–17 (2014)

12. Diamond, A., Ling, D.S.: Conclusions about interventions, programs, and approaches for improving executive functions that appear justified and those that, despite much hype, do not. Dev. Cogn. Neurosci. **18**, 34–48 (2016)
13. Díaz-Lauzurica, B., Moreno-Salinas, D.: Computational thinking and robotics: a teaching experience in compulsory secondary education with students with high degree of apathy and demotivation. Sustainability **11**(18), 5109 (2019)
14. Gerosa, A., Koleszar, V., Tejera, G., Gómez-Sena, L., Carboni, A.: Cognitive abilities and computational thinking at age 5: evidence for associations to sequencing and symbolic number comparison. Comput. Educ. Open **2**, 100043 (2021)
15. Gioia, G.A., Isquith, P.K., Guy, S.C., Kenworthy, L.: Behavior rating inventory of executive function. Child Neuropsychol. **6**, 235–238 (2000)
16. Lazar, M.: Working memory: how important is white matter? Neuroscientist **23**(2), 197–210 (2017)
17. Misirli, A., Komis, V., Ravanis, K.: The construction of spatial awareness in early childhood: the effect of an educational scenario-based programming environment. Rev. Sci., Math. ICT Educ. **103**(1), 111–124 (2019)
18. Papadakis, S.: Robots and robotics kits for early childhood and first school age. Int. Assoc. Online Eng. (2020)
19. Shute, V.J., Sun, C., Asbell-Clarke, J.: Demystifying computational thinking. Educ. Res. Rev. **22**, 142–158 (2017)
20. Scherer, R., Siddiq, F., Sánchez Viveros, B.: The cognitive benefits of learning computer programming: a meta-analysis of transfer effects. J. Educ. Psychol. **111**(5), 764–792 (2019)
21. Sisman, B., Kucuk, S., Yaman, Y.: The effects of robotics training on children's spatial ability and attitude toward STEM. Int. J. Soc. Robot. **13**, 379–389 (2021)
22. Verdine, B.N., Golinkoff, R.M., Hirsh-Pasek, K., Newcombe, N.S.I.: Spatial skills, their development, and their links to mathematics. Monogr. Soc. Res. Child Dev. **82**(1), 7–30 (2017)
23. Voogt, J., Fisser, P., Good, J., Mishra, P., Yadav, A.: Computational thinking in compulsory education: towards an agenda for research and practice. Educ. Inf. Technol. **20**, 715–728 (2015)
24. Wang, A.H., Firmender, J.M., Power, J.R., Byrnes, J.P.: Understanding the program effectiveness of early mathematics interventions for prekindergarten and kindergarten environments: a meta-analytic review. Early Educ. Dev. **27**(5), 692–713 (2016)

Assessment of Pupils in Educational Robotics—Preliminary Results

Jakub Krcho and Karolína Miková

Abstract We have been working on educational robotics for several years and our master work has involved cognitive and psychomotor taxonomies for educational robotics, resulting in our own taxonomy for the cognitive domain and a taxonomy for the psychomotor domain. Throughout the research, we were constantly asked questions about the assessment of the stated learning objectives for educational robotics in the classroom and the assessment of the pupils themselves working with the robotic kits. After these questions were insufficiently answered, we decided to choose the problem of assessment of pupils working with educational robotics on the teaching process as our main research problem, thus providing a deeper insight into this issue. Assessment is one of the most powerful interventions found in the educational research literature that can improve learning and teaching. However, identifying models and tools to assess pupils in relatively new educational activities like robotics is a question whether current approaches to assessing pupils are sufficiently appropriate for such a relatively new educational activity as robotics. Assessment is a compulsory part of the Slovak teaching process. Usually, teachers use grades to assess their pupils, but if we want robotics to be taken seriously, we need to come up with a good assessment method that will motivate pupils and fulfill the basic functions of assessment. In this paper we will try to present the preliminary results of our research.

Keywords Educational robotics · Assessment · Pupils · Learning process

---

J. Krcho (✉) · K. Miková
Comenius University in Bratislava, Mlynská dolina F1, 842 48 Bratislava, Slovakia
e-mail: krcho@fmph.uniba.sk

K. Miková
e-mail: mikova@fmph.uniba.sk

R. Balogh et al. (eds.), *Robotics in Education*, Lecture Notes in Networks and Systems 747, https://doi.org/10.1007/978-3-031-38454-7_18

1 Introduction

Robotics has been finding its place in the teaching process at all levels of education for years. However, there are questions that are still being answered. How to adequately set learning objectives for the use of robotic kits in classes? How to properly assess pupils working with educational robotics? Due to many years of experience in this field, we can say that the use of robotic kit provides many competences and skills to pupils. All our experience comes from our active learning through robotics kits, from leading a robotics interest activity for primary and secondary school pupils, being involved in various robotics national and international competitions (as an organizer, jury) and finally, we also act as teachers of computer science and programming at high school and university, where we teach with mentioned kits. Within our master thesis we have addressed taxonomies for setting educational goals in educational robotics. I am currently in my first year of PhD studies and I have chosen the topic of learner assessment in educational robotics as my dissertation topic because it is a topic that needs to be understood more deeply and is also a natural outgrowth of the thesis.

2 Assessment and Educational Robotics

In the teaching process, curriculum, learning process, learning objectives and assessments need to be defined and established [1]. On the other hand, according to some research studies [2], assessment is an underestimated factor and is not given enough attention. However, when we use assessment in a formative perspective (feedback, in particular), it is one of the most powerful interventions found in the educational research literature that can improve learning and teaching and therefore the entire school system [3]. However, identifying models and tools to assess relatively new educational activities like robotics, coding, making is an additional critical issue, because of a lack of models [4].

Assessment in educational robotics is challenging [5] because ER consists not only of programming but also of the technical part, which includes work with the construction of the robot. By ER we mostly mean constructionist activity, where pupils can acquire deep knowledge based on active handling of objects (learning by doing—Papert) [6]. It often leads teachers to use unconventional teaching methods that contribute to the building of competences useful for life, e.g.: teamwork, problem solving and creativity [7], collaboration [8], communication and problem solving [9], or even technical and social skills [10]. This is the reason why the assessment of educational robotics is much more demanding than for other taught subjects.

We tried to find available assessment frameworks and tools, but we were not successful. There are some cases where teachers have used peer assessment [11] with educational robotics. Another article describes the use of rubrics to assess pupils [5] or discusses the need for assessment in educational robotics [4]. However, there are not many publications that address assessment in educational robotics in schools.

3 Methodology

The aim of our research is to investigate ways of assessing pupils when working with educational robotics in the classroom and, if necessary, to develop our own way of assessing pupils. Due to the nature of the research, we have chosen a qualitative approach, namely developmental research [12], which is conducted iteratively. Within this paper we present preliminary results from the first stage of our research. Thus, we naturally chose a qualitative type of research (due to the nature of this research as well as the possible number of participants in our research). In order to achieve our stated objectives, we stated the research questions:

1. How to assess pupils' progress in working with robotic building blocks post-teaching?
2. What aspects of assessing pupil's progression are key for us?

3.1 Data Collection

The first data of our research came from a questionnaire survey [12], which consisted of 8 questions and was conducted using an online form. In it, we asked computer science teachers whether they work with educational robotics and, if so, what type of pupil's assessment they use. We also asked the teachers about their experiences or recommendations for pupil's assessment when using educational robotics. We also decided to use a questionnaire survey due to the size of our country and the number of teachers who teach programming or programming via robotics kits.

The questionnaire survey is still active and the data collection for our research is still ongoing, hence these are preliminary results. To date, we have received 36 responses from teachers using educational robotics in the classroom across all levels of education (ISCED1, ISCED2, ISCED3) and all types of schools (private, vocational, bilingual, general education) that are established in Slovakia.

From the questionnaire survey, we approached those teachers who were willing to have a face-to-face guided interview with us, which allowed us to a deeper understanding of their choice in assessing pupils working with robotics. Three teachers from different types of schools and different levels of education participated in the guided interview (so far). These were a teacher at a vocational secondary school, a teacher at a second level of primary school and a teacher at a private grammar school.

The individual interviews with the teachers were conducted based on pre-designed specific questions that attempted to further seek answers to our research questions. The questions we asked them were incorporated under three main headings.

1. Reason for using educational robotics in the teaching process
2. Using and setting educational objectives
3. Monitoring the fulfilment of the educational objectives (pupil assessment)

After completing individual guided face-to-face interviews with teachers, we had collected data to gain a first insight into the issue of pupil assessment when working with educational robotics. The guided personal interview with teachers was recorded on a digital voice recorder and supplemented with our personal notes we took during the interview. The data thus prepared and collected were ready for subsequent analysis.

3.2 *Data Analysis*

We worked with verbatim transcripts of the recordings using an inductive approach to analyse the individual interviews. Prior to the actual analysis of the transcripts, main categories were created for each of the study areas (which we identified based on the research questions) to reflect the structure of the guided interview. The categories were elicited based on the questions we asked the teachers and described in the previous chapter. They were also defined using existing secondary data on the topic, information gathered through previous analysis of the issue and personal observations made during the interviews.

During the analysis, we used an open coding process, which we applied to all transcripts of the recordings created. After open coding, we proceeded to an axial coding process whereby we were able to identify relationships between the categories that emerged. In analysing the interview transcripts, the aim was to identify as broad a thematic spectrum as possible relating to a particular area and, on the basis of these, to create sub-categories that relate to the issue, explain its causes or context. The interview transcripts were segmented, and the different sections were assigned to the corresponding categories. Using constant comparisons, we continually compared the data and looked for commonalities and differences between them.

4 Preliminary Results

Based on the analysis of the collected data, we found that teachers try to implement education robotics in their teaching process as much as possible. Even though pupils enjoy working with robots more, teachers have shown us in our findings that it is much more challenging to teach using educational robotics. Every one of our respondents stated that the most challenging part of learning with a robotics kit is assessing the pupil and the question they ask themselves every time they go to assess a pupil is—"What is important for me to assess?". Assessing not only in terms of what the pupils have learned, what they know, but above all to assess the pupils in a certain way (grades, points, percentages, ...) and which part of working with the robot has greater importance (programming, construction, overall functionality, effort to work, ...).

4.1 How to Assess Pupils and What Should Be Assessed?

Interviews with teachers showed us that they most often use assessment through grades, mainly because this is the preferred form in their schools. However, when assessing in educational robotics using grades, they discovered several shortcomings. One of them was the lack of information for the pupils as to why they received the grade in question.

Another problem they face is what is most important for them to assess, as the robotic model needs to be programmed as well as constructed. The result of these difficulties has been that many times the teachers have slipped into a mould—if the pupil has a working model they get an A, if the model is not working they remained unmarked or receive F. We also tried to find out why they assessed pupils in this way and whether they considered other forms of assessment. Respondents told us that this type of assessment is more common in schools in Slovakia and they did not think at all to explore other forms of assessment that might be more suitable for this field.

When we interviewed teachers, we also explored what they put emphases on. What we found with the teachers (who were interviewed in person) is that they try to assess primarily the programming part, i.e., how well the robotic model was programmed. Whether they used the right programming concepts according to the task and whether the robotic model works. The pupils construct the robotic models according to the instructions and hence the robot is built mostly mechanically functional and therefore the teachers assess only the programming part. In the case that pupils build the model based on their creative activity and not on the instructions, the model part still remains unmarked. When we interviewed the teachers, we tried to find out why they do not assess the design part as well. All the respondents told us they are unable to assess the construction part objectively.

By examining the second research question, we found that teachers are aware of all aspects that influence a robotic model (properly prepared design, properly used programming concepts), but they are not sure what is more important in the assessment, moreover they are not sure how to assess the design part of the kit. Often teachers ask questions such as, "How to assess that the design is stable enough?", "How to assess a pupil whose robot works but the code is "ugly" written?", "How to assess pupils whose robots work, but one pupil has a better design, and the other pupil has a better program?" and many similar questions. This may also be due to the fact that programming is most often taught in computer science and thus they are naturally interested in programming.

5 Conclusion and Future Work

In the course of our research, we sought to find answers to these complex research questions:

1. How to assess pupils' progress in working with robotic kits after the teaching process?
2. What aspects of assessing pupil's progression are key for us?

In exploring the first research question, we found direction for the next stages of our research. Questions such as—"What knowledge do teachers have about possible assessment methods?", "How can we change the established trend of using grades as an assessment of pupil progress?" emerged. During our research we discovered that teachers try to apply grades as a way of assessment at any cost. Yet, they are aware that this assessment is not an appropriate choice. Especially because of the lack of informativeness of such assessment. In addition, teachers feel difficulties in assessing the construction of robotic models because they do not know how to assess whether the pupil has built the construction correctly, stably enough, reliably, and so on. The point is that each pupil, in his/her own creative activity, can come to different ways to build a reliable robotic model. Despite seeing these problems, teachers did not look for other forms of assessment to use.

By examining the second research question, we found that teachers are aware of all aspects that influence a robotic model (properly prepared design, properly used programming concepts), but they are not sure what is more important in the assessment.

As mentioned earlier in this article, data collection and analysis are an ongoing process and continually influences the development of our research. We have also come to realize that assessment is a process in which not only the teacher, but also the learner plays a role. For this reason, one of the next stages of our research will be to embed pupils as research participants and try to gain pupils' perspectives on assessment (how pupils perceive being assessed in a particular way).

The obtained results gave us an initial insight into the issue of pupil assessment when working with educational robotics and we found that this topic needs to be addressed because teachers in Slovakia do not know how to appropriately assess their pupils in lessons with educational robotics, which results in a lack of feedback on what the pupils have mastered.

Acknowledgements We would like to thank referees for their comments on this paper. We would like to express thanks to projects VEGA 1/0621/22, APVV-20-0353, FERTILE 2021-1-EL01-KA220-HED-000023361 and UK/389/2023 for the provided funding and have our results published.

References

1. Leighton, J., Gierl, M.: Cognitive Diagnostic Assessment for Education: Theory and Applications. Cambridge University Press (2007). Author, F., Author, S.: Title of a proceedings paper. In: Editor, F., Editor, S. (eds.) Conference 2016, LNCS, vol. 9999, pp. 1–13. Springer, Heidelberg (2016)
2. Black, P.: Research and the development of educational assessment. Oxf. Rev. Educ. **26**(3–4), 407–419 (2000)
3. Black, P., Wiliam, D.: Inside the Black Box. King's College, London (1998)
4. Tegon, R., Labbri, M.: Growing deeper learners. How to assess robotics, coding, making and tinkering activities for significant learning. In: Makers at School, Educational Robotics and Innovative Learning Environments: Research and Experiences from FabLearn Italy 2019, in the Italian Schools and Beyond. Springer International Publishing, Cham (2021)
5. Veselovská, M., Mayerová, K.: Assessment of lower secondary school pupils' work at educational robotics classes. In: Educational Robotics in the Makers Era 1. Springer International Publishing (2017)
6. Papert, S.: What is Logo? Who needs it. In: Logo Philosophy and Implementation, pp. 4–16 (1999)
7. Kabátová, M., Pekárová, J.: Learning how to teach robotics. In: Constructionism 2010 Conference (2010)
8. Eguchi, A.: RoboCupJunior for promoting STEM education, 21st century skills, and technological advancement through robotics competition. Robot. Auton. Syst. **75**, 692–699 (2016)
9. Usart, M., et al.: Are 21st century skills assessed in robotics competitions? The case of first LEGO league competition. In: Proceedings of the 11th International Conference on Computer Supported Education (CSEDU 2019), pp. 445–452 (2019)
10. Kandlhofer, M., Steinbauer, G.: Assessing the impact of educational robotics on pupils' technical and social skills and science related attitudes. Robot. Auton. Syst. **75**, 679–685 (2016)
11. Anderson, W.L., et al.: A Taxonomy for Learning, Teaching, and Assessing: A Revision of Bloom's, 1st edn. Pearson Education Limited, Harlow (2014)
12. Creswell, J.: Educational Research: Planning, Conducting, and Assessing Quantitative and Qualitative Research. Pearson Education, New Jersey (2008)

Technologies for Educational Robotics

esieabot: A Low-Cost, Open-Source, Modular Robot Platform Used in an Engineering Curriculum

Gauthier Heiss, Elodie Tiran Queney, Pierre Courbin, and Alexandre Briere

Abstract This paper is about the design and the pedagogical applications of a modular robot called *esieabot*. The robot has been used as part of an engineering curriculum and has already been distributed to more than a thousand students. The authors explain the technological choices and compare with other available platforms used in education. Uses of the robot in different educational activities are discussed. The results show that when each student has their own robot, it encourages the development of ideas outside the classroom and stimulates creativity. It has also been observed that the manipulation of this tool facilitates multidisciplinary approaches in the classroom. Future work will focus on setting up a systematic measurement of the impacts of the robot's application in educational activities. It will also include an analysis of the obstacles to the adoption of this educational tool, particularly in terms of maintenance and teacher training.

Keywords Educational robots · STEM · Low cost robotics · Active learning · Interdisciplinary approaches · Teacher acceptance

1 Introduction

When building pedagogical content, whether at the curriculum level or for a specific course, the following questions arise: (1) What skills should students acquire? (2) In order to acquire these skills, what are the best educational activities? (3) How can one assess the level of skill acquired?

P. Courbin (✉) · A. Briere
Learning, Data and Robotics Lab, ESIEA, Paris, France
e-mail: pierre.courbin@esiea.fr

A. Briere
e-mail: alexandre.briere@esiea.fr

G. Heiss (✉) · E. Tiran Queney
ESIEA, Paris, France
e-mail: gauthier.heiss@esiea.fr

E. Tiran Queney
e-mail: elodie.queney@esiea.fr

R. Balogh et al. (eds.), *Robotics in Education*, Lecture Notes in Networks and Systems 747, https://doi.org/10.1007/978-3-031-38454-7_19

In this paper, we focus on the second question. Many studies have shown much interest in an active pedagogy supported by the use of Science, Technology, Engineering and Mathematical (STEM) tools. The educational use of robotics in particular have shown to increase motivation, involvement and thus the acquisition of knowledge and skills by students [1–3]. Research has also shown that the use of robotics facilitates the development of collaboration and cooperation skills as well as communication skills [3, 4]. It can also help in the teaching of various non-scientific subjects and in generating interest from the students [2]. Robotics is also obviously used successfully for teaching computer science [5].

With these positive aspects in mind, 3 years ago, we started to look for solutions to meet four main objectives:

1. *Adaptive*—A robot was needed that could be used in all the computer science fields taught to students.
2. *Interdisciplinary*—The same platform should allow teachers of other disciplines such as maths and physics to use it in their lessons.
3. *Open*—The platform must be open to encourage each student to want to contribute to its improvement.
4. *Personal*—Each student must have their own robot to allow them to explore and implement their own ideas outside of class time.

Many alternatives were considered: including those developed whether for the general public (LEGO Mindstorms), developed with a DIY approach (InMoov) or dedicated to research and teaching (Poppy) [6]. All these platforms have the advantage of being well supported by companies or communities, but the high price would not allow us to provide a robot to each student.

We therefore decided to create an in-house robot kit to match our educational needs and budget. The reduced cost allowed us to provide every student with their own robot. It is interesting to note that other studies have also focused on the development of open platforms at lower cost, some examples being Otto DIY [7], FOSSbot [8], Thymio [9] and Hydra [10].

In this paper, we detail the platform which was created and has been used more extensively during the last year. After a description of the platform in Sect. 2, examples of activities where we explore the *Adaptive* and *Interdisciplinary* applications of the robot are presented in Sect. 3. Section 4 gives an overview of the way the students themselves worked on their own initiative on various projects, this is facilitated by the *Open* and *Personal* objectives of the platform. Finally, in Sect. 5, we explore different avenues for the future of this platform including its educational uses, and we also highlight certain obstacles observed.

2 Proposed Platform: *esieabot*

In this section, we detail the context as well as the hardware and software components of our robotic platform shown in Fig. 1.

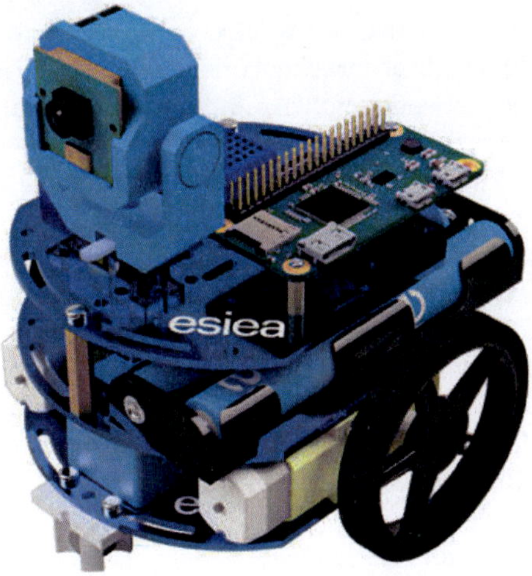

Fig. 1 Robot overview

2.1 Background

Back in 2014, the ancestor of our project was a simple dual-motor Raspberry Pi-based platform which was mainly focussed on our *Interdisciplinary* objective. The chassis was a single piece of transparent plastic; It was neither very modular nor solid but was satisfactory for a project which lasted a week. A toy USB rocket launcher had to be connected by the students to the robot in order to test the physics formulas learned during the semester.

In 2019, it was decided to create a solution which met all our four objectives (*Adaptive*, *Interdisciplinary*, *Open* and *Personal*). One of the main challenges was the *Personal* objective i.e. to create a kit to be given to every new student (about 300 new students each year) in order to inspire them and develop their creativity in computer science and electronics. We started to design a robot kit based on the one we had used earlier but which had a more robust, compact and modular form. The *esieabot* was born.

In the next sections, we give details of the hardware and software. Please, note that all the hardware and software used for the *esieabot* is opensource and published under GNU GPL v3 [11, 12].

2.2 Hardware

In Table 1, we list the main parts of the robot and the corresponding prices.

Focus on Raspberry Pi 0 WH

To satisfy our *Adaptive* objective, we wanted our students to use the robot in all computer-science fields taught at our school (including web and mobile development, cybersecurity, computer networks, etc.). An Arduino-based solution would have been too restrictive. Whereas with a Raspberry Pi, the robot can be used as a standalone computer.

We selected the *Raspberry Pi 0 WH model* as it is small (65 by 31mm), has a low power consumption (under 2W during stress-testing) and has built-in Wi-Fi and Bluetooth connectivity. Even if it's smaller than a regular full-size Raspberry Pi board, it has the same number of General-Purpose-Input-Output pins (40).

To satisfy our *Personal* objective, we had to be sure of supply and also be able to guarantee a fair price. Thanks to the *Raspberry Pi Foundation* with whom we developed a partnership, we now buy hundreds of *Raspberry Pi 0 WH* at the Manufacturer's Suggested Retail Price of 12€ every year.

Focus on the global price and alternatives

To satisfy our *Personal* objective, we had to give a robot to each student, and therefore keep its price quite low. Our robot was ultimately produced for the fairly low price of about 50€. This low price is in part due to pretty high volumes (more than 500 kits made every year, 300 for students and 200 others for teachers, maintenance,

Table 1 Main robot parts. More details in the online documentation [13]

Function	Parts	Details	Price
Computing	RaspberryPi 0 WH	Small form-factor, low power consumption, built-in Wi-Fi and Bluetooth connectivity	12€
Movement	2 motors L293D H-bridge	Generic 3-6V hobbyist motors	7€
Vision	5Mpx camera 2 servo motors	3D printed gimbal can move on 2-axis (pan and tilt)	8€
Connect and extend	Small breadboard	Allows students to extend functionalities	1€
Storage	8 GB SD-card	Stores the Operating System and student projects	5€
Controls	USB gamepad	Easily controls the robot	6€
Power	USB battery pack 4 AA batteries	Two sources to separate the power supply from the calculation part and the movement part	7€

giveaways, etc.) but also to the partnership with the *Raspberry Pi Foundation* and good deals with parts manufacturers in Asia.

We have looked at other existing opensource robots for education to compare with ours. The most striking difference between ours and other projects is the price. Even if our price is a cost and not a resale price, we are at the low end of the price range. Otto DIY, a walking robot, is sold for around 70€ [7]. Thymio, a dual-motor robot, costs around 100€ [9]. FOSSbot [8], a similar alternative to *esieabot* costs up to 120€. Hydra, a modular robot, costs around 35€ [10] but without power supply and with an Arduino.

2.3 Software

Our *Interdisciplinary* and *Personal* objectives, required us to make the robot as easy to use as possible for beginner students and for teachers. We built a custom GNU/Linux operating system, called *esieabot-os*, which is based on *Raspberry Pi OS*, the official operating system of Raspberry Pi. To satisfy our *Open* objective, its source code [12] and documentation [11] can be found online.

We created a few additional tools:

- *esieabot-manager*, the core software wich allows the robot to be controlled and managed remotely with ease. Moreover, to facilitate the addition of a program, a file structure is present on a FAT32 partition (so that it can be read from any operating system) and any file or program added to these folders is automatically copied and executed the next time the robot is started.
- A Wi-Fi access point, which is used to connect wirelessly to the robot and control it. It is set up automatically on first boot with a unique name based on the card's mac address. A password is also generated on first boot and can be retrieved from a simple text file stored on the SD-card. The only issue we have with this system is the Wi-Fi access point uses the same Wi-Fi channel as the main Wi-Fi connection. This makes the Wi-Fi access pretty unstable, especially when a lot of students are using their robots in the same room. That's why we recommend the access point only be used for configuring purposes.
- A Few demonstration programs are included and started on boot. Out of the box, the robot can be controlled with the included USB gamepad or with a smartphone thanks to a web application with camera control. Without any need to program, students can test and explore the main features of the robot.
- A simple remote API, to control the robot using HTTP requests for high-level usage.
- Built-in *pigpio* library [14] to control all the GPIO with a *C* or *Python* program. The *pigpio* library is easy to use and provides APIs for all GPIO functions, included servomotor control. To satisfy our *Open* objective, we created few documents to help students and teachers to build and program on the robot for development and usage purposes [11].

3 *Adaptive* and *Interdisciplinary* Objectives: Examples

In this section, we give details of the experiments carried out that allowed us to test the adaptability of our solution and the possibilities of using it in an interdisciplinary context. Our robot has been tested in two situations:

- In 2022, colleagues were able to use it during a summer school in Croatia. Students with various academic profiles had lessons on project management and an introduction to AI. At the end of the summer school, teachers suggested that students work on a project to allow the recognition of traffic signs by driverless cars using the *esieabot*. Teachers noted very positive reactions from the students who were able to understand the concrete practical applications of more theoretical subjects.
- Since 2014, first-year students of our school (18 years old on average) conclude their academic year with a one week end-of-year challenge. The aim of this challenge is to apply what they have learned during this first academic year in their main subjects: Electronics, Computer science, Mathematics, Physics and English.

In what follows we focus on the second experiment. We will look at the opportunities and difficulties encountered in using our robot with first-year students (Adaptability to novice students) and combining different academic subjects (Interdisciplinary with Electronics, Physics and Computer Science).

3.1 Context and Educational Objectives

The one week end-of-year challenge allows first-year students to combine multiple subjects in a playful final challenge. First-year students are gathered in teams of 3 and work on a challenge mixing Electronics, Physics and Computer Science during the last week of the academic year. The composition of teams is free to boost the motivation of the students. At the end of the entire week of work each team presents its technical achievements, respecting the specifications imposed by the subject, and compares it with the other teams of the class. A competition based on different criteria is organized, and the winner of each class takes part to an inter-class tournament. Students must also make a video, in English, detailing the work done during the whole week.

The *esieabot*—which every first year student receives and assembles at the start of the year—fits well into this multidisciplinary challenge.

In June 2022, the challenge aim has been to turn the students' *esieabot* into a "catapult robot" (Fig. 2a and c) able to throw ping-pong balls as far as possible. The catapult had to be made from cardboard or recycled materials and objects. At least one functional catapult robot by team was required at the end of the week.

The time spent by the students on this challenge is around 50 h spread over 5 d (Monday to Friday): 15 h with a multidisciplinary teaching team (computer science, electronics, physics, mathematics) and 35 h on their own.

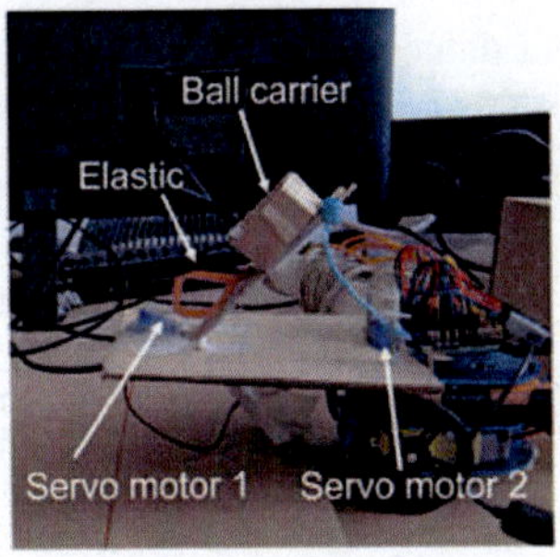

(a) Catapult robot with recycled materials (cardboard, string, elastic and paper clip...) and imposed electronic components: servomotors, LEDs and 7-segment display and decoder.

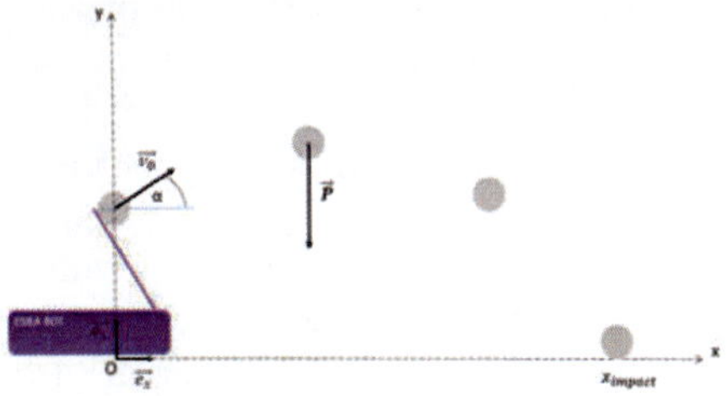

(b) Theoretical parabolic trajectory of the ping-pong ball. The students had to find out the theoretical impact distance of the ball with some approximations (no friction, initial position $y_0 = 0$): $x_{impact} = \frac{sin(2\alpha)v_0^2}{g}$.

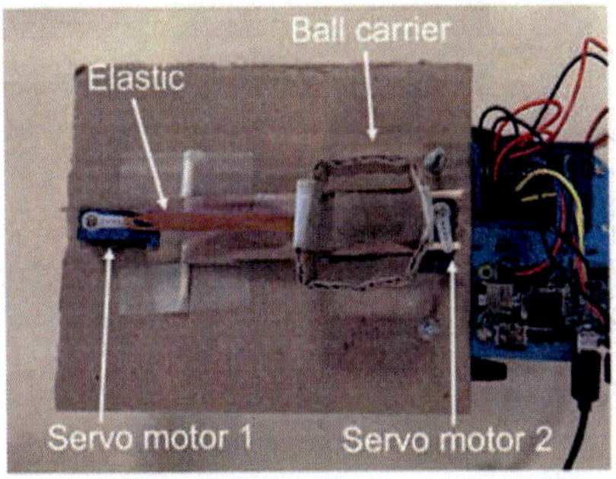

(c) Top view of the same catapult robot.

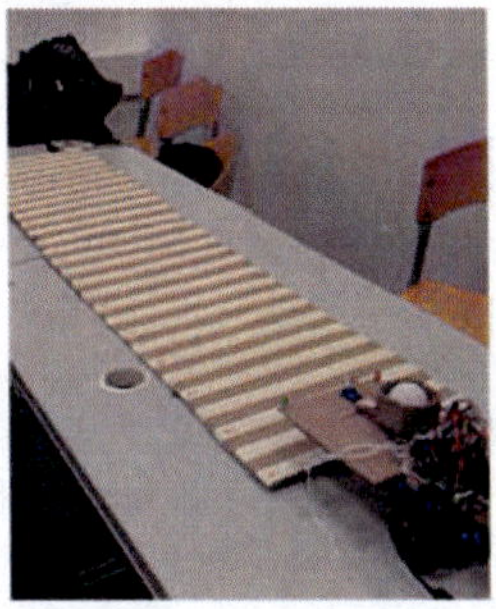

(d) Homemade measuring bench to validate their maximal impact distance after optimization of the catapult.

Fig. 2 End-of-year challenge

The work was divided into three interrelated parts:

1. **Electronic assembly**: assembly of the cardboard catapult and electronic components: servomotors, LEDs, 7-segment display with decoder.
2. **Computer programming**: control of the electronic components via a *C* language program using *pigpio*.
3. **Physical modeling and parameters optimization**: computations and measurements of the ball trajectory (Fig. 2b), optimization of physical parameters to launch the ball as far as possible.

Electronic assembly: The assembly of the robot itself had already been done by the students at the beginning of the year following the detailed guide [11]. An other detailed and step-by-step assembly guide was provided to obtain a functional but intentionally not optimized cardboard catapult model. Then, students carried out the electronic wiring of the various components they added on their basic robot. The electronic assembly had to be done first virtually on a free software for designing and editing printed circuits (Fritzing [15]) to avoid damaging the students' robot. Once validated by a teacher, it was physically assembled on their *esieabot*.

Computer programming: The programming target was to control all the components previously assembled using the joystick of the robot. Students connected their *esieabot* to Wi-Fi, created a code to allow the LEDs to light up as a turn signal, and activate the motors and servomotors with the joystick. The motors allowed the robot to move and the servomotors to control (1) the tension of the rubber band installed at the front of the catapult (servomotor 1 in Fig. 2a and c) and (2) the release of the catapult loaded for firing (servomotor 2 in Fig. 2a and c). The 7-segment display has to show the level of elastic tension. The students had been given a very basic example of *C* code to run one of the motors, as well as a document explaining how the motors are connected to the robot. The students had to understand how to enrich the basic code to control all the motors to move the robot correctly. For the joystick use, a fill-in-the-blank *C* code was also given to the students to let them understand what happens when a user presses a button.

Physics modeling and optimization of parameters: The physics part of the project was divided in two parts: theory and practice. In the theoretical part, students had to calculate the formula of the impact distance of the ping-pong ball according to the parameters of the catapult (angle of fire and initial speed - the latter being connected to the tension of the rubber band at the front of the catapult) (Fig. 2b). Then, the students had to use the theoretical result to optimize their catapult and launch the ball as far as possible. The teacher then measured the maximum impact distance of the ball on a homemade measuring bench (Fig. 2d).

Final Contest: In the final contest, two teams faced in a game arena. During two minutes, the robots had to throw the maximum number of balls into scoring bins: the farther the bins were from the robots, the more points they won. A robot hit by any opponent's ball was frozen for 30 s.

3.2 Results

3.2.1 Quantitative Results

About half the teams managed to get a fully functional catapult robot, allowing them to take part in the challenge within their class. Up to 20% of the teams managed to make two functional robots or more during the week.

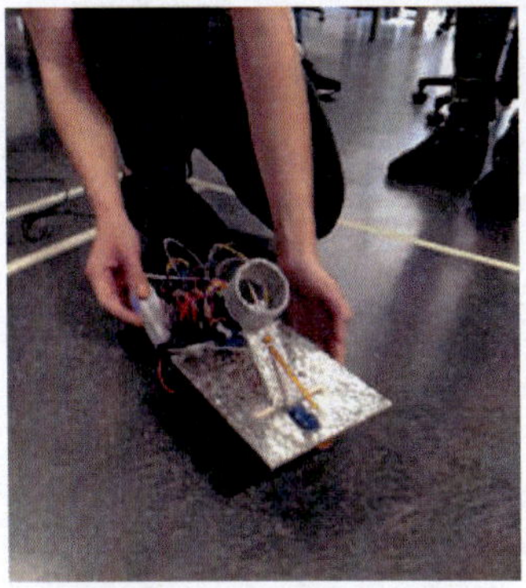

(a) A glittery cardboard catapult robot.

(b) A cardboard catapult robot on the theme of "Playmobil" and castle.

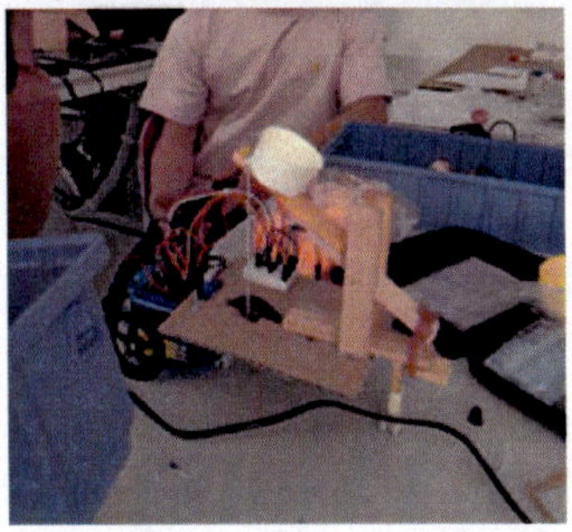

(c) A wooden catapult robot made from "Kapla"

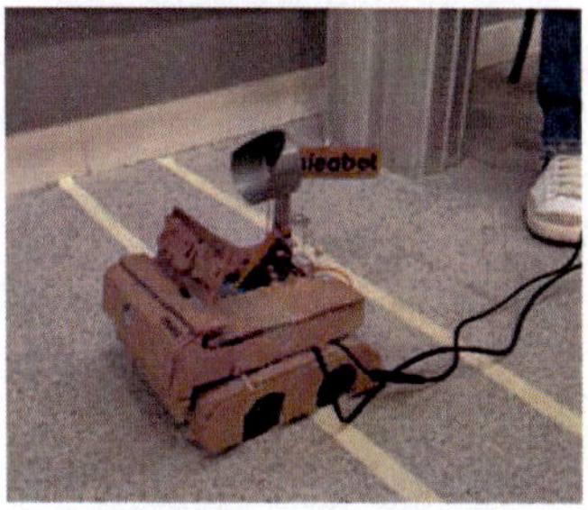

(d) A tank catapult made from two combined robots.

Fig. 3 Example of creative catapult robots made by the students

The main challenge faced by the students was *C* programming of the catapult to control the electronic components with the joystick. As the challenge was done with first-year students, they didn't have a lot of experience with systems and computer networks. It was quite difficult for them to get started by connecting their laptop to their robot and then their robot to the Internet.

As longer than expected shooting distances were achieved (vs first tests), size of the playgrounds had to be increased (1.20 m on average against 1 m during preparation tests). Some students have cleverly used recycled materials or objects other than the cardboard initially provided (wood, 3D printing, etc.) (Fig. 3) which allowed greater shooting performance (over 10 m).

3.2.2 Qualitative Result

A great creativity among the students has been observed, showing involvement and enthusiasm for the challenge (Fig. 3).

Some students, who perform poorly in theoretical exercises, revealed themselves in practice, happy to show their "know-how". Their freedom on this challenge highlighted the potential of this type of student.

3.3 Conclusions and Perspectives About This End-of-Year Challenge Using esieabot

The final competition seems to have been a great source of motivation for the students. Next year, it could be interesting to record and broadcast the competition in order to further increase the playful spirit.

Each team having its own robots is an additional source of involvement; it pushes students to develop their creativity. Their final catapult robots are customized and they keep it at the end of the challenge as a proof of the work done.

The first-year "physics of movement" course does not have dedicated practical work hours, so it was interesting for the students to see an actual application of some concepts learned in the first semester. From physics teachers' point of view, it was interesting to see the student's practical skills and difficulties.

The teacher supervision rate was a weak point for this 2022 session. First-year students are not used to work independently without being guided step by step by a teacher continuously. We were unable to have one teacher per topic (Physics, Electronics, Computer Science) at any time in all the classrooms. Some classes found themselves without teacher for several hours or with a teacher not specialised in the topic they needed at that time. More Computer Science teachers would have been needed because main challenges faced by students were on this topic. Putting all the first-year students in a large common workspace would partly limit the problem of the supervision rate, because all the supervisors would be in the same place, available to everyone. Above all, training of the teachers to the use of the robot has to be improved as it will become an increasingly used pedagogic tool in the training of our students.

Students' difficulties during this challenge concerning the computer programming part could be reduced by adding a few days of work on this theme earlier in the year, as the challenge would then not be their first encounter with the programming of the robot. The main difficulty of the challenge would then become making a link between the different topics (Physics, Electronics, Computer Science) around their familiar *esieabot*.

This end-of-year challenge has allowed students to discover the potential of their robot and to raise their curiosity and interest in robotics, electronics or programming. This interest can be seen in the slight increase of the number of second-year "scientific and technical projects" involving a robot in 2022–2023: 9 groups chose a scientific and technical project on the theme of robotics, including 4 using their *esieabot*, compared to an average of 6 in previous years (since 2018).

4 *Personal* and *Open* Objectives: Examples

4.1 Projects Proposed by Students

The students from the 2nd to the 5th year conduct technical one-year projects in groups of 3–5. The topics of these projects are open and proposed by the students themselves in 2nd and 3rd years, and by companies and laboratories in the 4th and 5th years.

Associated with our *Personal* objective, since we gave each 1st year student a robot, we have noticed a continuous increase in the number of projects using it. Therefore, we can cite various examples of projects proposed and realized by students using their robot:

- adding a LiDAR (Light Detection And Ranging) module to the robot in order to be able to autonomously explore a building and perform a 3D mapping of it
- transforming the robot into a flying drone
- building a semi-autonomous delivery robot able to transport objects (in adequacy with its size naturally) in an already known environment
- creating an elevator dedicated to our robot allowing it to move from one floor to another since it is not able to climb stairs, at least not in its basic form
- making a wristband with an accelerometer to substitute the joystick for the control of the robot

Another interesting fact we noticed is students do not hesitate to propose projects requiring skills that will be addressed later in their curriculum, which forces them to be trained either by themselves or with students from higher years who are members of the school's technical clubs.

More conventionally, students also turn to teachers, especially in electronics and computer science, for help with their projects. For these teachers it is an opportunity to see if the skills they taught to the students are acquired or not. It is also an opportunity to better identify skills that have not yet been covered in the curriculum. This feedback is therefore a chance for teachers to improve their teaching practices by targeting the topics which are the most difficult to learn for students.

4.2 Contribution of Students

Associated with our *Open* objective, thanks to their experience in assembling and using their robot, some students propose improvements of the basic version that we provide them. For example, one of them realized a PCB (Printed Circuit Board) to replace the connections composed of cables and a breadboard (Fig. 4). Therefore, students will have more space on their breadboard to implement more components. On top of that, the add-on PCB adds a Analog to Digital Converter to make possible the implementation of analog sensors directly on the robot.

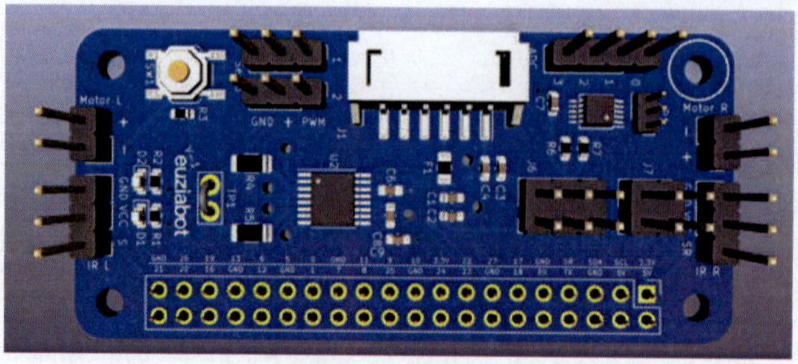

Fig. 4 Add-on PCB designed by a student

5 Conclusion and Perspectives

In this paper, we have shown that the platform developed allowed students to more easily approach other subjects while at the same time developing their creativity and motivation, changing the way that teachers can view student skills. Future work will include the development of a systematic measurement of the impacts of using *esieabot* in educational activities. Finally, the open aspect of the platform generated several ideas from students themselves who proposed their own projects or developed upgrades to the robot. Some areas for improvement are detailed below.

Opensource and user support For nearly a year now, *esieabot* has been 100% open-source. Everything from hardware to software is published under GNU General Public License v3. Every part of the robot can be used, edited and shared under GNU GPL. This has motivated other people, especially students, to contribute to the project. We are aware that this could generate a need for more maintenance and for increased user support.

Training teachers The robot can be used in other subjects where useful. We believe that using a robot in academic courses with few hours of practical work and in which students are generally more difficult to interest could engage them more efficiently. For this to be possible we need to train more of our colleagues. Even if we made the platform as easy as possible to develop and use for the teaching of STEM, it is nevertheless important to do as much as possible to communicate and support so that the robot can be used as a real learning platform and not just as a toy.

Use in secondary education In the future, we want to deploy our robot in high schools. To do so, we want to develop a block-programming interface to make it easier to program. A Scratch-like API implementation could be the way to do this. Our goal is the same: introduce more and more people to computer science and robotics.

Use in other curricula, abroad and also with non-technical students We want to continue to test the use of our robot in lessons for students with non-technical profiles. As presented in Sect. 3, colleagues have already been able to experience its successful use during a summer school in Croatia in 2022. We wish to broaden our understanding of the constraints and advantages of these uses with students of different profiles.

Acknowledgements This work was supported by ESIEA engineering school and Raspberry Pi Foundation. We are particularly grateful to Nadjim BENALI, Jeremy COCKS, Vincent GUYOT, Bassem HAIDAR, Siba HAIDAR, Loic ROUSSEL and Pierre TOUCHARD.

References

1. Alimisis, D.: Educational robotics: open questions and new challenges. Themes Sci. Technol. Educ. **6**, 63–71 (2013)
2. Benitti, F.: Exploring the educational potential of robotics in schools: a systematic review. Comput. Educ. **58**, 978–988 (2012)
3. Sapounidis, T., Alimisis, D.: Educational Robotics for STEM: A Review of Technologies and Some Educational Considerations, pp. 167–190. Nova Science (2020)
4. Eguchi, A.: Educational robotics for promoting 21st century skills. J. Autom., Mob. Robot. Intell. Syst. **8**, 5–11 (2014)
5. Major, L., Kyriacou, T., Brereton, P.: Systematic literature review: teaching novices programming using robots. Softw. IET **6**, 502–513 (2012)
6. Lapeyre, M.: Poppy: open-source, 3D printed and fully-modular robotic platform for science, art and education. Theses, Université de Bordeaux (2014)
7. Otto DIY store. https://www.ottodiy.com/store
8. Chronis, C., Varlamis, I.: Fossbot: An open source and open design educational robot. Electronics **11**, 2606 (2022)
9. Riedo, F., Chevalier, M., Magnenat, S., Mondada, F.: Thymio II, a robot that grows wiser with children. In: 2013 IEEE Workshop on Advanced Robotics and Its Social Impacts, pp. 187–193. IEEE (2013)
10. Tsalmpouris, G., Tsinarakis, G., Gertsakis, N., Chatzichristofis, S., Doitsidis, L.: Hydra: introducing a low-cost framework for stem education using open tools. Electronics **10**, 3056 (2021)
11. ESIEA: documentation of esieabot (2023). https://esieabot.readthedocs.io
12. ESIEA: esieabot git's repository (2023). https://gitlab.esiea.fr/esieabot
13. ESIEA: esieabot's bom and manual (2022). https://esieabot.esiea.fr/bom
14. Documentation of pigpio library. https://abyz.me.uk/rpi/pigpio/
15. Fritzing. https://fritzing.org/

Introductory Activities for Teaching Robotics with SmartMotors

Milan Dahal, Lydia Kresin, and Chris Rogers

Abstract In this research paper we describe the activities designed for a 5 day engineering design workshop using an educational robotics tool called SmartMotors. SmartMotors are low cost solutions to teach elementary and middle school students about robotics and Artificial Intelligence in under-resourced classrooms. In a few simple steps these motors can be trained by the users to run to various states corresponding to different sensor inputs. We believe that the low cost and trainable aspects of SmartMotors will reduce barriers of entry for both teachers and students in introducing robotics in classrooms and increase student access to robotics and engineering. In the summer of 2022, we used one of our prototypes to run a usability study in a workshop with ten middle school students aged 12–15. The students participated in an hour-long engineering design workshop for five days, and each day they received different prompts along with necessary scaffolding. The participating students had limited prior exposure to robotics and AI. By the end of the workshop, the students were able to train robots in group projects that reflected their individual interests. In this paper, we talk about the students' journey, starting with building simple projects and subsequently gaining the skills and confidence to showcase diverse and complex designs. We will then discuss the affordances of the tool and explore the opportunities and limitations of SmartMotors in an engineering design workshop.

Keywords Educational robotics · Workshops · SmartMotors

M. Dahal (✉) · L. Kresin · C. Rogers
Center for Engineering Education and Outreach, Tufts University, Medford, MA 02155, USA
e-mail: milan.dahal@tufts.edu

L. Kresin
e-mail: lydia.kresin@tufts.edu

C. Rogers
e-mail: chris.rogers@tufts.edu

R. Balogh et al. (eds.), *Robotics in Education*, Lecture Notes in Networks and Systems 747, https://doi.org/10.1007/978-3-031-38454-7_20

1 Introduction

Integrating robotics in the classroom can be an effective way to teach STEM concepts to students [1–3]. It can also enhance the development of skills like collaboration, interpersonal communication, creativity [4, 5], critical thinking, and problem solving [5] and inquiry [6]. Properly designed robots can actively engage even younger children in programming and engineering [7]. The idea of using robots in education is not recent. It is rooted in the concepts of constructivism [8, 9] and constructionism [10]. With the advancement of technology in recent years educational robotics has found its way into many classrooms. However, many of these technologies require technological expertise and access to computers, which presents issues of equitable access for students whose schools may already lack sufficient facilities and funding to incorporate hands-on STEM activities into their curricula.

LEGO robotic kits have been used all around the world to teach robotics [11]. However, their high cost prevents many schools from incorporating them into their classrooms. Online tools like Teachable Machine [12] and Cognimates [13] require internet connection and computers, and technical expertise on the part of teachers may be necessary for them to be used effectively. Tools like Paperbots [14] are low cost solution to teaching robotics using papers but you still need a computer to code the bots. In order to lower these barriers to entry to robotics, we have devised an idea for an educational robotics toolkit called SmartMotors. SmartMotors are built using easily accessible low-cost materials and do not require computers or access to the internet to program. In a few steps these contraptions can be trained by the users, including students, to build responsive robots. SmartMotors are a low-cost way to bring engineering and robotics into classrooms, making them more accessible to students from a diverse range of backgrounds and from around the world.

2 Background

The central concept of SmartMotors is to train robots instead of coding them. We hypothesize that this shift from programming to training makes SmartMotors approachable and accessible for both STEM and non STEM educators, as well as beginner students. Additionally, the elimination of the need for computers to engage in robotics activities makes this system accessible to users in low-resourced environments. The use of locally available tools and open source instructions to build SmartMotors makes it accessible for educators working with a wide range of materials. Through carefully designed instructions and activities, we can ensure that students have rich meaningful learning experiences using SmartMotors.

As alluded in the previous paragraph, the concept of SmartMotors can be implemented on different existing robotics platforms, as well as microcontrollers. Prior research has shown that many teachers have found ways to build their own robotics kits [15]. To offer these innovative instructors more flexibility, we have documented and made available the instructions for building SmartMotors using different microcontroller boards on a variety of platforms, including Arduino Uno, ESP8266, Seeeduino Xiao, and LEGO SPIKE Prime [16]. We are also developing low cost kits under $20 for teachers who prefer off-the-shelf kits. For the study described in this paper, we used a prototype built using Wio Terminal, a microcontroller from Seeed Studio and a hobby micro servo motor. The Wio Terminal was coded with the SmartMotors algorithm with a friendly user interface.

We envision the use of SmartMotors in an engineering design activity as a tool that enables students to engage in quick builds that showcase complex ideas. SmartMotors allows users to think creatively about the issue at hand without worrying about the technical details. This allows the students to invest more time on the brainstorming, building, and sharing aspects of the engineering design process. The system also provides layered complexity to support learners at different stages. As teachers and students become comfortable with SmartMotors and want to add more complex behaviors to the motors, they can reprogram and modify the behavior of the system. In other contexts, when the focus is on generating ideas for discussions and creating innovative solutions, the system provides plenty of opportunities to foster creativity.

3 Description of the System

Wio Terminal SmartMotors is built using Wio Terminal microcontroller, a hobby servo motor and a Grove rotary encoder. Wio Terminal is an ATSAMD51-based microcontroller developed by Seeed Studio [17]. The Terminal has a wide array of sensors and features like light sensor, microphone, Inertial Measurement Unit (IMU), as well as 40-pin GPIO for expansion. From a pilot study performed with teachers, we learned that the users wanted the system to inform them of its various states. They described that it would be helpful for the system to communicate different sensor and motor values through a graphical representation. The Wio Terminal has a 2.4" LCD color screen on which we designed the user interface. As it is not a touch screen, the buttons were used for user input. The total cost of this robotics kit is under $50.

3.1 Building Instructions

A computer is needed only once to upload the SmartMotors code to the Wio Terminal. The SmartMotors code is publicly accessible through the QR code on Fig. 1 or from the website [16]. Once the Wio Terminal is put into bootloader mode the firmware.uf2 file can be loaded on the Wio Terminal. The steps are outlined in the graphics in detail (Fig. 1).

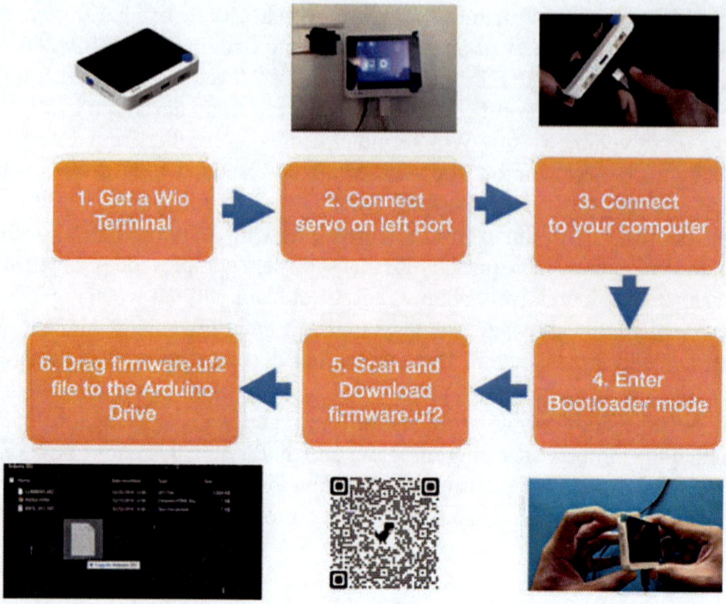

Fig. 1 The six steps to build Wio Terminal SmartMotors

3.2 *Features*

The tilt sensor is pre-selected as the default sensor in Wio Terminal SmartMotors. From the Home page (Fig. 2), the users can navigate to the Sensor Select page to choose a different sensor, if needed. On the Sensor Select page (Fig. 3a), the users have three different sensors to choose from—namely, light sensor, tilt sensor, and rotary encoder. The Home page screen shows the icon of the selected sensor as well as its reading with an interactive bar. The motor can be turned in either clockwise

Fig. 2 Homepage of the Wio Terminal SmartMotors user interface

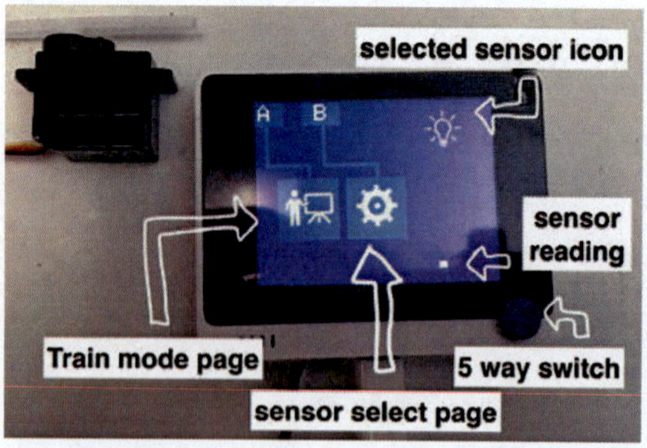

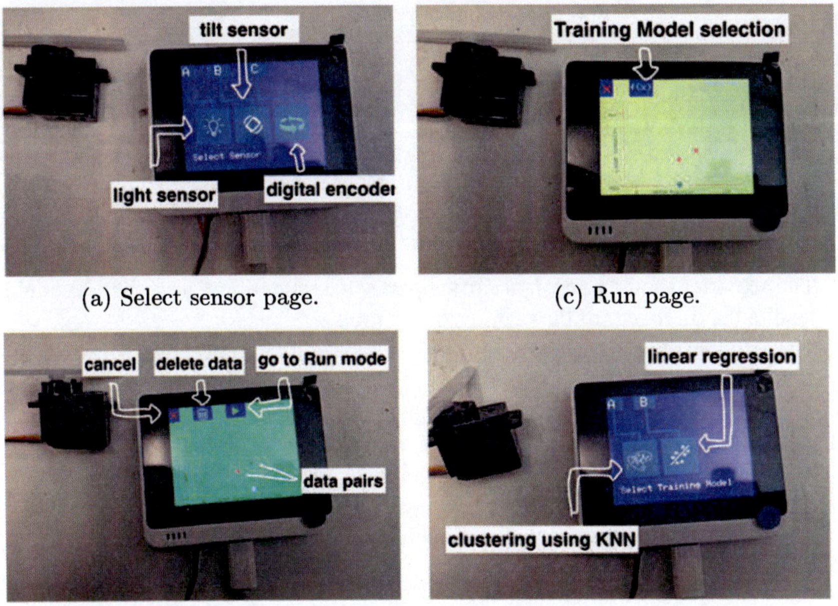

(a) Select sensor page.

(c) Run page.

(b) Train page.

(d) Train Model page.

Fig. 3 Other screens on Wio Terminal SmartMotors user interface

or anti-clockwise direction by pressing the 5-way switch left or right, respectively. From the Home page the users can move on to the Train page by clicking the Train icon. In the Train page (Fig. 3b), they can begin pairing sensor values and motor positions by moving the motor to the desired position and pressing the 5-way switch while setting the sensor at a desired reading. Once the user inputs all the desired sensor reading and motor position pairs to the system, they can enter Run mode by clicking the Run icon. In this mode, the motor responds according to the sensor values inputted in the Train mode. The Run page screen (Fig. 3c) shows the current sensor reading and motor position overlaid on a graph of inputted data points. On the Train Model page (Fig. 3d), users can choose to use a categorization model using k-nearest neighbor algorithm or a linear regression algorithm on the training data. The data is not saved on the system, so upon exiting the Run mode page, users must retrain the system with new data.

4 Method

4.1 Activity Prompt Design

SmartMotors are designed to support both teachers and students with little to no experience with robotics. We considered several criteria to build complexity one step at a time while designing the activities. Some of the criteria are listed below:

1. The activities should be designed to support self exploration. The activities should lead to the discovery of the features of the tools.
2. The activities should encourage group work and collaboration between team members.
3. The activities should inspire the students to think critically.
4. The activities should support students with different comfort and skill levels. They should lower the barriers to engagement, as well as raise the ceiling of their potential accomplishment.
5. The materials used in the activities should be low cost or easily found around the household.
6. The items that are used in the workshop should be sustainable. Use of recyclable materials like paper and cardboard is encouraged.
7. The activities should be completed within an hour including sharing and feedback sessions.

4.2 Workshop Design

Ten students from grades six to nine participated in the five day workshop. For one hour each day, they worked in small groups, in which they followed the engineering design principle—planning, designing, creating, testing and repeating. Activities were directed using instruction placemats [18]. Placemats are two-sided instructional sheets used in robotics activities. The prompts, challenge questions, examples, and instructions are put in specific order to elicit solution diversity and support self exploration [19]. The activities listed on Table 1 were adapted to the placemat design [16]. Each session began with an introduction to the prompt, planning in small groups, followed by 30–40 min of building time. The students were given cardboard, scissors, paper, tape, colored pencils, and hot glue guns to build their projects. At the end of each session, the students shared their project with the larger group. The other student groups gave feedback on the design and contents of the project. The activity was videotaped and the artifacts were photographed for documentation and analysis.

Table 1 Wio Terminal SmartMotors activities for robotics

Day	Activity title	Learning goals (Students will be able to...)
1	Learn to train and use the Wio SmartMotor to animate your creations	train their SmartMotors
2	Use the Wio SmartMotor to tell a story of what you did yesterday using the light sensor	use the light sensor to train the SmartMotor and use it in their project
3	Use the Wio SmartMotor and tilt sensor to make something relating to the environment	use the tilt sensor to train the SmartMotor and use it in their project
4	Use the Wio SmartMotor and digital encoder to make a game	use the digital encoder to train the SmartMotor and use it in their project
5	Choose a sensor and use what you have learned about Wio SmartMotors to tell a story	use SmartMotors to build a project that tells a story that they have read or composed

5 Observation and Discussion

5.1 Day 1: Learn to Train and Use the Wio SmartMotor to Animate Your Creations!

The first activity was designed for students to learn how to use Wio Terminal SmartMotors. Most of the students were doing robotics activities for the first time, and only one student said they had prior programming experience. The example projects on the placemat (Fig. 4) were simple and easy to build. The second page of the placemat was dedicated to instructing how to train the SmartMotors.

The students worked in groups of two and three. All the groups built simple arm-like attachments to the motors that moved from side to side when the sensor was triggered. In this activity, there was limited solution diversity. The older students seemed to have easier time building with the materials and working in groups. Every group was able to train their SmartMotor by the end of the first day.

5.2 Day 2: Use the Wio SmartMotor to Tell a Story of What You Did Yesterday Using the Light Sensor!

The prompt for the second day was to use the light sensor to train the SmartMotor and build an artifact to share an event from their personal life. The examples on the placemat (Fig. 5a) showed waking up, brushing teeth and reading a book as sample

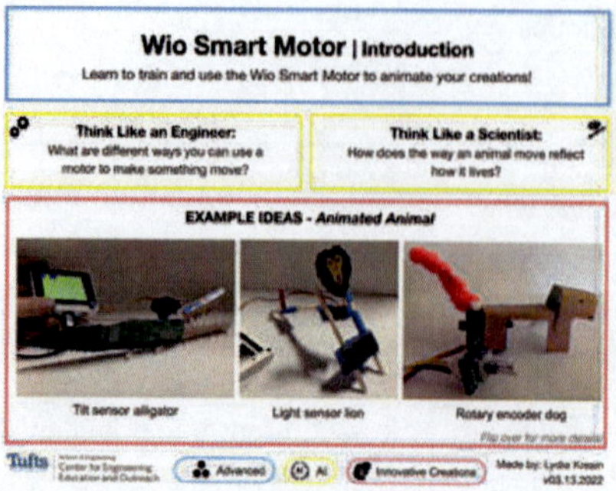

(a) Front page

(b) Back Page

Fig. 4 Placemat for day 1

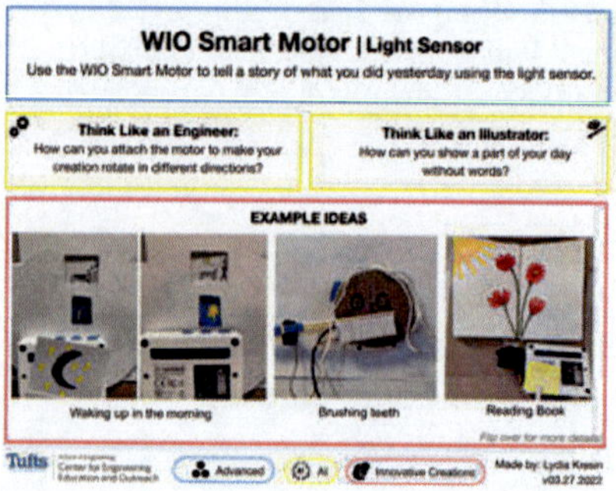

(a) Front page of day 2 placemat

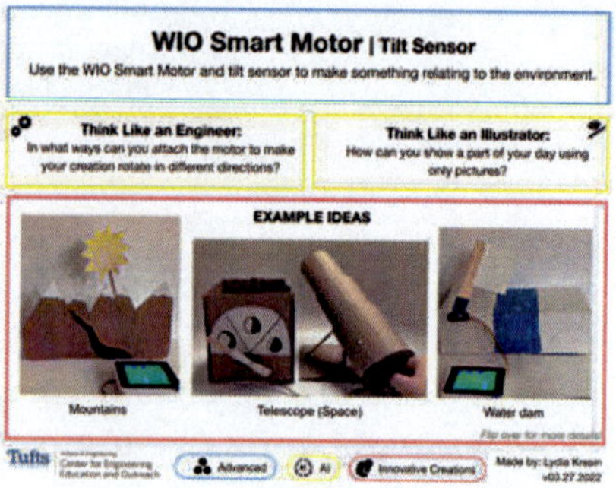

(b) Front page of day 3 placemat

Fig. 5 Front page of placemats for day 2 and 3

projects. On the back page of the placemat were detailed instructions on how to attach the motor to the cardboard using tape and how to use the light sensor to build interactive projects.

The students continued to work in their respective groups. Since they had already used SmartMotors the previous day, they were able to dedicate more time to building. On the second day, the diversity of projects was evident (Fig. 6). One project demonstrated how they put sunglasses on when they went outdoors. Another group built an artistic representation of the breakfast they had eaten that morning (Fig. 7a). Another group was inspired by the first group's idea of a sunny day and built a fan that turned on when the sun came out. Finally, the last group built a leg with two degrees of freedom to show how they played football in the morning (Fig. 7b). They used two SmartMotors to actuate two sections of the limb and coordinated to kick the ball. The overall complexity of the projects was significantly higher than that of the first day.

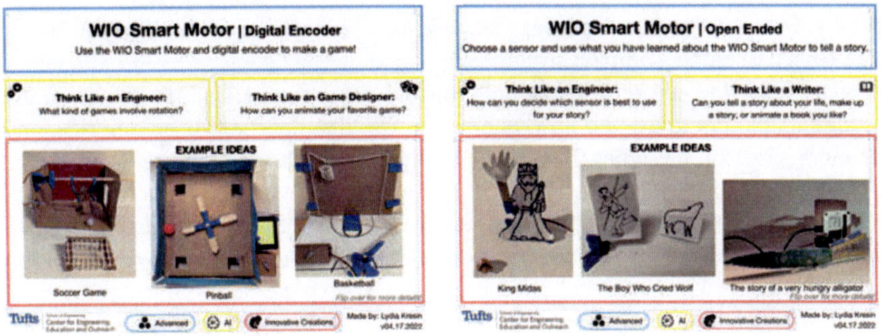

(a) Front page of day 4 placemat (b) Front page of day 5 placemat

Fig. 6 Front page of placemats for day 4 and 5

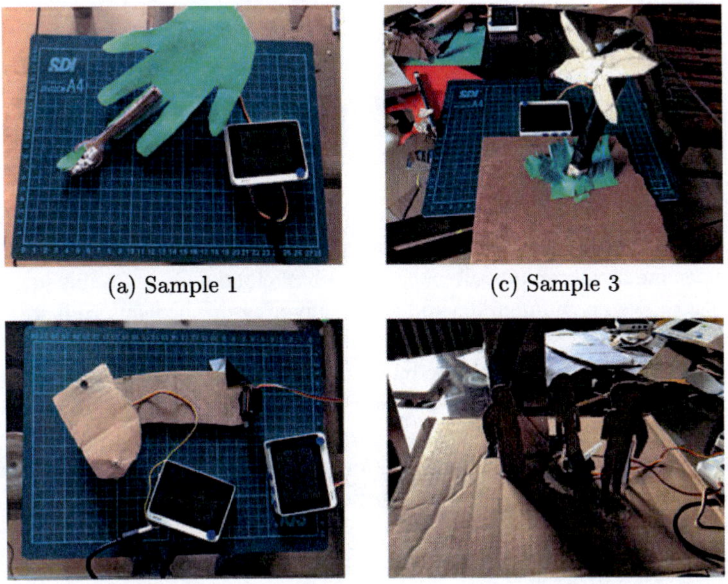

(a) Sample 1 (c) Sample 3

(b) Sample 2 (d) Sample 4

Fig. 7 Sample student projects from the workshop

5.3 *Day 3: Use the Wio SmartMotor and Tilt Sensor to Make Something Related to the Environment*

For the third activity, we asked the students to simulate an activity relating to the environment. The placemat (Fig. 5b) showed examples of the sun setting down behind the mountains, a telescope showing different moon phases, and a gate on a water dam moving up and down. The back page showed detailed instructions on how cardboard can be used to create complex structures.

Inspired by the sun setting example, one of the groups added the moon on the same shaft and demonstrated how when the earth tilts, the sun sets on one side, and the moon rises on the other. Another group wanted to bring attention to the issue of deforestation. They built a tree trunk and attached a cardboard axe to the motor shaft, which swiveled around and cut the tree trunk in half. They also wanted to use another motor to lift the tree back up to extend the message that deforestation should be stopped, but they needed more time to complete it. The third group built a windmill to bring up the topic of renewable energy. This group wanted the motor to move continuously; however, the servo could only sweep between 0 and 180°C (Fig. 7c). The final group built an assistive device to push the trash into a trash can. During this activity, the students worked on a wide range of project ideas meaningful to them and were able to get their message across using the SmartMotors.

5.4 Day 4: Use the Wio SmartMotor and Digital Encoder to Make a Game!

The prompt for the fourth day was to build a game using a digital encoder. The digital encoder is coded to work as a rotation sensor. The examples on placemat (Fig. 6a) showed a game of soccer, pinball game, and a basketball with a movable hoop. The back page showed how to build a sliding and pivoting mechanisms with string and pins.

The students were very excited to build the games and each pitched several ideas to their groups. Only two groups were able to complete their projects on time. One of the groups made a goalkeeper game, in which the objective for one player was to score a goal and for the other to move the goalie, controlled by the SmartMotor. Another team build an air hockey-type game with two controllers. The other teams who were not able to complete their projects cited lack of proper cardboard and glue guns as their reasons. They also indicated that they did not have enough time to realize their projects. It was noted that the groups who could not complete their projects had ambitious project ideas.

5.5 Day 5: Choose a Sensor and Use What You Have Learned About Wio SmartMotor to Tell a Story

On the final day of the workshop, the students were asked to brainstorm in their groups a story they wanted to share with the larger group through SmartMotors. They were free to use a sensor of their choice. The examples on the placemat (Fig. 6b) showed classic children's stories, such as King Midas and The Boy Who Cried Wolf. The back page of the placemat showed all the materials they could use in the project. It also informed how they could use a linear regression model instead of a categorization model on the training data collected with the SmartMotors.

One of the groups built a project based on their own story involving a devil, monster, and shields, in which, interestingly, the main character lost the final battle. Another group shared the story of a woodcutter and an angel. In short, the angel who lives in a lake is delighted by the honesty of the woodcutter and rewards him with gold and silver axes, in addition to his iron axe that had fallen in the lake. The group attached the angel to the motor to mimic her rising from the lake surface with the different axes as they told the story. The third group retold the story of "This Ends With Us," with figures attached to two motors that swiveled and turned around to demonstrate how the events unfolded and the characters moved on with their lives (Fig. 7d). Finally, the last group told a miner's story who mined a diamond using a pick axe.

5.6 Summary

In this 5-day workshop, we observed that SmartMotors enable students to construct unique and creative projects. The students generated all the project ideas based on the prompts from the placemats. The facilitator asked questions during the build to encourage discussion and critical thinking. The availability of different sensors allowed them to combine different actions to trigger motor movements. Since the students were all new to the robotics and engineering design process, there was a lack of solution diversity on their first project. However, the primary goal of the first day of the workshop—to teach them how to use SmartMotors—was accomplished.

Beginning on the second day, we saw the students pushing the limits of materials to develop exciting project ideas. The students demonstrated outside-the-box thinking, from modifying the example on the placemat to using two motors to actuate different leg joints. Throughout the workshop, students worked in teams and challenged each other to build sophisticated projects. Some students focused on mechanical movements, while others focused on stories they wanted to share. One of the groups was focused more on the design and aesthetics part of the project, which was evident by how meticulously they built and decorated their projects. Advanced system features were introduced to groups that were already comfortable with SmartMotors. On the final day, all students demonstrated creativity and imagination by telling a story through projects built using cardboard and the SmartMotor system.

We envision SmartMotors as a tool to lower the barriers of entry to robotics for beginners and students with limited access to computers, and the students' thinking about the features and further enhancements is an essential step in the direction of guiding students to become agents of creation than just consumers of technology [20]. The students encountered and highlighted some of the system's limitations—including the lack of continuous motors and the inability to control multiple motors at once. These issues can be opportunities to let the students explore more independently and creatively using the tools. Some issues, like the lack of speed control, can be easily solved by swapping the standard servo with a continuous servo. Issues of using multiple sensors or multiple motors can be solved by editing the code, which can be found in our website. However, the barriers of cost of the Wio Terminal based SmartMotors can only be lowered by either building a ecosystem where teachers and advanced students can build their own SmartMotors using the tools they already have, or building a cheaper version of SmartMotors. We have guides to help teachers and students to assist in this process in our website [16].

6 Conclusion

In this workshop, we demonstrated that students without prior experience in robotics can quickly build meaningful projects using SmartMotors. They may take some time to get comfortable with the new system [21], but with regular engagement they can learn how to navigate the system. The carefully designed activities can teach important robotics concepts, such as sensing, processing, and actuation, as well as the ideas of data collection and training. The use of placemats meant that minimum teacher intervention was required. The workshop demonstrated that students can move independently through the processes of discovery and self-exploration and that teachers without prior robotics experience can run these activities effectively.

Through SmartMotors and the corresponding activities outlined in placemats, we can engage users to think creatively and critically. The progression of complexity of the students' projects throughout the five-day workshop demonstrated that the SmartMotor system is adaptable to users with different levels of comfort with technology and robotics. SmartMotors also enabled users to meaningfully interact and learn from one another. The next step in the development of the system is to connect these activities to curricula and standards [22] for teachers, testing them in workshops in a variety of educational contexts. This version of SmartMotors uses Wio Terminal and the cost can be prohibiting for many users. The learning from this workshop will be used to design an accessible custom SmartMotors.

References

1. Khanlari, A., Mansourkiaie, F.: Using robotics for stem education in primary/elementary schools: teachers' perceptions. In: 2015 10th International Conference on Computer Science & Education (ICCSE), pp. 3–7. IEEE (2015)
2. Williams, D.C., Ma, Y., Prejean, L., Ford, M.J., Lai, G.: Acquisition of physics content knowledge and scientific inquiry skills in a robotics summer camp. J. Res. Technol. Educ. **40**(2), 201–216 (2007)
3. Ortiz, A.M.: Fifth grade students' understanding of ratio and proportion in an engineering robotics program. Tufts University (2010)

4. Sahin, A., Ayar, M.C., Adiguzel, T.: Stem related after-school program activities and associated outcomes on student learning. Educ. Sci.: Theory Pract. **14**(1), 309–322 (2014)
5. Okita, S.Y.: The relative merits of transparency: investigating situations that support the use of robotics in developing student learning adaptability across virtual and physical computing platforms. Br. J. Educ. Technol. **45**(5), 844–862 (2014)
6. Ganesh, T., Thieken, J., Baker, D., Krause, S., Roberts, C., Elser, M., Taylor, W., Golden, J., Middleton, J., Kurpius, S.R.: Learning through engineering design and practice: implementation and impact of a middle school engineering education program. In: 2010 Annual Conference & Exposition, pp. 15–837 (2010)
7. Bers, M.U.: The tangiblek robotics program: applied computational thinking for young children. Early Child. Res. Pract. **12**(2) (2010)
8. Bruner, J.: Celebrating divergence: Piaget and vygotsky. Hum. Dev. **40**(2), 63–73 (1997)
9. Piaget, J.: Genetic epistemology. trans. e. duckworth. 1970
10. Papert, S.A.: Mindstorms: Children, Computers, and Powerful Ideas. Basic Books (2020)
11. Butterworth, D.T.: Teaching c/c++ programming with lego mindstorms. In: Proceedings 3rd International Conferences on Robotics In Education (RiE2012) (2012)
12. Teachable Machine. https://teachablemachine.withgoogle.com. Accessed 31 Jan 2023
13. Cognimates. http://cognimates.me/home/. Accessed 31 Jan 2023
14. O'Connell, B.: The development of the paperbots robotics kit for inexpensive robotics education activities for elementary students. Ph.D. thesis, Tufts University (2013)
15. García-Saura, C., González-Gómez, J.: Low cost educational platform for robotics, using open-source 3d printers and open-source hardware. In: ICERI2012 Proceedings, pp. 2699–2706. IATED (2012)
16. Wio Terminal SmartMotors Placemats. https://smartmotors.notion.site. Accessed 31 Jan 2023
17. Seeed Studio. Get Started with Wio Terminal (2021). https://wiki.seeedstudio.com/Wio-Terminal-Getting-Started/. Accessed 31 Jan 2023
18. Willner-Giwerc, S., Hsin, R., Mody, S., Rogers, C.: Placemat instructions. Sci. Child. **60**(3) (2023)
19. Willner-Giwerc, S., Danahy, E., Rogers, C.: Placemat instructions for open-ended robotics challenges. In: Robotics in Education: Methodologies and Technologies, pp. 234–244. Springer (2021)
20. Blikstein, P.: Digital fabrication and 'making' in education: the democratization of invention. FabLabs: Mach. Makers Inven. **4**(1), 1–21 (2013)
21. Buchner, R., Wurhofer, D., Weiss, A., Tscheligi, M.: Robots in time: how user experience in human-robot interaction changes over time. In: Social Robotics: 5th International Conference, ICSR 2013, Bristol, UK, October 27-29, 2013, Proceedings 5, pp. 138–147. Springer (2013)
22. Eguchi, A.: Bringing robotics in classrooms. In: Robotics in STEM Education: Redesigning the learning Experience, pp. 3–31 (2017)

Collaborative Construction of a Multi-Robot Remote Laboratory: Description and Experience

Lía García-Pérez, Jesús Chacón Sombría, Alejandro Gutierrez Fontán, and Juan Francisco Jiménez Castellanos

Abstract In this paper the Robotarium-UCM is presented, a low-cost multi-robot remote laboratory for teaching robotics, control and distributed robotics. Robotarium-UCM is a physical laboratory capable of being remotely operated that is part of the remote laboratories of the research group to which the authors belong. The software that supports the Robotarium-UCM is composed of three pillars: Robot Firmware, Agent and Hub. A student multi-robot rendezvous experiment using the Robotarium-UCM is described as well as a collaborative construction experience.

Keywords Remote-laboratories · Multi-Robot · Control

1 Introduction

In Science and Engineering subjects, the importance of practical work with real devices is well known. Specifically, in recent years, improvements in computing, sensors, actuators, communication and power have helped the development of small, low-cost robots. These robots are used assiduously in laboratories not only in robotics but also in other areas such as control, signal processing, sensors [1, 2].

Multi-agent systems and distributed control are a priority focus in the area of control and systems. Despite this, there are still few practices that students of this type of disciplines can do with real robots, since maintaining a multi-robot practice system is not easy. For this reason, many educators use simulation environments that are also very simplified since realistic simulation requires great computational power that increases as the number of agents and interactions grows. However, these simplified simulation environments elude many of the most deterministic and important issues of the algorithms to be studied [2]. A different and particularly suitable option is to

This work was supported in part by the State Research Agency under project PID2021-127648OB-C33 and in part by the UCM University under project Innova-Docencia 2022-23 313.

L. García-Pérez (✉) · J. C. Sombría · A. G. Fontán · J. F. J. Castellanos
Facultad de C.C. Físicas, Universidad Complutense, Plaza de Las Ciencias 1, 28040 Madrid, Spain
e-mail: liagar05@ucm.es
URL: https://www.ucm.es/dacya

R. Balogh et al. (eds.), *Robotics in Education*, Lecture Notes in Networks and Systems 747, https://doi.org/10.1007/978-3-031-38454-7_21

have a multi-robot robotic laboratory, where users program and execute their code on real robots through a web platform that also allows them to observe the execution.

Two well-known and successful examples of multi-robot systems are Robotarium and Duckietown. Robotarium is a Georgia Tech Universty's project, originally intended to provide robotics and control researchers with a remote testing platform. As described in [3] although its use was initially intended for researchers, most of the requests to experiment with Robotarium came from students and teachers.

Duckietown is also a learning and research-oriented multi-robot platform. It is a scalable platform that can become a remote laboratory.

An operational swarm of small robots is a really interesting option, allowing novel and innovative exercises with students [4], both in distributed control and in robot coordination and collaboration [5]. These practices could be very valuable for our students, giving them real experience with actual field problems in areas like multi-robot systems and distributed algorithms.

The deployment of a laboratory with the desired characteristics requires a high economic and personal effort. On the one hand, although there are low-cost solutions especially focused on educational robotics [6], having a fleet of robots capable of collaborating can represent a significant economic investment. On the other hand, the deployment and maintenance of the necessary infrastructure to support the proposed activities entail a high level of dedication, due to the distributed nature of the swarm, multiple technological factors involved, security issues, etc.

Due to the high cost, limiting the use of the laboratory to a few practice sessions represents an underutilization of resources. In this sense, the development of remote laboratories has been found to be a very appropriate solution for maximizing the availability of experimental hardware, since it allows (1) a return on the investment in time and money made to set up the practices and (2) a simple solution to unforeseen situations (as we have seen with the COVID 19 pandemic) that prevent or reduce attendance.

This paper describes the implementation of Robotarium-UCM, a multi-robot system with remote access that allows students and researchers to carry out practices and experiments with real robots that would otherwise be very expensive or unfeasible. Remote programming, through Internet, makes the system very flexible and highly reusable. An important contribution of this work is that it presents the experience of building a robotarium with students, thus facilitating the replication of an interesting teaching and research resource in other centers in a simple way and with the participation of the entire educational community.

The paper has been arranged as follows: The first section describes the objectives to be achieved with the Robotarium-UCM as well as the requirements needed to achieve them. Section 2 describes the hardware and software used. Section 3 describes the remote laboratory infrastructure in which Robotarium-UCM is integrated. Section 4 shows the results of an experiment performed using the Robotarium-UCM. Section 5 describes an event with students organized on the occasion of Computer Science Week to replicate the robotarium in another location. Finally, Sect. 6 is devoted to conclusions and future work.

2 Robotarium-UCM Requirements

Robotarium-UCM project is part of ISA (*Ingeniería de Sistemas y Automática*) remote laboratory system [7] at the Faculty of Physics of Complutense University of Madrid. The shared objective of all these remote laboratories is to allow students to carry out practical work remotely, in order to optimize the use of scarce and expensive resources and to deal with unforeseen situations, as it was the case during the COVID-19 pandemic. The objective of Robotarium-UCM is to have a system with multiple robots (10 initially) that can be programmed remotely so that undergraduate and graduate students can practice distributed control systems or distributed robotics.

For this Robotarium-UCM has to meet the following minimum requirements:

- To have at least 3 physical robots.
- To have a delimited Robot-Arena where the robots are going to move. Obviously this Robot-Arena must be sized according to the number and size of the robots in the Robotarium.
- To have a localization mechanism that allows obtaining the position of each robot. Multi-robot algorithms require the robots to know its neighbours' relative pose (position and orientation) using local sensors or a central mechanism. We have decided to implement a centralized location system for reasons of simplicity as we have few robots.
- To have a server that on the one hand receives the code from the students through the platform of the remote laboratories and on the other hand is able to send to the robots the instructions and/or the code that they have to execute.
- To have a visualization system that allows students to see the result of their algorithms.
- Robots are endowed with encoders. Additional sensors could be added later.

In addition we have the following desirable requirements:

- Low cost robots are preferred. For education and outreach programs, low cost is a critical requirement, as robot system has to fit within the budget [8]
- The hardware of the robots must be flexible enough to be able to incorporate new functionalities to the robots.
- Different kinds of robots are allowed.

3 Robotarium-UCM: The Robots

3.1 Hardware

The robots of the Robotarium-UCM are holonomic robots, with two driving wheels and an idler wheel. We have two different robot models in terms of size and chassis materials. The initial model, type A, is larger and has a wooden and methacrylate

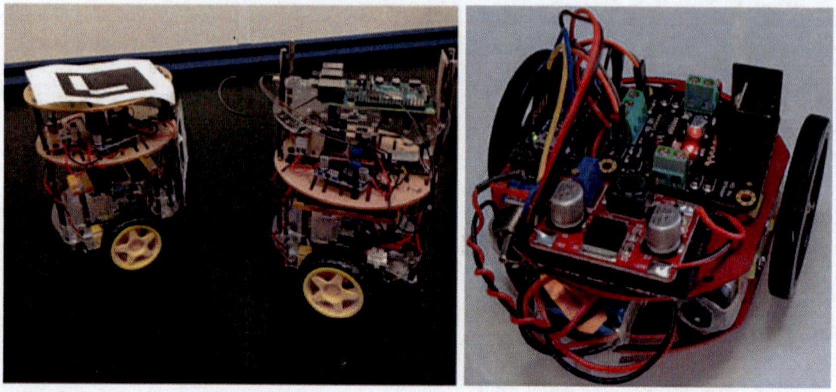

(a) Robot (type A). (b) Robot (type B).

Fig. 1 Robots of the Robotarium-UCM

Table 1 Size of both robot types

Robot model	Robot diameter (cm)	Robot height (cm)
Robot A	15	20
Robot B	10	10

chassis (see Fig. 1a). The other model, type B, has a metal chassis and is smaller (Fig. 1b). Sizes of both models are detailed in Table 1.

The robots have two *Pulse-Width Modulation* (PWM)-controlled DC motors as their only actuators, as is common in educational robots. At present the only sensor that the robots have is an encoder attached to the axle of each of the driving wheels with a resolution of 20 steps per turn.

Both, the activation of the motors (calculation of the appropriate PWM signal) and the reading of the encoder are done from an Arduino Nano. This Arduino Nano communicates via serial port with a Raspberry Pi, which is in charge of communicating with the server. Each robot-agent carries a unique ArUCO code [9], (Fig. 1a) which allows the sever to identify and locate the robot-agent, using an overhead camera.

The total cost of the assembled parts (chassis, circuit boards, motors, encoders and battery) is around 150€ for both robot models.

3.2 *Software*

The software that supports the Robotarium-UCM is composed of three pillars: Robot Firmware, Agent and Hub. The conceptual division in three subsystems is convenient and helps decoupling the design and maintenance, nevertheless the mapping between each subsystem and the hardware where it is actually deployed might depend on the setup. In particular, it might have sense either that the firmware and agent run in the same physical device (when the robot only ships an Arduino Nano), or in different devices (if the robot is also equipped with a Raspberry Pi).

Robot firmware. The firmware is the code that runs the Arduino Nano, which is responsible for controlling the basic functions of the robot. It reads from a text file the parameters needed to configure each robot (agent identification, IP, controller constants...). It is responsible for performing the following tasks:

- Read the robot's encoder and integrate the sensor readings to estimate the robot's position and orientation.
- Calculate PWM values to drive motors according to received speed and direction commands.
- Communicate with Raspberry via serial communication to receive speed and orientation commands and send position estimates.

Agent software. The agent is responsible for managing the communications between the hub and the robot firmware. On the one hand, the agent receives the messages originated in the robot firmware and forwards them to the Hub. These messages usually contain information about the state of the robot such as the velocity of the wheels as measured by the encoders. On the other hand, it takes the messages coming from the Hub and forwards them to the firmware. These messages, which can be originated in the Hub as well as from other external entities that use the API provided by it, usually contain commands that the robot must follow or information such as the position or velocity estimated by the positioning system.

The rationale behind the separation of the firmware and agent is that the communications with the hub might not be managed by the embedded control board (i.e. the Arduino Nano) but rather by an onboard computer such as Raspberry PI or NVIDIA Jetson Nano. This also opens the possibility of remotely programming the Arduino Nano board, therefore providing even greater flexibility to the activities that can be carried out with the robots.

Hub software. The Hub is a piece of software that (1) keeps track of the different agents that are present in the robotarium, (2) gathers data from the agents and (3) provides an API for clients to interact with the Robotarium-UCM. It can be thought of as a middleware that glues the agents (robots, cameras, etc. that make up the Robotarium-UCM experimental platform) and other services that are external or not directly related to it. ReNoLabs is an example of the latter, it is the software that provides a platform for remote experimentation, and it exists as an independent project.

The Hub keeps a list of agents which allows for a dynamic and, to a certain extent, autodiscoverable topology where new agents can be easily added. Every new agent that comes into the system must ask the hub to be registered, so the Hub can register the agent and establish a new link. In this way, the Hub start to receive data from the agent and it is able to send commands or other messages as discussed before.

Besides, the existence of the Hub is convenient because it provides flexibility in the topology, a centralized log, facilitates the redefinition of virtual topologies, or the simulation of problems in communications.

This software is still under heavy development, but is freely available for download in https://github.com/UCM-237/robotarium-hub.

4 Robotarium-UCM as a Part of the RemoteLabs

As pointed out before, an important feature of the Robotarium-UCM is that it offers students the ability to carry out their experiments with the mobile robots remotely, with the only requirement of having a stable Internet connection and any personal device that is able to run a modern web browser.

Figure 2 depicts the whole software/hardware infrastructure that enables remote access to the laboratory and supports the experimentation with the mobile robots and any other device that might be available in the Robotarium-UCM. The architecture is based on *ReNoLabs* ([10]), a Remote Laboratory Framework (RLF) developed by the authors. As discussed in the cited work, the creation and putting in production of a remote laboratory are usually non-trivial, with many technical aspects that have to be addressed. RLF simplifies and rationalizes that process.

Focusing on our use case, the experimentation with the mobile robots requires a software platform that has to provide different services, including:

- Authentication and Authorization.
- Schedule the access to laboratory resources.
- Web User Interface (WebUI).
  - Monitor the state of the laboratory.
  - Code Edition and Robot Programming.

Some of these services are already provided by ReNoLabs, such as authentication and authorization, web hosting, and a framework that allows to building and putting in production remote laboratory applications, including other capabilities.

The ReNoLabs RLMS provides a WebUI that students can access with their respective users. An important feature of the system is that, since the remote laboratory is a web application, it can be accessed from any device with a web browser, which includes any personal device such as cell phones, tablets or laptops. Within the web application, students can visualize the Robotarium-UCM through several webcams located in the lab (which have a twofold function, being the other one the localization of the robots using the ArUCO markers). Besides, the students can also take control of the robots in two different modes:

Manual Control In this mode, low-level commands are directly sent to the actuators (e.g. *move left wheel ten steps forward*).

Automatic Control In this mode, the student provides high-level commands that the robot will follow (e.g. *move to point* (x, y)).

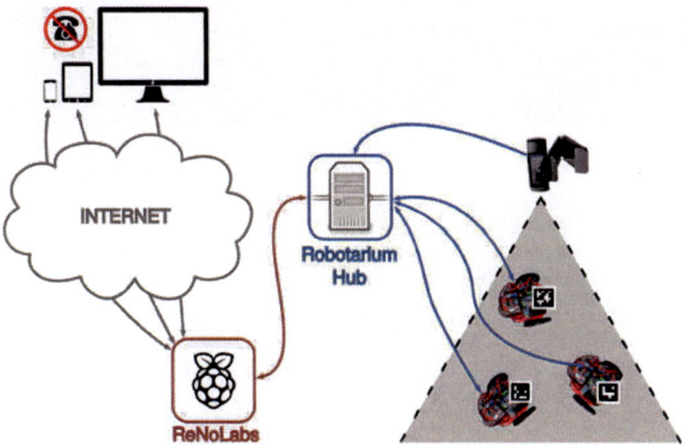

Fig. 2 An overview of the architecture of the Robotarium-UCM

Moreover, in *Automatic Control* mode, three different levels of granularity can be selected depending on the activity to be carried out.

Built-in Control The robot will be controlled by a built-in control algorithm that ensures a robust and safe performance.

Custom The student provides high-level commands, but the robot is still responsible for guaranteeing basic functions such as obstacle avoidance, or direct control of the motors.

Override This is the most open working mode, in which one or more essential functions can be exposed to the students. This mode should be used only in advanced courses and in a controlled environment, to avoid harm to the Robotarium-UCM.

5 Results: A Rendezvous Experiment Using Robotarium-UCM

In robotics, the rendezvous problem, also called an agreement or consensus problem can be stated as *get n mobile robots to move to a common location using distributed control* [11]. The rendezvous problem is an entry point to the more complex problem of distributed control in robotics. Therefore, it is a good first example to demonstrate Robotarium-UCM.

For the robots to move to a specific point in the Robotarium arena, a rendezvous algorithm has been programmed. The algorithm orients the robots to the desired point in an approximate way. Once they are oriented, they move to the point and correct their course by following the instructions provided by the server.

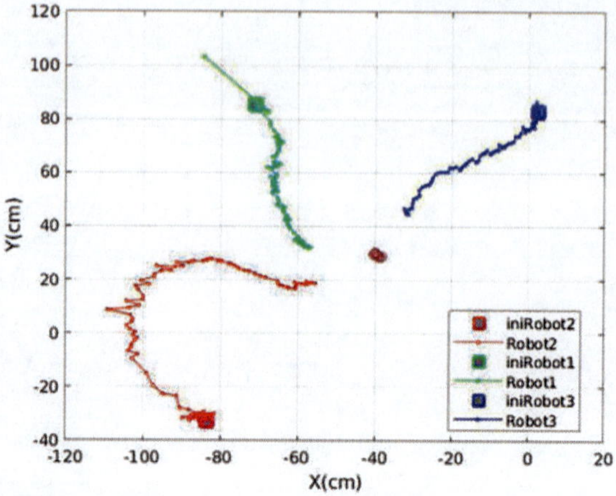

Fig. 3 Robot navigation trajectories to the target point (square)

The algorithm is as follows:

1. Each robot knows its position $\mathbf{p_R}$ and the goal position $\mathbf{p_G}$, provided by the server.
2. Using this positions, the difference vector is computed $\mathbf{v_D} = \mathbf{p_G} - \mathbf{p_R}$ and its angle to the x-axis θ_D.
3. Orientation error is defined as $\theta_e = \theta_D - \theta_R$.
4. Robot is controlled to correct this error using a PI controller.
5. Once the orientation of the robot is correct, the robot moves in a straight line towards the target, correcting the orientation if necessary. A fixed heading speed $V_R = 27.47$ (cm/s) is set and the robot is guided to a distance of 25 (cm) from the target point, then the robot stops.

Figure 3 shows the navigation that the robots have carried out to reach the desired point. A square indicates the starting point. Robot2 start point (-81, -35) has travelled further than either Robot1 start point (-71, 88) or Robot3 start point (5, 82) as can be seen in the figure.

Figure 4 presents a frame from the recording of the test, showing the initial and final position of the robots. An additional static marker to establish the meeting point for the robots is shown in the center of both figures.

6 Robotarium-UCM Clone Event

On the occasion of the Computer Science Week, a robotarium assembly and start-up event (2023/02/08) has been planned at the Computer Science Faculty of the Complutense University of Madrid. Teams of students will have to assemble and program a robot and perform a series of tests. These robots will be the first to be part of the robotarium of the Faculty of Computer Science.

Fig. 4 Robot initial (left) and final (right) positions

6.1 Event Objectives

The objectives of this event are varied. On the one hand, we are looking for a fun workshop, where students have a good time, learn and awake their interest in robotics. It is also intended to promote during the week of computer science, these knowledge areas that are closer to the hardware. But in addition, another objective is that the robots that the students build and implement, will be the germ for a robotarium as described in this work. In this way, the benefits of the workshop will last over time and the students will be able to have a multi-robot experimentation and learning platform. This new robotarium will be a replica of the Faculty of physic's robotarium, also with the aim to stimulate cooperation among students of both faculties.

6.2 Organizing the Event

Each team will be provided with all the materials needed to be able to assemble a robot and they will have to bring a laptop, with the Arduino IDE installed (there is the possibility of getting a laptop on loan from the university).

The workshop will start with a brief explanation, where the students will be told the general objective of the workshop and the partial goals. For each partial goal the teams will get points. At the end of the workshop, the team with most points will get a robot as a prize. The partial goals are: (1) assemble the robot and make it to move the two wheels; (2) program the robot to travel in a straight line a preset distance; (3) program the robot to move on a circular trajectory of a defined radius; (4) program the robot to move along the perimeter of a square of a given side.

To achieve these goals, students will have access to GitHub pages where the workshop content, the assembly steps and the necessary code (which they will have to complete to perform the tasks) are supplied.

6.3 Results

In order to know whether the results of the workshop were satisfactory, two different aspects were evaluated: (1) the extent to which the teams of students managed to complete the proposed tasks and (2) the degree of involvement and satisfaction of the students who participated in the workshop.

Although initially only 11 students signed up for the robotics workshop, another 7 joined in during the workshop, as they passed by the area where the workshop was being held, became interested and asked questions. In the end, 18 students attended, working in 5 groups, three groups of 4 people, two groups of 3 people. One of the groups of 4 and one of 3 joined in the middle of the workshop. All the groups, except the two groups that joined late, managed to assemble the robot, but only one of them managed to make the robot move in a straight line and to read the encoders.

An anonymous survey was used to measure the degree of satisfaction of the participants. Unfortunately, only 5 of the participants answered the survey. The survey first asked about the appropriateness of the materials, information, difficulty and help provided during the workshop. All respondents considered the materials, information, difficulty and help provided during the workshop to be adequate Fig. 5.

Secondly, they were asked about their previous knowledge of robotics and whether they felt they had learned anything during the workshop. Except for one of the cases none of the respondents claimed to have previous knowledge of robotics and all of them considered that they had learned a lot in the workshop, Fig. 6. Finally, all of them indicated that they would be willing to participate in similar events. Here are two free comments left by the respondents: "I loved the experience of tinkering with Arduino and being able to assemble the mini car piece by piece and see it move forward, even if it is only 10cm on the track XD. I will be waiting for the second edition of the robotarium! Thank you very much :)" "I liked it a lot".

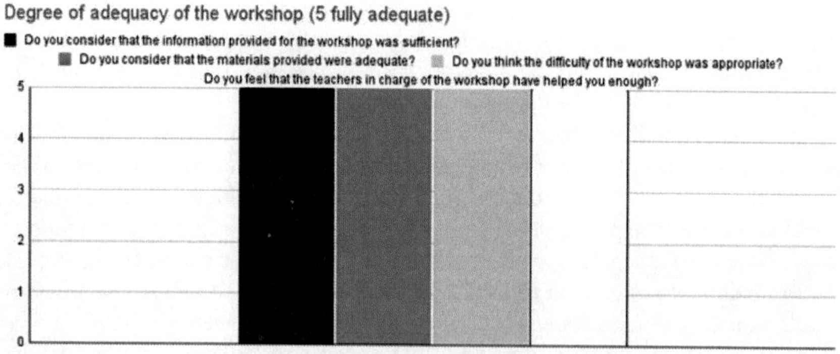

Fig. 5 Results of the participant survey. Answers to the questions about the suitability of different aspects of the workshop (5 means very suitable 1 not suitable at all)

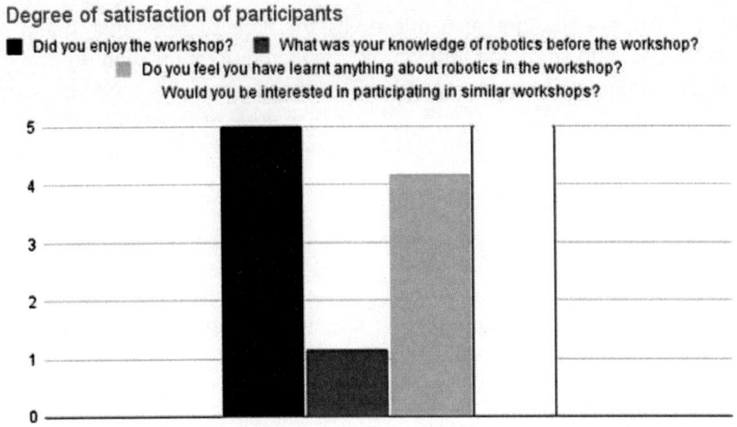

Fig. 6 Results of the participant survey. Answers to the questions about satisfaction and knowledge of robotics (5 means very much 1 means not at all)

7 Conclusions and Future Work

There are many good reasons to use multi-robot system in college's robotic class. Robotic swarms have many potential uses but many algorithm and applications are tested only in simulation. Optimizing the budget and effort devoted to build and maintain a multi-robot educational resource, enabling its remote use, is an excellent option. It is possible to increase both the time of use, which is no longer limited to class time, and the number of users, who can perform their experiments from outside the university laboratory.

With these two premises in mind we have started an educational project, called Robotarium-UCM, in the ISA lab at UCM, where we have 10 low-cost robots connected in a remote lab available to our students of different subjects. We have also presented the organization of a hackathon type event to start the assembly of another Robotarium in the Computer Science of the University. We consider the experience to be very positive, since the main objective of involving the students in the Robotarium project is being achieved. With a view to organizing other events, we have learned that the objectives of these events have to be much more limited, since for many of the students it will be their first contact with robotics.

Much work remains to be done along different lines. On the one hand, to favor the use of the laboratory by students both, by increasing the number of exercises and practices carried out with multi-robot systems and by encouraging undergraduate and graduate students to make use of the remote laboratory for their own projects.

On the other hand, the use of remote laboratories allows us to have data about how students interact with the laboratory, how much time they dedicate to the tasks, at what times they do it, which exercises require more dedication... This data can be of great help for teachers to design of the educational activities of the different subjects.

And last but not least, it also allows us to increase the capabilities of the Robotarium-UCM, to study the limits of centralized communication, to implement distributed communication methods, to increase the perceptual capacity of the robots and improving aspects such as safety, automatic loading or risk management.

References

1. Yu, J., Han, S.D., Tang, W.N., Rus, D.: A portable, 3D-printing enabled multi-vehicle platform for robotics research and education. In: 2017 IEEE International Conference on Robotics and Automation (ICRA), pp. 1475–1480 (2017). https://doi.org/10.1109/ICRA.2017.7989176
2. Wilson, Sean, et al.: The robotarium: globally impactful opportunities, challenges, and lessons learned in remote-access, distributed control of multirobot systems. IEEE Control Syst. Mag. **40**(1), 26–44 (2020)
3. Wilson, S., Egerstedt, M.: The robotarium: a remotely-accessible, multi-robot testbed for control research and education. IEEE Open J. Control. Syst. **2**, 12–23 (2023). https://doi.org/10.1109/OJCSYS.2022.3231523
4. Prorok, A., Malencia, M., Carlone, L., Sukhatme, G.S., Sadler, B.M., Kumar, V.: Beyond Robustness: A Taxonomy of Approaches Towards Resilient Multi-Robot Systems (2021). arXiv:2109.12343
5. Paull, L., Tani, J., Ahn, H., Alonso-Mora, J., Carlone, L., Cap, M., Chen, Y.F., Choi, C., Dusek, J., Fang, Y., Hoehener, D., Liu, S.Y., Novitzky, M., Okuyama, I.F., Pazis, J., Rosman, G., Varricchio, V., Wang, H.C., Yershov, D., Censi, A., et al.: Duckietown: an open, inexpensive and flexible platform for autonomy education and research. In: ICRA 2017-IEEE International Conference on Robotics and Automation, pp. 1497–1504 (2017)
6. Wilson, Sean, Gameros, Ruben, Sheely, Michael, Lin, Matthew, Dover, Kathryn, Gevorkyan, Robert, Haberland, Matt, Bertozzi, Andrea, Berman, Spring: Pheeno, a versatile swarm robotic research and education platform. IEEE Robot. Autom. Lett. **1**(2), 884–891 (2016). https://doi.org/10.1109/LRA.2016.2524987
7. Chacon, J., Besada, E., Carazo, G., Lopez-Orozco, J.A.: Enhancing EJsS with extension plugins. Electronics **1**(5) (2021)
8. McLurkin, J., Rykowski, J., John, M., Kaseman, Q., Lynch, A.J.: Using multi-robot systems for engineering education: teaching and outreach with large numbers of an advanced, low-cost robot. IEEE Trans. Educ. **56**(1), 24–33 (2013)
9. Garrido-Jurado, S., Muñoz-Salinas, R., Madrid-Cuevas, F.J., Marín-Jiménez, M.J.: Automatic generation and detection of highly reliable fiducial markers under occlusion. Pattern Recognit. **47**(6), 2280–2292
10. Chacon, J., Besada, E., Garcia, L., Lopez-Orozco, J.A.: Efficient deployment of remote laboratories with TwinCAT-PLCs and EjsS Plugins. IFAC PapersOnline **55**(17), 326–331 (2022). ISSN 2405-8963
11. Francis, B.F., Maggiore, M.: Flocking and Rendevouz in Distributed Robotics. Springer, Heidelberg (2016)

Simulators and Software

Environment for UAV Education

Martin Sedláček, Eduard Mráz, Matej Rajchl, and Jozef Rodina

Abstract Simulation tools are often used in robotics education to transfer theoretical knowledge into practical experience. The correlation between practical and real-world experience depends on the quality of the simulation environment and its ability to serve as a replacement for the real world. In this study an environment for UAV education is proposed, where the transfer of theoretical knowledge into practical experience is achieved through the implementation of both simulation and real UAV deployment. Initially, this environment allows students to develop their applications in the simulation. Subsequently, in the later stages, they are able to easily deploy them into the educational UAV and test the solution in the drone laboratory. Compatibility between simulation and physical UAV is achieved by combining the appropriate components in both the simulation and physical UAV. The software technologies used include ROS, ArduPilot, and MavLink, while the hardware platform for the educational UAV was chosen as DroneCore.Suite by Airvolute s.r.o. These technologies are not only suitable for the proposed environment but are also commonly used across the robotics field, which enhances relevant students competencies. Later in the article, an example assignment is presented and described how to use the environment for UAV education in a classroom setting.

Keywords UAV · Simulation · Education

This article was written thanks to the generous support under the Operational Program Integrated Infrastructure for the project: "Research and development of the applicability of autonomous flying vehicles in the fight against the pandemic caused by COVID-19", Project no. 313011ATR9, co-financed by the European Regional Development Fund. This publication was also supported by the project APVV-21-0352—"A navigation stack for autonomous drones in an industrial environment" and by the company Airvolute s.r.o.

M. Sedláček (✉) · E. Mráz · M. Rajchl · J. Rodina
Slovak University of Technology in Bratislava, Vazovova 5, Bratislava, Slovakia
e-mail: martin.sedlacek@stuba.sk

E. Mráz
e-mail: eduard.mraz@stuba.sk

M. Rajchl
e-mail: matej.rajchl@stuba.sk

R. Balogh et al. (eds.), *Robotics in Education*, Lecture Notes in Networks and Systems 747, https://doi.org/10.1007/978-3-031-38454-7_22

1 Introduction

Robotics as a field of study is a very broad topic, which incorporates quite a large range of subtopics e.g. computer science, mechatronics, control systems and a lot more. Given the complexity and multidisciplinary nature of robotics, it can be challenging to teach this subject effectively. It is often very difficult to apply students' theoretical knowledge into complex robotics problems which simulates real life practical examples. One of the obstacles is the mentioned complexity of a robotics system itself, which can lead to failures during the practical testing phase of student assignment. These failures are a natural part of the educational process. However, they can reduce the time effectiveness of students and increase the cost of materials used in the class. To overcome these obstacles a simulation environment is commonly used [1]. Depending on the quality of the simulation environment the real practical experience gain can be reduced for the students. In the environment for UAV education, the impact of failures is intended to be reduced by utilizing the software in the loop (SITL) capabilities of the autopilot used in a real UAV. Additionally, a virtual environment was created to serve as a digital copy of an existing drone laboratory. The use of UAV as a platform is becoming increasingly common in higher education as well as in various industries. This trend is reflected in the growing number of applications of UAVs in various fields [2, 3]. This has also been observed in studies such as [4]. The proposed environment for UAV education is an implication of this trend and is expected to provide students with easier access to this growing industry.

One of the goals of the environment for UAV education is to create a virtual environment that simulates reality as realistically as possible. But the main goal is to create an environment that enables the development of easily deployable student projects in a real drone laboratory. The simulation part of the environment use the same technologies as those utilized in real hardware. By using the same technologies in both the simulation and the real hardware, it is possible to achieve that a solution developed in the simulation will also be able to run seamlessly in the real environment. The UAV used in this study is intended to fly indoors, which poses some new challenges for the UAV education environment. Those will be addressed in detail further in the article. Indoor UAV is a great platform to teach robotics across the whole spectrum of the students' years. The platform can be used by students in the early years of their studies to gain general knowledge about robotics, UAVs, and related topics through projects and education. The environment for UAV education can abstract students from critical low-level tasks such as localization and control. Students in the final years of their studies can use the UAV environment to develop much more complex projects. This environment will not only provide students with a simulation for rapid testing of their solution, but also enable them to test their solution in the real environment. Therefore, provide them with the motivation for the project, final thesis or a course and real world experiences.

As mentioned before, robotics is a very broad topic. The environment for UAV education also reflects this. It encapsulates many technologies that are used together to achieve working solution. In the selection of these technologies, the aim was to choose those commonly used in the field of robotics, with the intention of making the knowledge easily transferable to other subjects in the student curriculum [5]. Furthermore, technologies were chosen to be compatible with the capabilities of the UAV. In summary, the environment for UAV education includes several technologies such as ROS [6], ArduPilot [7], Gazebo [8], MAVLink [9] and the operating system Ubuntu [10]. Similar technologies are used by other researchers and technical teams to simulate UAV behaviour in various applications [11, 12]. The proposed environment differs from these works by providing a possibility to seamless transfer of the solution between simulation and real world as well as creating environment focused on teaching robotics.

Although the proposed solution is hoped to be viable for general use as described, certain limitations are expected to be encountered in achieving a seamless port between simulation and reality. These limitations are largely attributed to the unique precision of localization and other low-level systems for each UAV. As a result, extra safety or compatibility parameters may need to be incorporated into the simulation solution, which could make the porting process more complicated and require additional parameter adjustments.

2 Used Technologies

2.1 Education UAV

To ensure high quality of education while maintaining accessibility of UAV systems in school environments, the requirements for the UAV are somewhat contradictory. The system needs to be robust, complex enough to demonstrate real world tasks and able to fly in an indoor environments, but on the other hand it should be affordable and also not too complicated. The requirements may be met by some of the commercial solutions, but their platforms are usually closed (proprietary) or protected by licenses, which makes it impossible to use them in the environment for UAV education. Therefore, custom solution had to be developed (Fig. 1).

2.2 Architecture of Environmnet for UAV Education

2.2.1 Electronics

The UAV system from company Airvolute s.r.o. is used as the core of the custom UAV. This system (DroneCore.Suite [13]) combines UAV autopilot (ArduPilot) with the Nvidia Jetson Xavier NX system on a module [14]. DroneCore.Suite provides rich connectivity for various sensors, however for the custom UAV, the most relevant sensors are range finders and stereo camera pairs. This system is also capable of

Fig. 1 Prototype of education UAV

handling more complex and computationally expensive tasks such as reactive navigation or real time mapping elevating Nvidia Jetson Xavier NX coprocessors (GPU and other). DroneCore.Suite also includes DroneCore.Power, which is an electronic speed controller capable of working with ArduPilot.

Using DroneCore.Suite it was possible to build a relatively simple UAV capable of flying indoors, which is compatible with the environment for UAV education. The drone configuration consists of a DroneCore.Suite minimal viable configuration (see from the diagram in Fig. 2) and a camera stereo pair. Software compatibility is ensured by incorporating Nvidia Jetson Xavier NX that is powered by Nvidia JetPack SDK. Jetpack SDK is a collection of software components, which are compatible with Nvidia Jetson Xavier NX, the operating system of Jetpack SDK is Ubuntu. This operating system is used also in the environment for UAV education. This way we are able to port the application from the simulation environment to the real UAV comfortably and therefore reduce the overhead for the students.

2.2.2 Construction

To ensure stiff construction which can be easily replaced a set of carbon plates has been used, which creates the main chassis of the UAV (see Fig. 1). It encapsulates all of the electronics in a safe robust covering whilst maintaining a very good weight ratio of the construction to the power of the motors. The bottom plate also seamlessly extends outwards to create four arms of the UAV. This enables to cut the whole body out of 3–5 mm carbon plates using a simple hobby grade CNC.

Main attributes of the whole construction is that it is very light-weight and very stiff. It can resist many of the student mishaps such as collision with other objects, ground crashes or high strain forces due to aerodynamic forces caused by for example

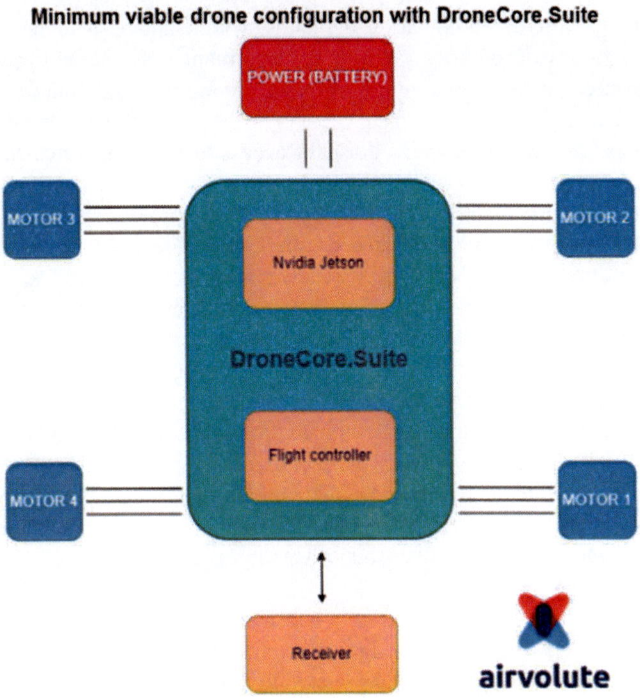

Fig. 2 Minimum viable drone configuration with DroneCore.Suite [13]

regulator instability. Also the light-weight nature of the whole UAV is suitable for educational environment since the amount of kinetic energy which is produced by a drone is much lower compared to heavier drones. Thus chances of injury during collision is lowered.

2.2.3 Simulation

In the simulation part of the proposed solution, an autopilot (ArduPilot) in Software-in-the-loop (SITL) configuration is used, which functions similarly to the actual autopilot. However, some sensors are either simulated or obtained from a simulation environment. As an alternative, autopilot PX4 could be used. Since our laboratories dispose of UAV's equipped with ArduPilot solution stack, ArduPilot is the preferred autopilot. The great advantage of ArduPilot is a very active development team and its opensource licensing. ArduPilot in SITL configuration is connected with a graphical simulator.

In the proposed environment for UAV education, Gazebo is currently utilized as the graphical simulator. One of the key advantage of Gazebo is, that it is relative lightweight in terms of computation demands, which allows it to run on lower-specification hardware without sacrificing performance. Additionally, Gazebo is well documented, with a large and active user community providing support and resources to users. These factors make Gazebo an excellent choice for use in educational settings, where students may not have access to high-performance hardware or the resources to troubleshoot complicated software. While Gazebo is an excellent choice for our task, there are other simulation software options available. One such option is AirSim [15], which is based on the Unreal Engine 4 and offers photo-realistic environments. While this may be advantageous for certain use cases, such as testing autonomous vehicle perception systems, there are significant drawbacks to using AirSim. One major disadvantage of AirSim is its computational complexity. The photo-realistic environment require significantly more computational power to simulate, which can limit performance of the simulation and in the worst case make the simulation invalid. Another issue with AirSim is that it is not fully compatible with ArduPilot SITL. While most of the features are supported, some have not yet been implemented. This is problematic as it can hinder the quality of the simulated flight, resulting in a deviation from real flight. These issues were considered critical for the given use case. Therefore, AirSim was not used, and instead, Gazebo was adopted.

Mavros ROS package [16] is used as an interface to ArduPilot SITL and graphical simulation environment. Mavros bridges and translates MAVLink protocol to ROS topics and provides other functionality related to ArduPilot.

Students applications are then written inside the ROS environment as a ROS nodes. This approach is considered favorable as it is one of the conventional methods of developing UAV control in the real world. The whole architecture of the simulation part of the environment of UAV education can be viewed on the Fig. 3. The ROS nodes created by students can access the ArduPilot SITL simulated UAV in the Gazebo graphical simulation via standard communication tools inside the ROS, such as topics and services. To communicate with ArduPilot SITL from ROS in a standard way, one can utilize Mavros (a ROS node) provided services and topics. These services and topic are converted and sent further via MavLink directly into ArduPilot SITL. So the ROS nodes created by students have access to autopilot and therefore are able to control the UAV. In addition to communicating with the autopilot, students are expected to work with other ROS topics that contain crucial data, such as the current position of the UAV in the simulation. By using this mechanism the simulation part of the environment for UAV education is capable to control the whole simulation using only a ROS node. By having all important data accessible through ROS topics, the complexity of the simulation environment is significantly reduced for students. This enables them to concentrate solely on the ROS node and handle all relevant tasks from there.

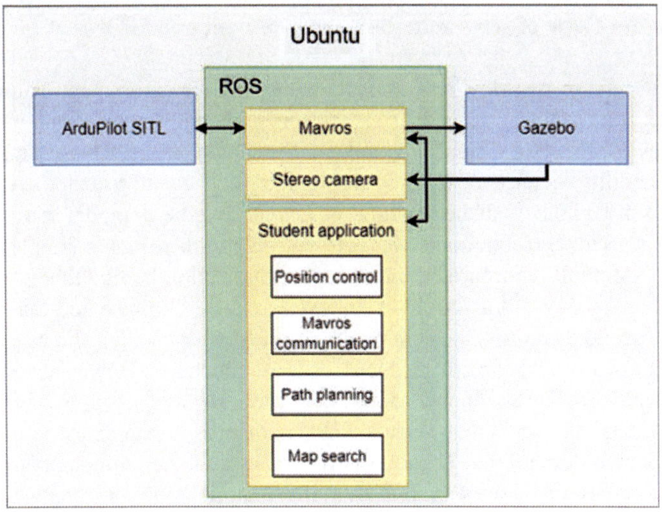

Fig. 3 Overview of the environment for UAV education

2.2.4 Simulation and UAV System

The software core of the environment for UAV education is the Ubuntu operating system both in the simulation and in the UAV.

This allows to use ROS and Mavros in the education UAV. Thus, it is possible to directly copy ROS node developed for the simulation to the board computer of the UAV, make minor modifications for subscribed topics names, compile it and launch it. This process is as seamless as possible, given the complexity of the UAV system. By utilizing ROS, Mavros, ArduPilot and Gazebo, the students are able to develop and test their application in a simulated environment, and then easily transfer it to the actual UAV. This approach greatly simplifies the development process and reduces the overhead for the students.

2.2.5 UAV System

The DroneCore.Suite from Airvolute s.r.o. provides localization from stereo camera pair and other lower level tasks, which are launched automatically after the drone starts. This greatly simplify the development process for the students or the teachers, as they do not need to focus on implementing these tasks themselves. Instead, they can rely on the pre-existing capabilities provided by DroneCore.Suite and only use the data provided by ROS topics. But, if it is needed the system of DroneCore.Suite is open and it is possible to make changes.

2.2.6 Digital Copy of the Simulation Environment and UAV

Simulation scene or world, where students work on their assignment, is one to one scale digital copy of the existing drone laboratory at the university. Blender was used to model environment for Unreal Engine 4 (AirSim) and Gazebo. This approach gives us some flexibility to reuse model in some other projects due to Blender's rich export options. To use photo realistic textures in Unreal Engine 4 model, Substance3D software was used. Gazebo model was textured only with colors or simple textures, which are by default incorporated into Gazebo, motivation behind this was to keep this model as lightweight as possible. In this section, it is presented that a digital copy of the drone laboratory was created for both AirSim and Gazebo as part of the development process for the environment for UAV education. However, it should be noted that only Gazebo is adopted in the latest version.

In this virtual environment, a virtual UAV is placed. A simple static model can be placed almost directly into Gazebo or AirSim with the help of export from CAD software. It is essential to check that the virtual model's dimensions are as close to the real model as possible. This ensures feasibility of the proposed trajectories both in the simulation environment and in the drone laboratory. Virtual drone is also equipped with stereo camera pair, which needs to be placed in the right position with the right orientation (see Fig. 4).

3 Environment in Education

The environment for UAV education encapsulates all the technologies mentioned in Sect. 2. It should be noted that not only virtual simulation is present, but also a drone laboratory and the education UAV are included. All these parts together ensure compatibility between simulation and reality.

The prerequisites for students to take full advantage of this environment include basic Linux skills (such as using bash, installing packages, and similar tasks), computer science skills (including knowledge of ROS, Python or C++ programming languages), and general robotics knowledge (such as understanding algorithms and basic control systems). Most of these skills are already present in a lot of robotics curriculum. This environment can also be used by students with less knowledge. For example, one can select such assignments, that will not require computer science or they can provided code templates. Our team believes, that even for starting students this environment can be a great motivation to further improve their skills and promote individual interest in robotics and UAVs.

Fig. 4 Digital copy of drone laboratory. Up—Unreal Engine 4, down—Gazebo

3.1 Deployment of the Environment for UAV Education

To integrate the environment for UAV education in a course, it is important to make sure, that students are at least a bit familiar with each used technology. In our class, we attempted to achieve this by not providing pre-built Linux images or similar system distributions for desktops that already have the environment installed. Instead, students were given with an installation guide to set up the environment on their own. The guide is divided into several parts, each of these parts include control points. The control points do not only indicate the correct setup, but also demonstrate some basic functionality. This approach takes away some of the time in the class, which could be used to improve other talents. Despite this, the advantage of students getting familiar with the environment is more important to us. Therefore, it was decided to use this approach in our course. In general, we would recommend using the installation guide approach with students in higher years. It is expected that using the installation guide with new students would take too much time to be effective. In this case, each of the technologies used can be explained in the form of a presentation.

To deploy compatible system in the education UAV, the system image derived from Airvolute s.r.o. system version of Nvidia Jetpack for Nvidia Xavier NX was used.

This system image, as mentioned in Sect. 2.1, includes pre-installed software and features supporting hardware DroneCore.Suite, which the education UAV is running on. As already mentioned, students deploy their application into ROS workspace, so their contact with the whole UAV system is limited. Therefore, some faults can be avoided. In the case that some additional access to the system is needed, students are able to alter almost all the parts of the system. If some unrecoverable error occurs the DroneCore.Suite can be flashed again with the default system image in a relatively short time.

In addition to the application computer on DroneCore.Suite, where the students' application will run alongside ROS, Mavros, and other software components prepared by Airvolute s.r.o. The DroneCore.Suite also includes an autopilot hardware CubePilot running on ArduPilot software. It is possible to preset this autopilot with tuned values for regulators and parameters for other systems. These parameters needs to be tuned to work well. Due to the good mechanical properties of the education UAV, they can be identified on one drone and can be transferred to other UAVs of the same type. This is very beneficial, because a relatively large fleet of the educational UAVs can be serviced with ease, which is ideal for robots in education. In some more advanced and profiled courses students are able to tune autopilot by themselves as a part of an assignment. It can be achieved by changing parameters of the autopilot, which is supported by the education UAV.

3.1.1 Demands on Drone Laboratory

To ensure good functionality and ease of use of the educational drones, it is necessary to enforce some properties of the drone laboratory. The dimensions of the laboratory should be enough to fly the drone freely. Good dimensions starts from at least $100\,m^2$ with a square or rectangle footprint. Height of the ceiling is sometimes overlooked factor, it is recommended to be at least 3.5 m. Space with these dimensions allows to plan a lot of types of automatic or autonomous UAV missions. At the same time it allows to fly UAV in manual mode, which can be really important during tuning. Flying drone, even in this size, can be dangerous. Therefore, the space should be equipped with safety nets or at least contain some protected space, where workspace for students can be set up. The lighting of the space is also important, good illumination should be present at all times, because of the use of visual odometry. In general, it is good to avoid some types of LEDs due to flickering in camera images. Visual odometry can be swapped for some other type of localization usable indoor, for example the education UAV is ready to be used with motion capture system, which can ease some demands on the lightning and improve stability of the localization.

3.2 Proposed Assignment—Simple Automatic Mission

To better illustrate how the environment for UAV education can be used in courses, an exemplary practical assignment was created. In this assignment students have to apply a wide spectrum of theoretical knowledge about robotics to solve a simple automatic mission.

In this assignment, UAV have to fly on predefined trajectory (set of waypoints). Trajectory waypoints are given in a coordinate system of the simulation environment, each waypoint has predefined accuracy of reach and some waypoints have special tasks assigned to them. This task can be assigned to students using similar table to Table 1.

To solve this assignment, multiple subtasks have to be solved. If the assignment is divided the right way, there will be some continuity between subtasks. Therefore, each subtask can be developed relatively standalone. This makes the assignment valid as a group project, which does not only promote teamwork, but also simulates real world technical problem solving. One possible way in which a simple automatic mission could be split is as follows: map-related tasks (including map generation and map handling), trajectory planning (involving pathfinding and trajectory planning), a ROS drone control node (providing an interface to the autopilot, managing mission tasks and implementing simple position control). After this division, it becomes apparent that the simple automatic mission involves a wide range of topics from robotics. It is possible, that some of the students have already solved similar tasks as a part of projects, final theses or other courses. To avoid unnecessary implementation of the same algorithms, students are allowed to use already existing solutions in our course. The focus of the work of these students has shifted a little bit to integration with UAV environment. In our experience, this has led to increased motivation for the class and better overall solutions and class performance.

This assignment can be easily deployed from digital environment into the real one (if most of the suggestions for drone laboratory and digital copy has been satisfied). The process is not different as presented in Sect. 2.2.4. To avoid potential issues during the deployment, waypoints from Table 1 should be defined with enough clearance from the walls and other obstacles.

Table 1 Simple automatic mission

Waypoint	x[m]	y[m]	z[m]	Task	Precision
1	×0	y0	z0	Takeoff	–
2	×1	y1	z1	–	Soft
3	×2	y2	z2	Land and takeoff	Hard
4	×3	y3	z3	–	Soft

4 Conclusion

This study presents a detailed simulation environment for UAV education. It is worth noting that the environment for UAV education includes not only a collection of software components and a simulator, but also a real-world UAV and a drone laboratory. The main advantage and benefit of this solution is that it bridges configuration in simulation environment into the real-world with relative ease. This helps students to transfer theoretical or simulator experience into practical experience. Therefore it helps them prepare better for future in robotics. In addition to practical experience, this solution supports students' motivation by enabling them to deploy their solutions and test them in conditions close to the real world.

In addition to the pointed out educational benefits, we believe that our environment can contribute to the trend of ever-increasing use of UAVs by preparing future engineers to face challenges related to UAV deployment and development.

Although it is possible to port the working solutions of the simulated UAVs to the real environment rapidly. This aspect of the work was tested only by a team of teachers and researches. Therefore, the complete integration of the simulation environment for UAV education into the courses remains as a part of future work. To be exact, the simulation part of this solution is already integrated into a university course. The part of the solution, where students deploy application into the real world is currently being tested by a team of students within their semestral project. After this test, the feedback will be integrated into the environment and the complete environment for UAV education will be incorporated into the course.

Thus far, the feedback from the students regarding the simulation part of this environment has been very positive. One of the most praised aspects is that students are able to integrate multiple skills from different areas and create a functional solution. It seems, that even the simulation part of the environment was motivational enough for the students.

Not any big or critical issue arisen during the integration of the environment into the course. Most of the issues stemmed from insufficient familiarity with ROS and computer science concepts.

In the future work, it is planned to continuously gather feedback from university students and apply relevant suggestions into the environment. It is also planned to add more exemplary assignments, which should target specific areas of robotics.

References

1. Fernandes, D., Pinheiro, F., Dias, A., et al.: Teaching robotics with a simulator environment developed for the autonomous driving competition. In: Robotics in Education, pp. 387–399 (2019). https://doi.org/10.1007/978-3-030-26945-6_35
2. Ahmed, F., Mohanta, J.C., Keshari, A., Yadav, P.S.: Recent advances in unmanned aerial vehicles: a review. Arab. J. Sci. Eng. **47**, 7963–7984 (2022). https://doi.org/10.1007/s13369-022-06738-0

3. Maghazei, O., Lewis, M.A., Netland, T.H.: Emerging technologies and the use case: a multi-year study of drone adoption. J. Oper. Manag. **68**, 560–591 (2022). https://doi.org/10.1002/joom.1196
4. Bolick, M.M., Mikhailova, E.A., Post, C.J.: Teaching innovation in STEM education using an unmanned aerial vehicle (UAV). Educ. Sci. **12**, 224 (2022). https://doi.org/10.3390/educsci12030224
5. Zhang, L., Merrifield, R., Deguet, A., Yang, G.-Z.: Powering the world's robots–10 years of Ros. Sci. Robot. (2017). https://doi.org/10.1126/scirobotics.aar1868
6. Quigley, M. et al.: ROS: an open-source Robot Operating System. In: ICRA Workshop on Open Source Software (2009)
7. ArduPilot. https://ardupilot.org/. Accessed 22 Jan 2023
8. Koenig, N., Howard, A.: Design and use paradigms for gazebo, an open-source multi-robot simulator. In: 2004 IEEE/RSJ International Conference on Intelligent Robots and Systems (IROS) (IEEE Cat No04CH37566). https://doi.org/10.1109/iros.2004.1389727
9. Koubaa, A., Allouch, A., Alajlan, M., et al.: Micro air vehicle link (MAVlink) in a nutshell: a survey. IEEE Access **7**, 87658–87680 (2019). https://doi.org/10.1109/access.2019.2924410
10. Ubuntu. https://ubuntu.com/. Accessed 6 Jan 2023
11. Hentati, A.I., Krichen, L., Fourati, M., Fourati, L.C.: Simulation tools, environments and frameworks for UAV systems performance analysis. In: 14th International Wireless Communications amp. Mobile Computing Conference (IWCMC) (2018). https://doi.org/10.1109/iwcmc.2018.8450505
12. Chen, S., Zhou, W., Yang, A.-S., et al.: An end-to-end UAV simulation platform for visual slam and navigation. Aerospace **9**, 48 (2022). https://doi.org/10.3390/aerospace9020048
13. DroneCore.suite-autopilot solution for autonomous drones. https://www.airvolute.com/product/dronecore/. Accessed 21 Dec 2022
14. Jetson Xavier NX Series. https://www.nvidia.com/en-gb/autonomous-machines/embedded-systems/jetson-xavier-nx/. Accessed 21 Dec 2022
15. Shah, S., Dey, D., Lovett, C., Kapoor, A.: AirSim: high-fidelity visual and physical simulation for autonomous vehicles. In: Field and Service Robotics, pp. 621–635 (2017). https://doi.org/10.1007/978-3-319-67361-5_40
16. Mavros. http://wiki.ros.org/mavros. Accessed 23 Jan 2023

A Short Curriculum on Robotics with Hands-On Experiments in Classroom Using Low-Cost Drones

Sylvain Bertrand, Chiraz Trabelsi, and Lionel Prevost

Abstract This paper presents some feedback on a curriculum on robotics developed for Master's level students at ESIEA, a graduate school of engineering in France. The main particularity of this curriculum is that it is composed of only three short modules (control and estimation, computer vision, ROS), with a small number of hours (18hrs each), but proposes hands-on experiments with drones to students. Experiments are done in the classrooms with low-cost drones as exercises integrated into the practical work sessions. The detail of the curriculum, the pedagogic approach, and examples of experiments proposed to the students are presented in this paper.

Keywords Robotics curriculum · Experiments in classrooms · Low-cost drones · ROS

1 Introduction

Drones are very motivating platforms for students. They are also of huge interest to teachers from a pedagogical point of view and are now widely used for teaching Science, Technology, Engineering, and Mathematics (STEM), at different levels, ranging from young kids to graduate students [1]. More and more academic curricula devoted to robotics now integrate one or several modules focusing on drones [2]. Similarly, more and more projects realized by students during their curriculum are also oriented to drone design or applications. It is also worth noticing that open resources for education on drones are now easily accessible, see e.g. [3, 4].

S. Bertrand (✉)
Université Paris-Saclay, ONERA, Traitement de l' Information et Systèmes,
91123 Palaiseau, France
e-mail: sylvain.bertrand@onera.fr

C. Trabelsi · L. Prevost
Learning, Data and Robotics Lab, ESIEA, Paris, France
e-mail: chiraz.trabelsi@esiea.fr

L. Prevost
e-mail: lionel.prevost@esiea.fr

R. Balogh et al. (eds.), *Robotics in Education*, Lecture Notes in Networks and Systems 747, https://doi.org/10.1007/978-3-031-38454-7_23

As low-cost hardware and off-the-shelf components or platforms are now available and affordable, practical work and experiments with drones usually complete more theoretical lectures. Different types of hands-on experiments are usually proposed to students during academic modules. The first approach consists in making students develop a drone by designing and/or integrating parts. Most of the time, this type of approach is related to Project-Based Learning, with hours dedicated to practical work to build a *home-made* drone [5]. Another approach consists in providing students with a representative mockup (eg. a 2-DOF helicopter bench [6]) or an already existing and ready-to-fly vehicle. Some existing platforms that are commonly used for educational purposes are Parrot mini drones, Crazyflie, ARDrone, Beebop, or DJI Tello Edu. In this case, work of the students focuses on more specific technical topics, most of the time designing and implementing algorithms.

From a practical point of view, hands-on experiments with drones by students are usually performed in flight arenas, equipped with nets for safety purposes. When possible, motion capture systems are used for precise indoor localization [2]. Nevertheless, such systems are expensive, and disposing of free space large enough for a flight arena is not always possible in teaching facilities.

In this paper, feedback is proposed on a short curriculum on robotics at ESIEA, a graduate school of engineering in France, which aims at integrating hands-on experiments by students with drones in the classrooms.

The curriculum is dedicated to last-year students, at the Master's level. This is a short curriculum, only composed of three modules, 18 h each. A strong effort is put into practice since practical work sessions represent half of the hours for two of the modules and almost all the hours for the third one. This is a challenge since the curriculum is also open to students from different backgrounds and has been designed to require very few prerequisites except standard scientific and engineering backgrounds. Another challenge is that the third module also integrates some hands-on experiments with drones as exercises during practical work sessions. Experiments are done by the students in the classroom with low-cost drones and equipment. A link with student projects should also be made in the sense that some students are already working on drone projects at the beginning of this curriculum.

The paper is organized as follows. The next section introduces the context of this curriculum. Its content is then detailed in Sect. 3. Sections 4 and 5 respectively present the approach proposed to students to progress step-by-step in simulation first and then easily perform hands-on experiments. Examples of such experiments given as exercises to students are also described in Sect. 5. Before concluding remarks, Sect. 6 provides some feedback on student projects on drones and their links to this curriculum.

2 Context

The engineering curriculum at ESIEA is composed of three years. The objective of the first one is to teach students all the required scientific and technical background (mathematics, physics, computer science, etc.). The next two years, which correspond to Master's classes, are devoted to more specific curricula, chosen by the students themselves to develop their own expertise in some specific area. Some examples of these *technical majors* are embedded and autonomous systems, software engineering, cyber-security, virtual reality, AI & data science. During the last third year, students can personalize further their experience by choosing an additional technical or managerial short curriculum (*minor*), e.g. robotics, low tech, innovation and entrepreneurship, economic intelligence, etc. Each minor is composed of three modules of 18 h each and some conferences on additional topics, organized with professionals from the industry.

During the third year, a *technical minor* is devoted to robotics. It is composed of three modules of 18 h each: control and estimation for mobile robotics, perception and computer vision, and computer programming for robotics (ROS - Robot Operating System). It is open to students from several technical majors with very few specific requisites regarding robotics, automatic control, etc. The syllabus of this *minor* is detailed in the next section.

3 Content of the Curriculum

The first module is *Control and estimation for mobile robotics*. It consists of 9 h of lectures and 9 h of practical work sessions. The lectures cover the basis of dynamic modeling for mobile robots, pathfinding and trajectory generation, motion control and obstacle avoidance, state estimation, and information fusion for localization. Different fundamental algorithms such as Dijkstra, A*, PID control, potential fields, odometry, inertial navigation, Kalman Filter, etc. are presented to the students with applications to ground mobile robots (differential drive) and drones (quadrotors). Simulation results and videos of real experiments are presented to illustrate the algorithms. Some live demonstrations with real hardware equipment are also proposed during the lectures, mostly concerning sensor technologies (LIDAR, IMU, stereovision).

Three practical work sessions of 3 h each consist of direct applications of some of the algorithms presented during the lectures: pathfinding, position control with waypoint navigation and obstacle avoidance for a differential drive robot, and position estimation using Kalman filtering. Python codes that implement the algorithms are given to the students except for some parts to be completed during the sessions. When possible codes developed during different parts of the tutorials (eg. path finding and waypoint navigation) are mixed together to enable addressing more complex robotic missions.

The second module is dedicated to *Perception and computer vision for mobile robotics*. It also consists of 9 h of lectures and 9 h of practical work sessions. The lectures address the topics of image processing and geometric computer vision, visual odometry, Simultaneous Localization and Mapping (SLAM), environment modeling and cartography, and Machine Learning for computer vision. Students can discover different algorithms and approaches such as features extraction in computer vision (blobs, Harris, SIFT), optical flow (Lucas-Kanade), camera models and calibration, stereovision, point clouds processing, Octomaps, 3D meshes, supervised learning, SVM, neural networks (RNN, CNN), etc. Three tutorial sessions of 3 h each are dedicated to applying computer vision methods (features extraction, matching, and visual odometry from stereo vision) using Python and OpenCV library.

The third module *computer programming for robotics (ROS)* has been designed in a different way to promote practice as much as possible. The objective is to present the basis and usage of the Robot Operating System by introducing during a short 1.5 h lecture its basic notions (workspace, ROS master, nodes, topics, messages, services, etc.) and useful tools (rosbag, RViz, RQT, Gazebo simulator). Some examples are presented to the students. The rest of the module (16.5 h) is devoted to practical work sessions including hands-on experiments on low-cost drones.

The next sections propose a focus on the content of this third module.

4 Step-by-Step Learning in Simulation

Before experimenting with real drones, the first hour of practical work sessions is devoted to tutorials on ROS to learn its basics: developing a node in Python, publishing messages on topics, and using tools. It is done with a virtual machine provided to each student. After this first tutorial, students are invited to use the TUM ARDrone simulator[1] with Gazebo to pursue learning ROS basics and the development of robotic codes through a motivating drone application scenario. A pedagogical step-by-step approach has been developed to propose simple exercises to students in an incremental way, starting from very basic tests and finishing with the development of codes to fulfill a complete drone mission. This pedagogical approach is summarized by the flowchart presented in Fig. 1.

The first step is to familiarize students with ROS concepts, data visualization, message publication, etc. They are first invited to launch a teleoperation code and control the drone's motion with the keyboard. This is usually a motivating and funny part well appreciated by students. Nevertheless, they are asked to visualize and record sensor and localization messages. Recorded data can be replayed using the *rosbag* tool, and also post-processed by the students to produce time plots and trajectory plots. This first step aims simultaneously at safely familiarizing students with the drone they will use for experiments, and learning how to deal with the experimental dataset (recording, processing, visualizing).

In the next step, students focus on the development of a controller for the vertical motion of the drone. A ROS node developed in Python is provided to the students that have to complete the controller equations. Students usually develop a Proportional Controller to stabilize the drone at a given reference altitude, based on range measurements provided by a downward-facing range sensor. This code is tested and validated in simulation. It enables students to have a first glimpse at the structure of

[1] http://wiki.ros.org/tum_simulator.

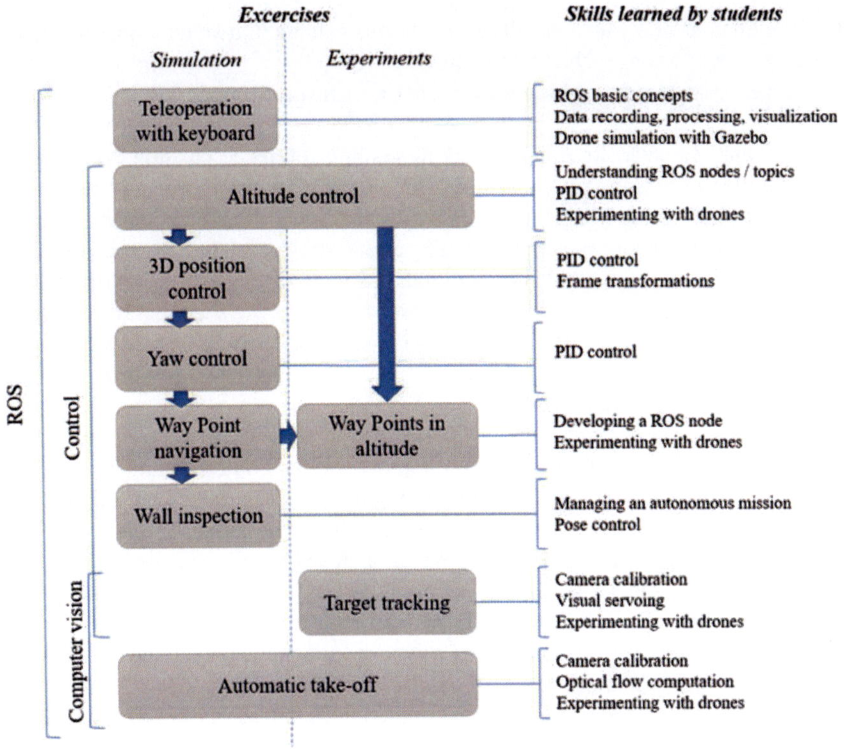

Fig. 1 Flowchart of the step-by-step pedagogical approach

a ROS node, and understand how to manage input and output data through topics and messages. Once validated in simulation, this altitude controller is then tested by the students on a real drone (see next section).

Pursuing the step-by-step approach, the problem of 3D position control of the drone is then investigated by the students. The first exercise considers a simplified problem with zero yaw assumption. Therefore no transformation between reference frames is required for the computation of the control inputs (velocity commands). Once equations are implemented for the control of the (x, y, z)-coordinates of the position, a second exercise is proposed to generalize the developed code to handle non-zero yaw for the drone.[2] Simple coordinates transformation is then performed by the students to express the control inputs, initially computed in the inertial reference frame used for localization, to the body frame attached to the drone and which orientation depends on the current yaw of the vehicle. This exercise is completed with

[2] Pitch and roll motions are neglected in the design of the controllers which aim at computing velocity commands (translational velocity and yaw rate) to be applied to the drone, taking profit of the existence on-board of inner-loop controllers.

the development of a yaw controller. This part is validated in simulation, providing a reference pose to be reached by the drone.

The last step before working on the final wall inspection mission is Way Point Navigation. Students are invited to develop a ROS node in Python that will act as a Way Point Manager. Its role consists in sending reference position and yaw to the controller, depending on the actual position and orientation of the drone, and a list of predefined poses (3D position + yaw) to be reached. This step makes students develop a full ROS node on their own, taking as a starting point the Python code of the controller node, to be adapted and modified. A simple distance criterion (including on yaw error) is implemented by the students to trigger a switch to the next reference pose. The code is validated in simulation by the students on a simple trajectory (eg. a square pattern with arbitrary yaw references), and in experiments considering vertical motion only (see next section).

At this stage of the practical work sessions, the students have developed a set of ROS nodes implementing in Python a simple drone controller and Way Point manager that can be used for autonomous flight. The final exercise of this part is to address a wall-inspection mission autonomously by the drone. Students dispose of a simulation environment in Gazebo which includes a wall structure (see Fig. 2). The objective is to make the drone take off, perform a visual inspection[3] of the two sides of the wall, and then return and land at the initial position. The mission must be realized in a fully autonomous way. Students are let free to decide how to parameterize this mission. Some of them use the teleoperated mode to control the drone with the keyboard to find adequate pose coordinates to be given to the Way Point manager. Other students proceed iteratively with tests and trials to get good coordinates directly using the simulation with the autonomous drone. Some groups of students defined the inspection trajectory with a rectangular shape around the wall at some constant altitude. Distance to the wall is sometimes adapted by some students to get the wall height inside the camera field of view. Other groups of students try to define some snake-like inspection patterns at different altitudes to improve coverage of each side of the wall. An example of realization can be seen in the video available at https://tinyurl.com/2p9ed3fj.

This step-by-step approach in simulation is completed by some experiments with real drones, as detailed in the next section.

5 Hands-On Experiments with Drones

5.1 Drone Platforms

Making students perform experiments with real drones in classrooms implies different constraints. Safe platforms have to be used in a constrained and cluttered

[3] The video camera of the drone is required to provide images of the wall. But no specific requirement on coverage is specified in the exercise.

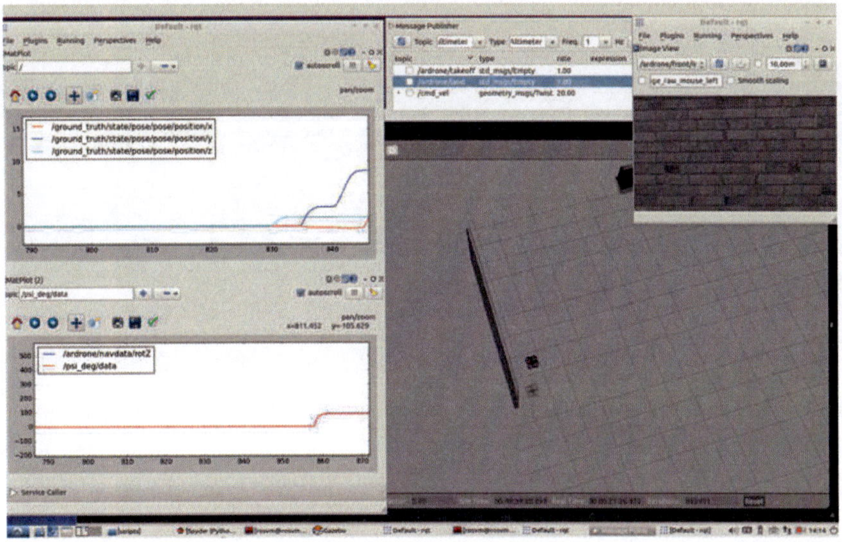

Fig. 2 Wall inspection mission in ROS Gazebo simulation. *Left*: time visualization of position and yaw, *Center*: Gazebo simulator, *Top-Right*: video from the simulated drone camera

environment with people nearby (a classroom with students). The drone and associated equipment (batteries, spare propellers, computer, etc.) should be low-cost, as crashes may occur during the experiments. Software compatibility should be ensured between the drivers of the drone and the software environment as well as the programming language used for the labs (Ubuntu+ROS+Python).

In previous years, the ARDrone has been chosen. The existence of the TUM ARDrone simulator and ARDrone Autonomy library [7] for driver and control motivated this choice, in addition to the fulfillment of the three aforementioned constraints. Thanks to these characteristics, this low-cost platform has been indeed widely used both for academic research and teaching. Nevertheless, it is now discontinued, being no longer sold nor maintained.

Therefore, a transition to another platform has been decided this year and the Robomaster Tello Tallent (RMTT) drone has been selected. It is an update of the Tello Edu drone which provides new features such as an extension module for some additional sensors or RGB led matrix + front-facing range sensor. For the Tello Edu, there exist open-source ROS drivers developed by the robotics community. Nevertheless, at the time of this year's lab sessions, no drivers were found for the RMTT, i.e. that include the extension module for ROS1. Therefore a specific driver has been developed, that provides a wrapping of the Python Robomaster SDK[4] from DJI and enables access via ROS topics to sensors, battery, control inputs, take-off and

[4] https://github.com/dji-sdk/RoboMaster-SDK.

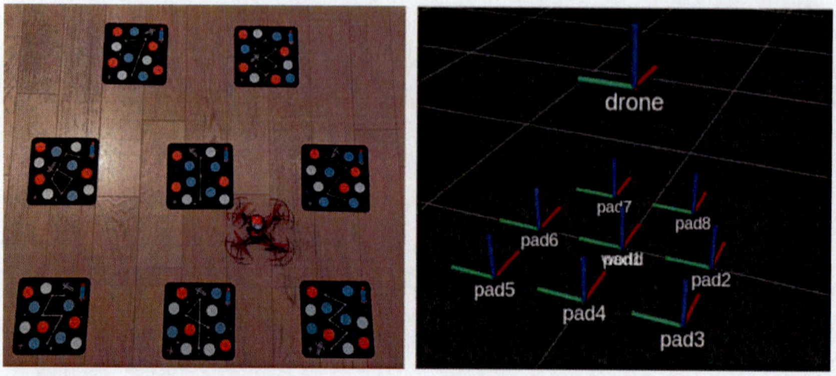

Fig. 3 *Left*: mission pads and drone before take-off, *Right*: Visualization in RViz of positions of pads and drone during flight

landing, mission pads (colored tags provided with the drone), and extension module (RGB LED, LED array, front-facing range sensor).

This driver is freely available[5] and work is currently being done to finalize the integration of a localization system based on the mission pads provided with the drone. The first step already done consists in using a *carpet* of mission pads with known relative locations to compute the 3D position of the drone during flight. An example is given in Fig. 3, showing the mission pads on the left and the localization of the pads and drone during the flight on the right. Taking one of the pads as a reference (pad no 1, at the center, in Fig. 3), the real-time position of the drone is computed in that reference frame. Using multiple pads enables to increase the volume where the drone can be localized.

Such a localization system will be useful to make students develop and experiment with control algorithms in the classrooms, without the need for an expensive motion-capture system or a specific setup with other types of markers [8] than the ones already provided with the drone. It will also be convenient for new exercises that will be proposed to the students in the close future, such as information fusion for localization (eg. Kalman filtering using accelerometers and position measurements from the pads) or Simultaneous Localization and Mapping (SLAM).

Classroom experiments already realized by students in the context of this curriculum are detailed in the next section.

5.2 *Experiments*

Several hands-on experiments are proposed to students as exercises as shown in the flowchart of the pedagogical approach of this curriculum (Fig. 1). These experiments are done in the classroom during the lab sessions. Students always enjoy testing

[5] https://github.com/bertrandsylv/rmtt_ros_driver.

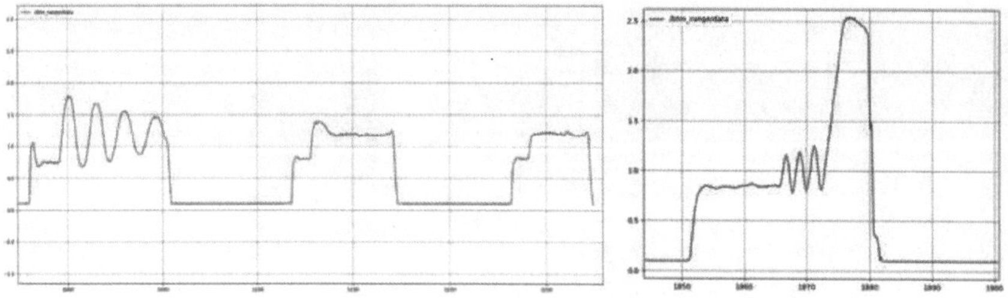

Fig. 4 Closed-loop vertical motion of RMTT drone for four different gain tunings of PI controller (flight experiments by the students)

their algorithms on real flying drones. It is very motivating for them, and this is the reason why experiments have been introduced at different steps of the practical work sessions. Students can record videos of their experiments and get experimental data as well.

The first and most simple experiment consists in controlling the vertical motion of the drone. Controllers developed by the students are tested and validated, each group using its own controller tuning. Figure 4 shows vertical trajectories (altitude recorded from the bottom range sensor) realized by different groups of students with the RMTT drone. The two plots at the center show a good closed-loop behavior of the drone, whereas the ones on the left and the right correspond to oscillating behaviors obtained for badly tuned controllers. For the plot on the right of Fig. 4, the drone oscillated before diverging to the ceiling of the classroom. Observing the drone behaviors during the flights made the students more easily understand the concepts behind automatic control (stability, damping, steady state error, etc.). After each group has performed its experiment, a debriefing with all the students has been done to compare and discuss together results and ways of improvement.

A direct extension of the previous experiment proposed to the students is altitude Way Point navigation, for which the drone has to reach and stabilize successively at points of different altitudes.

A third experiment proposed to the students concerns target tracking using image-based information. It has been realized in the past years with the ARDrone, taking profit from its tag recognition capability. A control law (visual servoing) is developed by the students to move the drone such that the detection of the target remains centered in the image provided by the bottom camera (see left part of Fig. 5). The same experiment will be done with the RMTT drone in the new version of the exercises.

Finally, to draw a link to the *Perception and computer vision for mobile robotics* module of the curriculum, the last part of the practical work sessions is devoted to an exercise involving computer vision. The objective is to develop a ROS node enabling automatic take-off of the drone when motion is detected in the image of its front camera. This could correspond to a surveillance scenario where a drone would take off if an intruder is detected. Optical flow monitoring is given as proposal of a solution for motion detection. A rosbag with a video record of someone walking

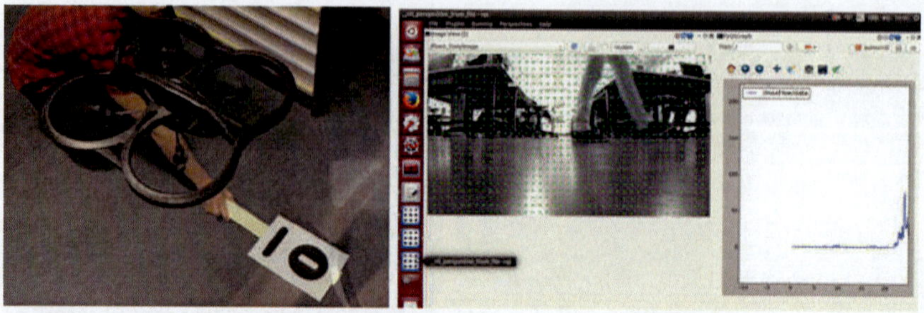

Fig. 5 *Left*: Visual servoing for tracking of a ground mobile target, *Right*: Optical flow computation for automatic take-off triggered by motion detection

in front of the landed drone is given to students for developing and testing their solution. Once tested, each group can validate by experimenting on the real drone. An example is given on the right part of Fig. 5 showing the video frame with optical flow visualization (left) along with a metric proposed by the students (maximum norm of the optical flow) for motion detection (right). The experiment can also be seen in the video available at https://tinyurl.com/2p9ed3fj. For time reasons, optical flow computation is realized using the ROS *opencv_apps* package.[6] A demonstration of camera calibration is also done in the classroom with all the students, before experiments.

6 Student Projects with Drones

In parallel to this curriculum, some groups of students also work on projects, such as the one in the context of the Cap Project initiative (see [9]). Some projects are dedicated to drones. Some examples over the past years are dot painting using single or multiple drones, designing a low-cost drone, participating in student drone contests, etc. These projects are managed by teachers and advisers in a complementary way to the curriculum to ensure that students can benefit from both.

An example of project is face and emotion recognition with drones. It aims to make the drone recognize the emotion of the person in front of it and interact accordingly through movements. The drone used in this project is the RMTT, with which we communicate using a Python interface that includes Tello libraries provided by the Tello SDK. Video frames captured by the drone are sent to the computer, which recognizes the emotion of the person in front of the drone and sends movement commands, so the drone reacts to the emotion. Social interaction in this project is based on four main tasks: face detection, face tracking, emotion recognition, and drone reaction.

---

[6] http://wiki.ros.org/opencv_apps.

Face detection is done using the OpenCV library. The computer receives the video stream sent by the drone and applies the OpenCV Haar Cascade classifier to detect faces in a video frame. Various face detection models are available in the literature such as OpenCV, SSD, MTCNN, and RetinaFace. In the context of this project, face detection has to be as fast as possible with acceptable accuracy. MTCNN and RetinaFace are very powerful face detectors in terms of accuracy. However, they require a long execution time, which makes them not suitable for real-time face detection. On the other hand, OpenCV and SSD face detectors proved to be faster at the cost of a slight accuracy degradation. This is why the OpenCV face detector was used in this project.

Drone movements are impacted by two main mechanisms: face tracking and reaction to the captured emotion. Face tracking consists in maintaining the detected face centered in the captured video frames. To do so, after capturing a video frame by the computer and detecting the face position in it, the difference between this position and the center of the frame is calculated and movement commands are sent to the drone so it slides horizontally (right/left) and/or vertically (up/down) so the next captured face position is the closest possible to the video frame center. This is done using PID (proportional-integral-derivative) speed controllers.

Emotion recognition is done using the open-source DeepFace Python library. This library offers several features such as face verification (comparing two faces), face recognition (finding a known face in a frame), and facial attribute analysis (age, gender classification, and emotion analysis). In this project, only the emotion analysis feature is used. Emotion recognition in DeepFace is based on a combination of convolutional and dense neural network layers. In [10], authors tested twelve emotion recognition models, all based on Deep Learning and Convolutional Neural Networks (CNN) using the CK+ (extended Cohn-Kanade) [11] and the Fer2013 (Facial Expression Recognition 2013) [12] datasets. They showed that the DeepFace algorithm was the most accurate in emotion recognition. Another interesting feature of DeepFace is that it is a lightweight system, which makes it suitable in the context of this project since emotion recognition time does not have a significant impact on the smoothness of the video transfers between the drone and the computer. The DeepFace system allowed for the recognition of six main emotions (anger, disgust, fear, happiness, sadness, and surprise) and neutrality.

As a reaction to the emotion of the person in front of it, the drone uses two mechanisms. First, it displays a smiley on its RGB-led matrix that mimics the recognized emotion. Figure 6 shows how emotions are displayed on the drone's RGB-LED matrix. Second, the drone makes some motions as a reaction to the emotion of the person in front of it. This reaction can be of two types: (1) showing empathy by making motions that interpret the detected emotion, and (2) making motions that try to change a negative emotion of the person.

Some of the students involved in this project also enrolled in the robotic curriculum presented in this paper. Using the same drone for the project as in the curriculum helps to make it profitable for students. As the project started before and finishes after the curriculum, students may also find a special interest in lectures, tutorials, and lab sessions, directly implied by practical concerns/issues raised from their project.

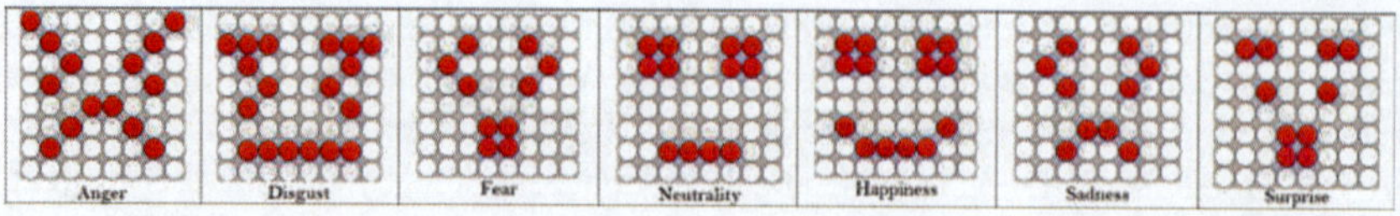

Fig. 6 Emotions displayed on the drone RGB-LED matrix

7 Conclusions

In this paper, some feedback has been presented on a short curriculum on robotics developed for last year's students at ESIEA, a graduate school of engineering. One of the characteristics of this curriculum is to propose hands-on experiments in classrooms during practical work sessions with low-cost drones.

The content of the curriculum has been presented, with a specific focus on one of its modules devoted to ROS and computer programming for robotics. Examples of experiments realized by the students as exercises integrated into the work sessions have been provided. This practical aspect is usually well appreciated by the students and helps to maintain their motivation and attention. Seeing, understanding, and analyzing practical results help them to have deeper looks at theory.

The next steps will concern the experimental setups, to be able to provide one drone per group of students with a localization system based on mission pads, as well as the development of new experiments and exercises regarding sensor fusion for localization and SLAM.

References

1. Yeppes, I., Barone, D.A.C., Porciuncula, C.M.D.: Use of drones as pedagogical technology in STEM disciplines. Inform. Educ. **21**(1), 201–233 (2022)
2. Beuchat, P.N., Sturz, Y.R., Lygeros, J.: A teaching system for hands-on quadcopter control. IFAC-PapersOnLine **52**(9), 36–41 (2019)
3. Canas, J.M., Martin-Martin, D., Arias, P., Vega, J., Roldan-Álvarez, D., Garcia-Pérez, L., Fernandez-Conde, J.: Open-source drone programming course for distance engineering education. Electronics **9**, 2163 (2020)
4. Bertrand, S., Marzat, J., Stoica Maniu, C., Makarov, M., Filliat, D., Manzanera, A.: DroMOOC: a massive open online course on drones and aerial multi robot systems. In: 12th UKACC International Conference on Control, Sheffield, UK (2018)
5. Brand, I., Roy, J., Ray, A., Oberlin, J., Tellex, S.: PiDrone: an autonomous educational drone using raspberry Pi and Python. In: IEEE/RSJ International Conference on Intelligent Robots and Systems, Madrid, Spain (2018)
6. Invernizzi, D., Panza, S., Giurato, M., Yang, G., Chen, K., Lovera, M., Parisini, T.: Integration of experimental activities into remote teaching using a quadrotor test-bed. In: IFAC Workshop on Aerospace Control Education, Milano, Italy (2021)
7. Monajjemi, M.: AR Drone Autonomy Library (2014). http://wiki.ros.org/tum_ardrone
8. Kayhani, N., Heins, A., Zhao, W., Nahangi, M., McCabe, B., Schoellig, A.P.: Improved tag-based indoor localization of UAVs using extended kalman filter. In: 36th International Symposium on Automation and Robotics in Construction (2019)

9. Bertrand, S., Prevost, L., Ionascu, F., Briere, A., Koskas, R., Taquet, R., Andrianantoandro, F., Kanzari, M.: Light painting with mobile robots as motivating projects for robotics and control education. In: 13th International Conference on Robotics in Education, Bratislava, Slovakia (2022)
10. Chiurco, A., Frangella, J., Longo, F., Nicoletti, L., Padovano, A., Solina, V., Mirabelli, G., Citraro, C.: Real-time detection of worker's emotions for advanced human-robot interaction during collaborative tasks in smart factories. Procedia Comput. Sci., 1875–1884 (2022)
11. Lucey, P., Cohn, J., Kanade, T., Saragih, J., Ambadar, Z. and Matthews, I.: The extended Cohn-Kanade dataset (CK+): a complete dataset for action unit and emotion-specified expression. In: 2010 IEEE Computer Society Conference On Computer Vision And Pattern Recognition-Workshops, pp. 94–101 (2010)
12. Goodfellow, I., Erhan, D., Carrier, P., Courville, A., Mirza, M., Hamner, B., Cukierski, W., Tang, Y., Thaler, D., Lee, D., et al.: Challenges in representation learning: a report on three machine learning contests. In: International Conference On Neural Information Processing, pp. 117–124 (2013)

A Beginner-Level MOOC on ROS Robotics Leveraging a Remote Web Lab for Programming Physical Robots

Sandra Schumann, Dāvis Krūmiņš, Veiko Vunder, Alvo Aabloo, Leo A. Siiman, and Karl Kruusamäe

Abstract With an increased demand for roboticists in the labor market, there is a growing interest in acquiring skills in ROS (Robot Operating System), one of the more popular robotics software development platforms. Most freely available resources for learning ROS rely on simulation or assume that people have a robotics platform at their disposal. In this study we extended professional robotics to self-guided learning using physical hardware by creating a massive open online course (MOOC) for learning ROS. Course participants could log into a web lab and remotely control robots located in a university classroom without the need to install anything locally on their computers. During the 7-week course participants of varying ages and backgrounds learned about the necessary skills to use ROS on a Linux machine through receiving an authentic ROS developing experience. The participants' weekly performance results and answers to a feedback questionnaire were subsequently analysed. The first iteration of this course was completed by 49 participants, with the biggest reason for dropping the course being lack of time. The course demonstrated the possibility of using a remote web lab to teach ROS. This paper summarizes our lessons learned.

Keywords Robotics · ROS · MOOC · Remote web lab

1 Introduction

With the prevalence of automation in the world constantly increasing, so does the need for people with knowledge in the field of robotics. One of the most popular software frameworks in robotics is ROS, an open-source middleware suite [7]. Due

Supported by Education and Youth Board of Estonia, European Social Fund via IT Academy programme, Estonian Centre of Excellence in IT (EXCITE) funded by the European Regional Development Fund, and AI & Robotics Estonia co-funded by the EU and Ministry of Economic Affairs and Communications in Estonia.

S. Schumann (✉) · D. Krūmiņš · V. Vunder · A. Aabloo · L. A. Siiman · K. Kruusamäe
University of Tartu, Tartu, Estonia
e-mail: sandra.schumann@ut.ee

R. Balogh et al. (eds.), *Robotics in Education*, Lecture Notes in Networks and Systems 747, https://doi.org/10.1007/978-3-031-38454-7_24

to its popularity and the fact that ROS is supported by a very large number of robotic platforms, knowledge of ROS is a good starting point for learning to operate a wide range of different robots.[1]

For learning ROS, many different resources are available, such as books or online tutorials [6, 8]. There are also various courses on typical educational platforms, on more expensive research-oriented hardware platforms or in simulation [2, 3, 10, 14]. Most online courses teaching ROS are taught on simulators.

However, online courses can also benefit from the use of physical hardware. Learning robotics with the help of physical robots is something that is deemed beneficial by a large number of robotics instructors [1]. There are a few resources that allow teaching online courses on physical hardware using remote web labs [2, 5, 12, 13]. However, to the best of our knowledge, there are no open online courses aimed at beginners teaching ROS using physical robots that can be controlled remotely and offer an authentic ROS developer experience.

We have developed an introductory online ROS course open to anybody, along with tools that allow the participants to complete practical tasks on physical mobile robots located in a university classroom. The participants could teleoperate the robots from afar and observe the behaviour through webcams. The course did not require participants to install anything extra on their computers, but rather allowed them to access an authentic developer environment on a Linux machine through a web browser. No prerequisite knowledge of Linux, ROS or robotics was required.

In the fall of 2022, we ran the first iteration of the course. Over the course of 7 weeks, participants completed six modules, each focusing on a different aspect of basic ROS and robotics skills. The course was in Estonian language, free and available for anybody interested. Most participants were recruited via advertising on YouTube, with the advert[2] gaining around 16,000 views. Participants included people from a wide range of different backgrounds, ages and levels of education.

In this paper we discuss the course design, the results achieved by the participants and their feedback. Our main research questions are as follows:

RQ1. What factors are relevant for the successful completion of an introductory ROS course?
RQ2. How did participants experience learning in this course?

The paper is organized as follows. Section 2 details the course structure from a pedagogical point of view and Sect. 3 the technical setup used in the course. In Sect. 4 we present the research methodology and Sect. 5 summarizes the results. We conclude in Sect. 6 with a discussion and ideas for future work.

[1] https://robots.ros.org.

[2] https://youtu.be/CCnGBWoKIYc.

2 Course Structure

The objective of the course is to give an introduction to programming robots using ROS to an interested participant regardless of level of background knowledge. The main learning outcome was for the participants to have the fundamental skills to engage in ROS development independently.

The 2 ECTS course consisted of six modules, released weekly. Each module consisted of theoretical material, practice tasks, an automatically graded test and a practical homework assignment that relied on the knowledge obtained from theoretical material and practical tasks. The topics covered together with the expected learning outcomes have been listed in Table 1. Altogether, the modules cover the necessary skills to use ROS on a Linux machine, the use of different ROS packages and messaging system, the way to describe the kinematics of a robot and to use the robot's actuators and sensors in one's custom code.

Table 1 Learning outcomes of course modules

Module	Learning outcomes	Homework assignment
M1	The learner navigates between directories, edits files and opens applications in Ubuntu; uses simple Linux command line commands	Teleoperating the robot via command line, motion visible by overhead camera
M2	The learner distinguishes ROS nodes, topics, messages; operates and constructs rosrun and roslaunch commands	Remapping nodes from several different pre-existing packages to make the robot autonomously drive between orange markers
M3	The learner designs a digital description of a robot using Unified Robot Description Format (URDF), modifies it using xacro	Creating a URDF description of a four-wheeled robot, moving it in RViz using keyboard teleoperation
M4	The learner prepares a new ROS package with an executable node; designs a program to make the robot to move using ROS publishers	Writing a publisher to make the robot move in a square, circular and Fig. 8 shaped static trajectory
M5	The learner extracts info from sensors; designs a ROS subscriber and a controller; constructs a launch file	Writing a simple controller that makes the robot turn between AR markers
M6	The learner performs simultaneous localization and mapping, uses ROS Navigation library	Mapping the environment, autonomously navigating a robot to a desired location

Table 2 Overview of course online environments

Environment	Purpose
Moodle	Links to other environments. Course timeline. Submission and feedback of assignments and tests. Course-wide announcements
Public website	Theoretical materials. Instructions for practical tasks
Slack	Instant messaging between course participants and instructors. Asking and giving assistance during the completion of practical tasks
Remrob	Access to machines with Ubuntu Linux and ROS installations. Booking system for reserving a time with a physical robot. Capability of interacting with a physical robot or a simulator. Viewing the image of a webcam pointed at the physical robot

The participants were given 7 weeks to complete all assignments. The course did not have hard deadlines for completing the tests and assignments other than the end date of the course, however, the participants were recommended to follow the pace of one module per week (with an extra week given as a buffer).

The final grade received in the course was either pass or fail. In order to receive a pass one had to pass all tests and practical homework assignments. The participants were given an unlimited number of attempts to pass the weekly automatically graded test, with passing threshold set to 90%. They also had a theoretically infinite number of attempts to resubmit a practical homework assignment, only limited by the tempo at which instructors could give feedback to the submissions. In practice, the homework assignments were mostly passed on the first submission attempt and never required more than two attempts.

3 Technical Setup of the MOOC

3.1 Online Environments

The course made use of four different online environments: the university installation of Moodle environment, a public website containing course material,[3] a Slack workspace and a newly developed remote web lab (Remrob), piloted in this course. Table 2 gives an overview of the purposes of the different environments.

[3] https://sisu.ut.ee/rosak.

3.2 Remrob Environment

The Remrob environment[4] offered each participant access to a workstation with Ubuntu Linux 20.04 [4]. The environment also included an online booking system for reserving a time slot to work with a workstation. The workstations included an installation of ROS Noetic Ninjemys, several ROS packages used for completing course assignments and Gazebo simulation software. A screenshot of the system is shown in Fig. 1. 7 workstations were configured to have a connection to actual physical robots located in a university classroom (shown in Fig. 2a), one robot per workstation. Additionally, those workstations could show a top view camera stream of the corresponding robot.

Another 9 workstations were provided in order to enable learning in simulation. They included the same software capabilities as other workstations, but instead of using a real robot, they allowed participants to test their code in simulation. This enabled learning even when access to robots was limited.

In order to use a workstation, a participant logs into the Remrob system and books a time slot to use either a workstation with a physical robot connection or a workstation meant for simulations. The participant could then access the workstation during their allotted time, with their personal files stored in a private folder accessible only to their user.

The time slots that participants could book were made available by the course instructors. The availability of simulation workstations was only limited by the server's capability of handling simultaneous connections and thus they were often available 24/7. The workstations with connections to physical robots, however, required that a human be present in the robotics lab in order to monitor the situation, resolve any technical problems and make sure the robots' batteries are sufficiently charged. Therefore, the workstations with robot connections were only made available at times when one of the course instructors could be present. Time slots (ranging from 30 to 90 minutes) were provided within a wide range of different times of day, in order to accommodate participants' various schedules.

3.3 Robotic Platform

The physical robots used by the participants of the course were the Robotont open-source omnidirectional mobile robots (Fig. 2b) [9]. A Robotont consists of a round base, three omniwheels for moving the base, motor controllers and other electronics necessary for operating them, an Intel NUC computer and an Intel RealSense depth camera. The robot is powered by a LiPo battery that provides autonomous driving experience in loaded conditions for more than an hour.

[4] https://github.com/unitartu-remrob/remrob-server.

Fig. 1 View of a workstation used by the participants in the course

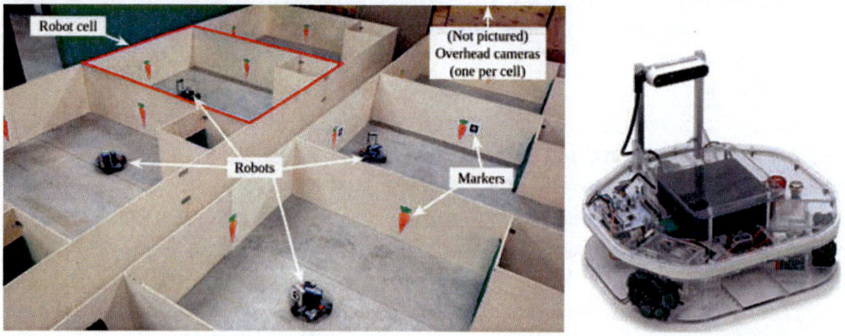

(a) Classroom (b) Robotont

Fig. 2 Classroom and the Robotont robot used in MOOC. Outlined in red is a single robot cell used by one learner at a time. The robot is free to move around within the cell (including into a smaller corner box through a curtain) and bounded by walls. The robot cells have colourful markers and AR markers

In addition to the physical robots, participants could also operate a digital twin of the Robotont robot. This was mostly used to allow participants test their solution without the need to book a time slot for using a physical robot, as simulation times were more available.

4 Methods

To answer the research questions, two approaches were adopted: learning analytics and feedback survey [11], analyzed using descriptive statistics.

After the final deadline for submitting course assignments, the learners' results in tests and homework assignments were gathered, along with their age and gender information. Additionally, for each participant the start and end times of their booked time slots on Remrob were extracted.

All participants, regardless of whether they passed the course or not, were asked to fill out an anonymous feedback survey. The survey consisted of 25 questions. The subsequent analysis relies on the 4 questions about demographic information, 6 questions about the participants' intentions and preparation before the start of the course and 2 questions about difficulty level of the assignments.

5 Results

In total, 191 participants signed up to take the course and 49 of them completed the course. 50 participants out of the initial 191 filled out the anonymous feedback survey. Among them 29 completed all the modules and 21 did not.

5.1 *Demographic Information*

5.1.1 Age and Education

Participants' age (based on registration information and feedback survey) and highest completed level of education (based on feedback survey) are shown in Figs. 3 and 4. The youngest person who signed up for the course was 14 years old and the oldest one 72. The youngest person to complete the course was 16 years old and the oldest one 72. Overall, people from the age group 40–49 were less likely than others to finish the course.

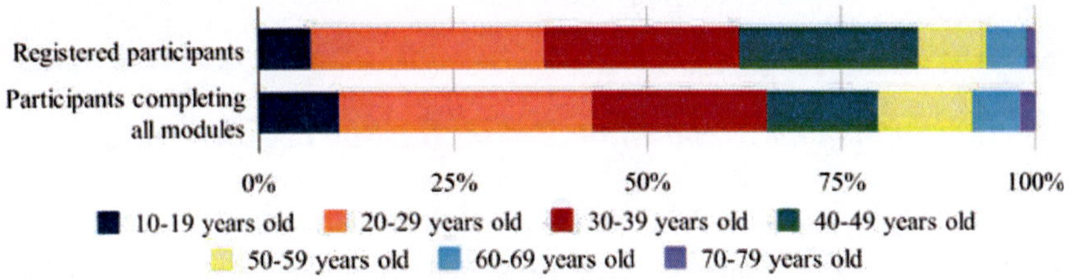

Fig. 3 The ages of all course participants ($N = 191$) and those completing the course ($N = 49$)

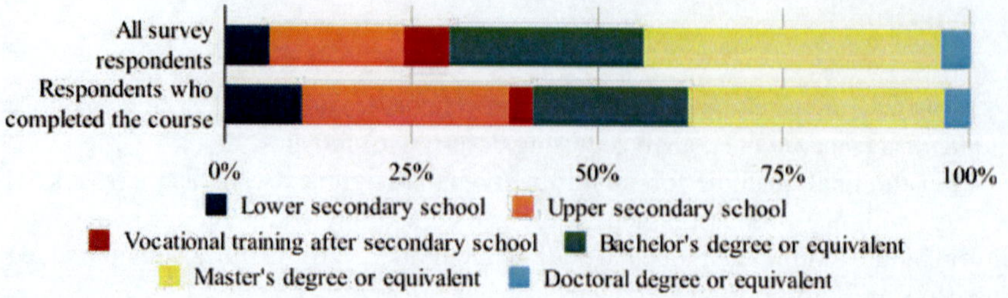

Fig. 4 Participants' highest obtained level of education based on feedback survey for all survey respondents ($N = 50$) and survey respondents who completed the course ($N = 29$)

5.1.2 Gender

Among the 191 signing up for the course, 27.2% were female and 72.8% male. Among those finishing all modules, however, only 18.4% were female and 81.6% male. Among those completing the feedback survey, 24.0% were female, 74.0% male and one respondent chose not to disclose their gender. This seems to indicate that men were more likely than women to complete the course.

5.1.3 Field of Speciality

Survey respondents were asked about their field of speciality and were allowed to pick more than one field in their response.

70% of the respondents reported being involved in IT or technology. 26% reported being involved with natural sciences. Overall, only 18% of participants did not indicate being involved in one of the previously listed fields, indicating that most participants had a background in technology or natural science.

There appears a difference in the self-reported completion rates between people of technology or natural sciences and everyone else. 90% of those who completed all modules reported being involved in the fields of technology or natural sciences. However, among those that did not complete all modules, only 71% reported being involved in those fields.

32% of the participants also reported being involved in the field of education, suggesting that the course was fairly popular among teachers of various kinds.

5.2 *Intentions and Preparation Before the Start of the Course*

Due to the voluntary nature of the MOOC, participants were asked about whether they were planning to complete the course or signed up without a clear intention of actually completing the course. 94% of the respondents indicated that they were

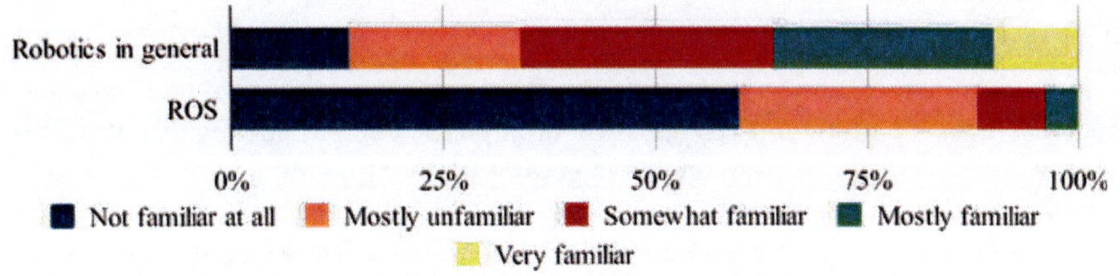

Fig. 5 Participants' reported level of familiarity with robotics in general and with ROS before the start of the course

planning to complete the course when signing up, with the rest being unsure or only intending to browse some course materials. 58% of the survey respondents also reported that they did actually complete the course, including two thirds of those who were initially unsure about their goals. This seems to suggest that dropping the course was not a planned activity from the start for the vast majority of the participants.

The survey also asked respondents about their prior experiences with ROS and robotics in general. On average the respondents had some familiarity with robotics in general, but not much prior knowledge of ROS (Fig. 5). There was no notable difference in the previous robotics or ROS knowledge between the survey respondents who completed the course and those who did not.

5.3 *Time Spent*

Participants were asked about the average number of hours they spent on the course every week in the feedback survey. Additionally, information about the time slots used by each person was extracted from the Remrob system.

Based on the feedback survey, those who completed all course modules report having spent, on average, 6.36 h per week on the course (standard deviation 3.86 $\frac{h}{week}$). Those that did not complete all modules report having spent 4.12 $\frac{h}{week}$ on average (standard deviation 3.40 $\frac{h}{week}$). Based on the Remrob usage data, those who completed all course modules used over the whole course (on average) 8.06 h on workstations with physical robots and 23.8 h on simulation workstations, for a total of 31.9 h on Remrob system over the entire course. However, these numbers vary a lot between participants. A few participants also opted to use their local installations of ROS, the usage of which is not included in the statistics.

5.4 Course Progression

Figure 6 depicts participants' progression through the course. The biggest obstacles to completing the course were related to getting started: logging into the system and completing the Moodle first test. There were a few learners that attempted using the Remrob system, but did not pass the first Moodle test. However, the majority of participants that dropped out before completing module 1 did not even attempt to book a time on the Remrob system. Because the nature of learning in the first module was largely by completing practical exercises, this suggests that these participants did not make a serious attempt at participating in the course.

After getting started, the biggest obstacle was completing the second homework exercise. The second module taught the basics of ROS topics, nodes and messages. This seems to indicate that the support offered by the course in covering those new topics might not have been sufficient for all beginners.

The participants that did not complete the course were also asked about their reasons for not doing so in the feedback survey. Participants could provide multiple reasons. Out of the 21 survey respondents that did not complete the course, 14 reported one of the reasons being lack of time. 8 reported that their progress was impeded by the fact that they could not use robots at suitable times. 4 deemed the course too difficult, 3 were experiencing technical difficulties and 2 reported that the content was too boring. These results indicate that the primary reason for not completing the course was the required time commitment.

The participants filling out the survey were also asked about the perceived level of difficulty of the automatically graded test on Moodle and the practical homework assignments (Fig. 7). The results show that most participants found the Moodle tests to be either at an appropriate level or slightly too easy, and the homework assignments to be at an appropriate level or slightly too hard.

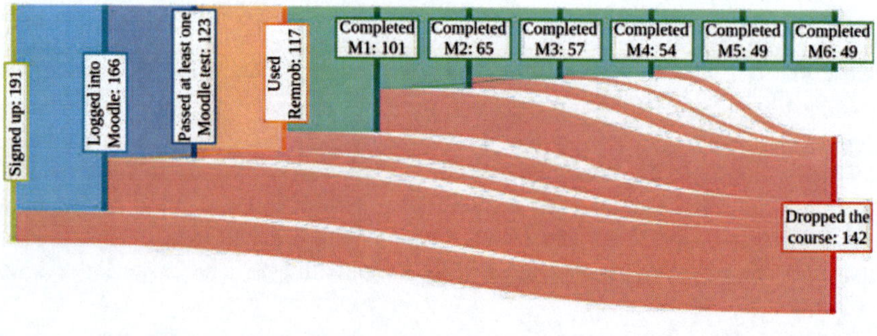

Fig. 6 Participants' progression in the course and path to completion or dropping out of the course

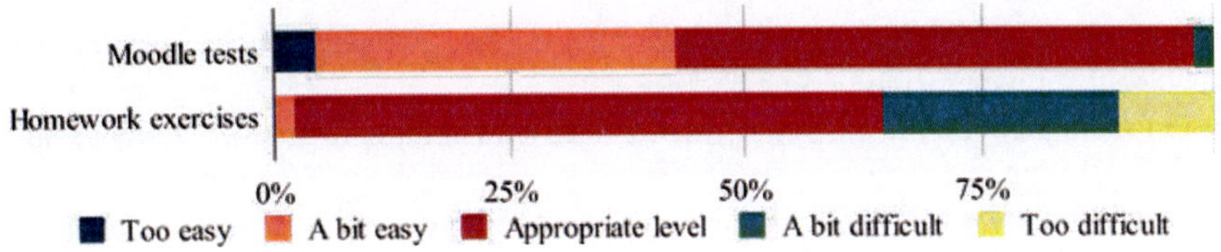

Fig. 7 Perceived level of difficulty of the Moodle tests and practical homework assignments

6 Discussion

6.1 *RQ1: What Factors are Relevant for the Successful Completion of an Introductory ROS Course?*

The first iteration of the course was successfully completed. The 49 participants completing all course modules formed 26% of those who originally signed up and 42% of those who used at least one time slot on the Remrob online system.

The most important factor identified predicting the successful completion of the course seems to be the amount of time spent on the course. Those that completed the course spent, on average, more time on the course per week than those that did not. Lack of time was also most commonly mentioned as a reason for not completing the course in the feedback survey.

Lack of time could also be one of the factors responsible for participants in the age group of 40–49 being less likely to complete the course: they are likely to be employed full-time and to have significant family obligations. The same effect could come into play with the level of education: younger participants, indicated as having completed lower or upper secondary school education, are more likely to have time that they can dedicate to the course.

Most participants listed their field of speciality as either being in IT, technology or natural sciences. Based on the limited number of responses of those who were not, it seems that background in one of the listed fields can be helpful for successfully completing the course. Interestingly, participants' self-reported previous experience with robotics or ROS specifically did not seem to have an effect on the likelihood of completing the course.

6.2 *RQ2: How Did Participants Experience Learning in This Course?*

Analysis of data in Fig. 6 shows that the most challenging step (other than getting started) was completing module 2, focusing on ROS nodes, topics and messages. Somewhat surprisingly there does not seem to be a big drop between reserving a time on the Remrob system and completing module 1, which included learning the basic skills of using Linux command line.

On average for all the modules, Moodle tests with automatic grading were considered easier than practical homework exercises in Remrob environment. This, however, is to be expected, as the homework exercises required higher order thinking skills than the Moodle tests. For both Moodle tests as well as homework exercises, most survey respondents indicated that the level of exercises was appropriate for them.

6.3 Limitations and Future Work

The course has so far only been offered once, giving a limited amount of data. Future iterations of the course can offer more data on participants' progression. Additionally, the feedback survey should be tailored to more accurately give data on the research questions. The feedback survey was also only sent out at the end of the course, which can affect answers to the questions querying the participants' preparation and intentions before the start of the course. In the future, this can be addressed by asking participants to fill out a pre-survey as well as a post-survey.

The course capacity was limited by the amount of hardware available. Currently, the robots used are not capable of autonomously charging their own batteries, which is the main reason for requiring the physical presence of an instructor. However, the capability is being developed, which should make the robots more accessible in the future. The availability of the 9 simulation-only workstations also helped relieve the capacity issues. Based on the number of hours the participants who completed the course reported having spent on the course each week and assuming that around 75% of the course work can be done in simulation, the current 9 simulation only machines available 24/7 would theoretically be sufficient for over 300 learners. The actual number will be different, as there are certain times during which there are more learners using the machines, such as evenings. Currently we were not able to host more than 9 simulation workstations simultaneously, but the capacity can be increased by making more workstations available, for example by adding a second server.

The course infrastructure proved itself to be mostly reliable. The video feed sent from robots to the server had to be either compressed or throttled or the image analysis done on the on-board computer in order to avoid congestion, all of which were viable solutions. No congestion was observed from users trying to access the Remrob system. The time delay between the users' computers and robots was minimal and in most cases did not hinder the progress. However, it was observed that the performance suffered significantly if the participants' own Internet connection was not reliable.

For the next iterations of the course, one can consider giving the participants more time to complete the same exercises (for example, with the average pace being two weeks per module instead of one). Additionally, it can be beneficial to pay special attention to module 2 covering the fundamentals of ROS nodes, topics and messages, as most participants that actually started with the course dropped out there, perhaps dividing the introduction of basics along more modules. The content of all other modules should also be revised based on performance and feedback. The future iterations of the course should also be conducted on ROS2. Since men were more likely than women to complete the course, a more thorough analysis into the methods to support women in the course could reduce the gender disparity. Lastly, the first iteration of the course was offered to Estonian-speaking audience. In the future, the course could also be offered in other languages, such as English.

References

1. Birk, A., Simunovic, D.: Robotics labs and other hands-on teaching during covid-19: change is here to stay? IEEE Robot. Autom. Mag. **28**(4), 92–102 (2021)
2. Cañas, J.M., Perdices, E., García-Pérez, L., Fernández-Conde, J.: A ROS-based open tool for intelligent robotics education. Appl. Sci. **10**(21), 7419 (2020)
3. Chen, H., et al.: Development of teaching material for robot operating system (ROS): creation and control of robots (2022)
4. Krūmiņš, D., Vunder, V., Schumann, S., Põlluäär, R., Laht, K., Raudmäe, R., Aabloo, A., Kruusamäe, K.: Open remote web lab for learning robotics and ROS with physical and simulated robots in an authentic developer environment. IEEE Trans. Learn. Technol. (May 2023)
5. Kulich, M., Chudoba, J., Kosnar, K., Krajnik, T., Faigl, J., Preucil, L.: Syrotek-distance teaching of mobile robotics. IEEE Trans. Educ. **56**(1), 18–23 (2012)
6. Pozzi, M., Prattichizzo, D., Malvezzi, M.: Accessible educational resources for teaching and learning robotics. Robotics **10**(1), 38 (2021)
7. Quigley, M., Conley, K., Gerkey, B., Faust, J., Foote, T., Leibs, J., Wheeler, R., Ng, A.Y., et al.: ROS: an open-source robot operating system. In: ICRA Workshop on Open Source Software, vol. 3, p. 5. Kobe, Japan (2009)
8. Quigley, M., Gerkey, B., Smart, W.D.: Programming Robots with ROS: A Practical Introduction to the Robot Operating System. O'Reilly Media, Inc. (2015)
9. Raudmäe, R., Schumann, S., Vunder, V., Oidekivi, M., Nigol, M.K., Valner, R., Masnavi, H., Singh, A.K., Aabloo, A., Kruusamäe, K.: Robotont-open-source and ROS-supported omnidirectional mobile robot for education and research. HardwareX (Mar 2023)
10. Roldán-Álvarez, D., Mahna, S., Canas, J.M.: A ROS-based open web platform for intelligent robotics education. In: Robotics in Education: RiE 2021 12, pp. 243–255. Springer (2022)
11. Schumann, S., Siiman, L.A., Kruusamäe, K.: Feedback Questionnaire used in the University of Tartu Introductory ROS MOOC. Zenodo (Jan 2023)
12. Tellez, R.: A thousand robots for each student: using cloud robot simulations to teach robotics. In: Robotics in Education: Research and Practices for Robotics in STEM Education, pp. 143–155. Springer (2017)
13. Wiedmeyer, W., Mende, M., Hartmann, D., Bischoff, R., Ledermann, C., Kroger, T.: Robotics education and research at scale: a remotely accessible robotics development platform. In: 2019 International Conference on Robotics and Automation (ICRA), pp. 3679–3685. IEEE (2019)
14. Wiesen, P., Engemann, H., Limpert, N., Kallweit, S., Schiffer, S., Ferrein, A., Bharatheesha, M., Corbato, C.: Learning by doing-mobile robotics in the FH Aachen ROS summer school. In: TRROS@ ERF, pp. 47–58 (2018)

Teaching Robotics with the Usage of Robot Operating System ROS

Miroslav Kohút, Marek Čornák, Michal Dobiš, and Andrej Babinec

Abstract This article presents an exercise framework for teaching robotics with Robot Operating System ROS. It is designed to teach the fundamental robotics principles in an interactive way, while using modern software tools that are commonly used in real-world practice. The exercises and assignments encourage students to implement the robotics theory into their own algorithms and test and compare them to ROS provided functionality, to better understand the development of robotic systems. This paper describes the framework in detail and explains the choice of ROS as the software tool for the framework. In the end, the results of an evaluation questionnaire are presented to assess the overall satisfaction and usefulness of the course.

Keywords Robot operating system · ROS · Education · Framework · Robotics · Course

1 Introduction

Modern technology has vastly transformed people's lives in the past decades, and current technological advances are accelerating this transformation more than ever before. Terms like robots, drones, intelligent systems, or artificial intelligence (AI) are gaining tremendous popularity, playing an important role in shaping the world of tomorrow. The vast field of robotics encompasses all these terms and applies them to practice. It is also clear that the world of tomorrow will need skilled and bright engineers and roboticists to properly use their knowledge to create innovative, productive, and effective solutions, hopefully for the benefit of all people.

This is also the main effort of the robotics and cybernetics study programme at the Slovak University of Technology (STU). The programme is designed so that graduates receive a high-quality engineering education based on knowledge of automatic system control, cyber modelling, control theory, computer science and design and development of robots [1]. The graduate's skill set includes the ability to apply

M. Kohút (✉) · M. Čornák · M. Dobiš · A. Babinec
Slovak University of Technology, 812 19 Bratislava, Slovakia
e-mail: miroslav.kohut@stuba.sk

R. Balogh et al. (eds.), *Robotics in Education*, Lecture Notes in Networks and Systems 747, https://doi.org/10.1007/978-3-031-38454-7_25

advanced knowledge of IT technologies, principles of electrical and mechanical engineering together with complex algorithms to create applicable robotic solutions. The university course Control of Industrial Robots (CIR) [2] teaches students how these classical fields are applied to solve the most common and basic robotic problems such as robot forward and inverse kinematics, spatial homogeneous transformations, Jacobians, Lagrange's equations of motion and polynomial trajectory generation. The subject focuses on applying these principles on robotic manipulators. However, they can be applied generally throughout other fields of robotics. The course's lectures provide students with necessary underlying theoretical knowledge and teaches them how to apply the principles into the practice. The traditional way for students to revise the theory and learn the principles used to be done by pen and paper assignments, where students would solve various problems using the equations and algorithms from the lectures. Authors find that this method is insufficient by today's standards. All modern robots are controlled by computers, thus it is necessary for good roboticists to transform their knowledge into the format that the computer can understand, such as a programming language. They must be able to use modern software tools to exploit all the potential for developing robots to produce powerful and effective solutions. We highly support a learn-by-doing method, where the students apply their expertise to solve practical examples.

In this paper, we present an educational framework using a Robot Operating System or ROS. It is designed to teach the theoretical principles of robotics, where students apply this knowledge to solve programming assignments. Thus, students learn robotics more interactively, and, using ROS, they are already learning the development process of real-world robotic applications. The authors believe that schools and educational courses often struggle to connect the presented topics with real world practice, so we find it important to teach tools, that students could widely use in their professional life and consequently, be more motivated to learn the concepts seeing how much the tools learnt can benefit them in the future. In addition, by programming algorithms and mathematical formulas, students also strengthen their algorithmic thinking and computer science knowledge. The main purpose of the framework is to present the students a set of assignments which they will solve by using ROS operating system. The framework is developed so that the students will not only learn to use the ROS tools, but most importantly, they learn the robotic principles in depth as they will program a lot the ROS functionality themselves and hopefully will get better understanding of the overall concepts and principles.

This paper presents an overview of the proposed framework, which currently is used as an accompanying practical part of the CIR subject. After completion, the graduate should understand basic concepts of kinematics, robot trajectory generation and control of robotic manipulators and be able to develop deployable robotic solutions. The knowledge obtained, can also be leveraged in other robotics fields such as mobile robotics, where kinematics, dynamics and trajectory generation also play crucial part in these applications. The framework consists of three blocks of assignments, each focused on a different topic. These assignments are completed using ROS framework; however, the learning of ROS is not the sole focus of the exercise framework. The main focus of our framework and the difference between our approach and other ROS focused courses is that we focus primarily to effectively teach the robotics principles using the framework, that we find is the most useful for the future career of the students in robotics in general. The choice of ROS for the framework will be discussed and, in the end, the results of a questionnaire collected from the graduates will be presented to evaluate their overall satisfaction with the course.

2 Related Work

The proposed exercise framework was originally developed as a practical part for the CIR course at STU, with the focus to interactively demonstrate main topics of the course. These topics and basic principles are, the forward and inverse kinematics of the robotic manipulator, the dynamics of the robot, the generation of polynomial trajectory, and robot's joint control. In this chapter, we summarise similar university courses, which were publicly available from online sources. We aimed to represent universities which are leading in robotics education as well as universities in the region, and we especially focused on practical teaching methods and the tools used in the teaching process. The Table 1 presents the university courses with the corresponding method or tool used in course practical lessons.

Table 1 A summary of similar university courses to CIR with its software tools for education

University	Course name	Tool used
Massachusetts Institute of Technology (MIT)	Robotic Manipulation [3]	Drake
California Institute of Technology (Caltech)	Robotic Systems [4]	ROS
ETH Zurich	Robot Dynamics/Programming for Robotics—Introduction to ROS [5]	Matlab RTB/ROS
Stanford	Introduction to Robotics [6]	Paper assignments
KTH Royal Institute of Technology	Introduction to Robotics [7]	ROS
Delft University of Technology	Robot Dynamics and Control [8]	Robopy/Matlab RTB
Polytechnic University of Milan	Control of Industrial Robots [9]	Matlab RTB
Technical University Vienna	Fundamentals of Robotics [10]	Matlab
Czech Technical University in Prague	Robotics [11]	Matlab RTB
Technical University of Kosice (TUKE)	Industrial Informatics and Robotics [12]	Matlab RTB

As the table shows, two of the most popular software for teaching advanced robotics are Matlab and its Robotic System Toolbox (RTB) [13] and ROS. According to Esposito [14] the most used software tools for robotics are Matlab (62%) followed by C (57%) and ROS (28%). It is also important to mention that one of the alternative complementary literature for the CIR course is the book Introduction to Robotics by Craig [15], which also provides practical exercises in Matlab RTB.

Matlab RTB provides the user with many powerful functions and tools for building and learning robots, such as simulation space, graphical interface, and plenty of examples with its vast ecosystem of additional libraries like neural networks and computer vision. Its scripting language like Python is easy and convenient to use, allowing the user to develop a variety of applications relatively quickly without the need to focus too much on the intricacies of the programming concepts. It also supports a few real-world robotic platforms such as Ginova Gen 3 and Universal robots. Matlab RTB thus provides a very good all-in-one, easy-to-use solution for programming and teaching robotic systems. However, its use in real-world applications is somewhat limited due to its relatively high licence fees and lack of hardware support.

ROS or a Robot operating system is a standalone open-source framework for developing robotic applications. ROS provides the user with a vast ecosystem of tools, libraries, and modules to build deployable robotic applications from scratch on a huge spectrum of supported hardware like KUKA, ABB, Universal Robots, Fanuc, etc. ROS mainly supports the C++ and Python programming languages, which are still immensely popular and relevant. The disadvantage of ROS is its relatively steep learning curve, as programming in ROS demands fairly advanced computer and programming knowledge. Nevertheless, ROS is still considered one of the most widely used tools for developing complex research and industrial applications, and its influence expected to be even stronger in the near future, as the second generation of ROS called ROS 2 is already being released, which is focused on the development of high-performance commercial grade robotic solutions. According to ABI Research [16] "nearly 55% of total commercial robots shipped in 2024, over 915,000 units, will have at least one ROS package installed", thus forecasting a need for ROS developers on the job market.

To sum up, Matlab and ROS appear to be the two best candidates for the implementation and building of robotic applications mainly due to their popularity in research, educational and commercial fields, and also their strong community support and vast documentation. However, it should be noted, that Matlab and ROS cannot be compared directly as Matlab was not primarily developed only for robotics. In real world, different applications require different functionalities and features, thus it is up to the developers to decide which framework provides them with better tools. Although the tools may be different, the main principles are the same, so it is much more important to understand the underlying theoretical knowledge than the syntax of the concrete framework, hence the focus of our approach to teach the principles, rather than the software tools. Fortunately for developers, Matlab and ROS can also work together through the Matlab ROS toolbox, so it is possible to combine them together to take advantage of both tools.

2.1 Why ROS

The biggest challenge in choosing the software framework and designing the assignments was finding the right balance between three main aspects:

1. The difficulty in understanding the framework itself.
2. The variety of additional tools supported by the framework, such as simulation and visualisation environments, supported libraries, available documentation, etc.
3. Real-world applicability in terms of supported hardware, programming practices, security, performance, and more.

In the Table 2, there is a comparison between ROS and Matlab frameworks regarding the mentioned aspects. It is important to mention that the comparison criteria are focused mainly on the field of robotics.

The next challenge was to fit the software framework into our education method and practices. As already mentioned, we are strong proponents of the active learning, or learn by doing method, where students actively solve practical (ideally real-world) problems so that the students perceive their new knowledge not only as useful but also necessary. We also wanted to present the topics and principles in an interactive and visually stimulating way, so the students get constant and useful feedback on their work. Both Matlab and ROS provide powerful tools for data visualisation and simulation. Regarding the difficulty of the frameworks, we believe that the students of the course should have sufficient technical background and computer science knowledge to understand ROS concepts and programming practices. ROS also supports the

Table 2 Comparison between Matlab and ROS

Software tool	Matlab	ROS
Installation difficulty	Easy	Difficult
Project development and overall use	Easy (use of Matlab scripts and functions)	Difficult, (needs understanding of ROS concepts, and code compilation and building)
Simulation/visualisation tools	Very good (Simulink, Matlab Graphing, good GUI)	Very good (Gazebo, Stage, Rviz, rqt graphing)
Documentation	Very good [17]	Very good [18]
Modularity	Restricted (by Matlab modules)	Very good (ROS package organisation provides access to vast number of open-source modules and libraries)
HW support	Poor (only a few robotic platforms and sensors)	Very good (wide spectrum of supported robots and other hardware)
Performance	Limited (using Matlab dynamically typed, higher level scripting language)	Very good (able to use, lower level, statically typed C++ language and libraries)

Python language, which is considerably easier to use compared to C++. However, assignments in ROS would likely require more preparation from the teacher's side, such as providing students with example pre-programmed templates to use, while in Matlab, programming from scratch is much easier.

The decisive factor for us was the real-world applicability of the framework. Although Matlab is easy to use and provides powerful tools for education, it lacks the hardware support and performance to apply it in deployable and production grade solutions. In our opinion, a big problem in schools in general is the lack of connection between the theory presented and its application in general practice. We consider that students should ideally learn methods and tools which have great potential in their future professional career. Due to a broad spectrum of supported hardware, emphasis on high performance and enormous modularity, ROS fulfils all the mentioned criteria. Moreover, thanks to a wide support of robotic platforms, it is more available for teachers and students to create and solve problems on real hardware. Another huge advantage is that open-source nature of ROS and its simulation tools enables to design, develop, and test robotic solutions completely for free. Additionally, this could impact the motivation to learn, as students could be discouraged to learn information that they do not feel they could use or need in their professional life. We believe that students will be more motivated to learn when they are shown how their expertise could be applied in the real world, even at the cost of the subject being more difficult. The support for this learning is provided in Fogg's behavioural model [19], in which he argues that ease of use and motivation to do a certain action do not go hand in hand. The third important factor of this model are triggers, which enable to perform the targeted behaviour in the first place. Similar approach for teaching ROS was also used by Cooney et al. [20].

ROS can be generally defined as a set of software libraries, tools, frameworks, APIs, and computing concepts that are running on the user's computer as a middleware. The core functionality of ROS consists of a distributed network of various (single purpose) processes which can run on one or multiple computers, able to communicate with each other asynchronously through so-called topics based on the publish-subscribe principle, or synchronously using services based on request-response communication model.

3 Framework Description

The proposed exercise framework is closely connected with the lectures of the CIR course and the theory explained in them. Our framework can be used as an independent tool for consolidating knowledge of robotics and acquiring practical skills in robot development.

There are 12 two-hour exercises during the course split into three assignment blocks. In each block, students are provided with pre-made ROS packages which are thoroughly explained. Students then proceed to solve the tasks on the packages. The basic framework structure is described in Fig. 1.

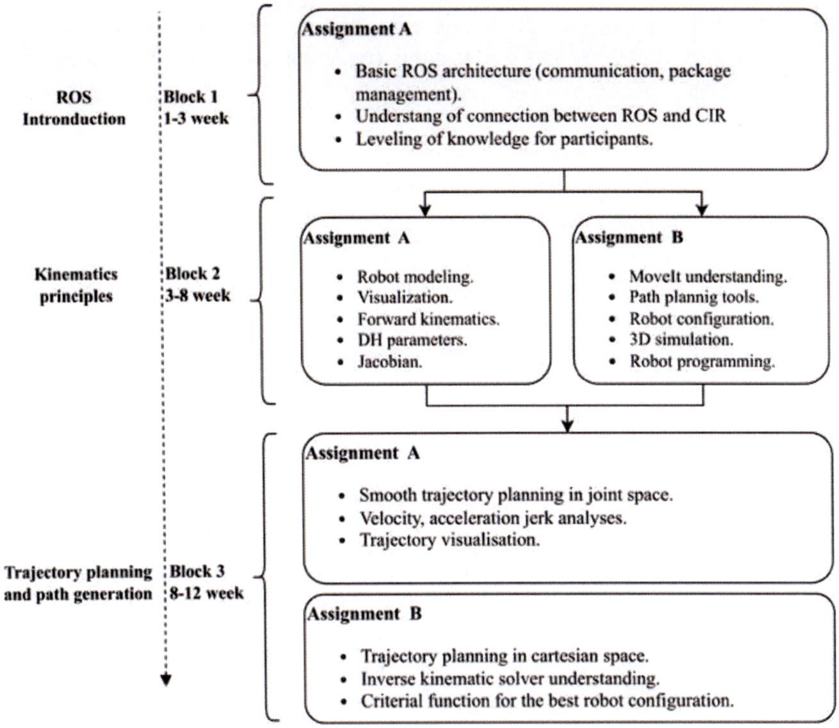

Fig. 1 Scheme of the assignment distribution during the whole course

Requirements necessary to complete the course are basics of Linux/Unix operating system, C++, CMake and Python. The fundamental basic knowledge of mathematics, physics and algebra is also required.

3.1 Block 1—ROS Introduction

In the first block students go through the beginner ROS tutorials and get basic knowledge of ROS standards (topics, services, parameter server etc.) [5]. This part was inherited from basic concepts of ROS introduction tutorials [18]. The difference is that we are providing practical information and program examples for faster understanding of the whole concept. There are three main goals for the first assignment:

- Understanding the basics of ROS and programming in C++.
- Understanding the correct configuration of the ROS workspace and package.

- Ability to program basic ROS concepts (topics, services, parameters, ROS configuration files), use basic ROS tools (rqt graphs, Rviz) and run their own processes.

3.2 Block 2—Kinematic Principles

The second part of the course is divided into two assignments. The first assignment of the second block is focused on understanding basic kinematic principles and modeling kinematic chain of the 3-DOF manipulator. The second assignment is about understanding basic planning principles like collision checking and trajectory planning with support of MoveIt! MoveIt! is a ROS package which provides specific build in functionalities for trajectory planning inverse kinematics and robot control.

After first block, students should already have enough experience to start implementing their own solutions in ROS. In the first assignment, students are shown how the basic kinematics principles are implemented in ROS and how to build a robot model using URDF format file.

After correct robot setup, using the example in our template package, students continue with analyses of existing robot structure. Using basic kinematic principles, they define transformation matrix for forward kinematic computation using both the classical "intuitive way" through elementary matrices and computation through DH parameters. They will do it using C++ Eigen library [21], which is commonly used for matrix operations in geometry, physics, robotics, and other applications. In the end they should produce a working kinematic structure based on the assignment, that they can then visualize using Rviz tool.

In the second part of exercise, students must analyze existing kinematic structure of a 6-dof manipulator ABB IRB 4600. They are instructed to create their own ROS robot configuration and understand the functionality of inverse kinematic solvers, trajectory planning and collision checking. This knowledge and understanding of the functionality will be required for the last part of the course.

Consequently, students should learn how is the real robot connected to ROS and how to control its movement through MoveIT! tool. Based on their obtained experience, they are instructed to implement their own application of a robot movement to simulate a simple task (applying glue, machining of a component).

3.3 Block 3—Trajectory Planning

In the last part students should already have a basic knowledge of the ROS system and MoveIt!, where they are able to plan and execute trajectory for industrial robot. The last part is focused on writing their own trajectory planner, where students will obtain basic principles of trajectory planning and understanding MoveIt backend.

The aim is to understand and implement algorithm for generation of smooth trajectories based on the polynomial path [22]. The first task is to determine simple polynomial functions (Eqs. 1–3), in which polynomial order depends on number of specific requirements e.g. initial and target position, velocities, accelerations or other required waypoints.

$$q(t) = a_0 + a_1 t + a_2 t^2 + \cdots + a_n t^n \tag{1}$$

$$\dot{q}(t) = a_1 + 2a_2 t + 3a_3 t^2 + \cdots + n a_n t^{n-1} \tag{2}$$

$$\ddot{q}(t) = 2a_2 + 6a_3 t + 12a_4 t^2 + \cdots + n(n-1) a_n t^{n-2} \tag{3}$$

$a_0, a_1, \ldots a_n$ are coefficients needed to be calculated in the second step. $q(t)$ is a joint value in time t and $\dot{q}(t)$, $\ddot{q}(t)$ are derivatives. The solution for Eqs. (1–3) could be found using C++ Eigen library, with which students have already been working in the second block.

The assignment is split into two parts and in both, trajectory is planned for standard 6-axis industrial manipulator. In the first part a trajectory is planned in the joint space and the goal is to create the trajectory shown in the Fig. 2, which contains motions of the first and the third joint.

The second part of the assignment is more difficult, as the goal is to determine polynomial functions in the cartesian space of robot's tool frame. The final trajectory is built from partial sequences of motions as Fig. 3 illustrates.

The first step is to determine polynomial functions for tool position $x(t)$, $y(t)$, $z(t)$ and rotations $rx(t)$, $ry(t)$, $rz(t)$.

The tool pose $P(t)$ with respect to the base coordinate system in cartesian space is given by Eq. (4), where T_{xyz} is homogenous transformation representing a pure translation and $Rx(t)$, $Ry(t)$ and $Rz(t)$. are homogenous transformations with pure rotations about x, y and z axis respectively.

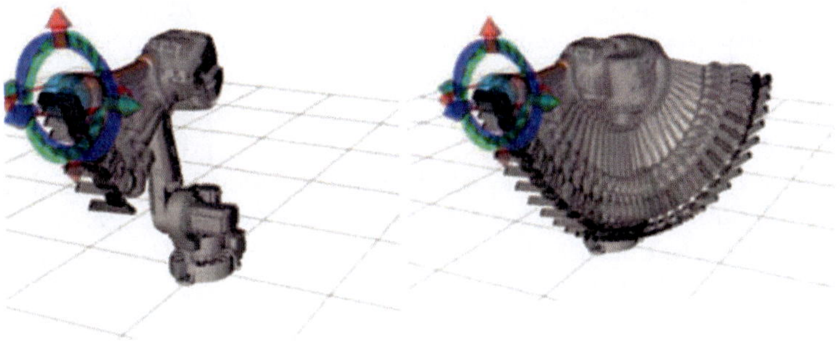

Fig. 2 The trajectory in the first part of the exercise

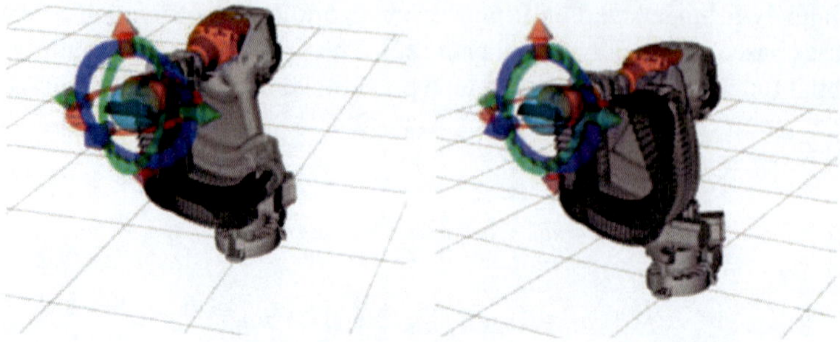

Fig. 3 The trajectory in the second part of the exercise

$$P(t) = T_{xyz}(x(t), y(t), z(t))R_x(rx(t))R_y(ry(t))R_z(rz(t)) \tag{4}$$

The second step is to convert the tool pose from the cartesian space to the joint space using inverse kinematics. The theory and principles of the inverse kinematics are explained in the lecture, but in the exercise, students are working with Ikfast library developed by Rosen Diankov in his dissertation thesis [23]. Ikfast outputs all valid robot joint configurations, and the last step is to determine the appropriate configuration depending on the first robot joint configuration. The most basic solution is to compute joint distance between the latest robot joint configuration and the inverse kinematics solution in a specific time sample (Eq. (5)). Where $q_{lj}(t_{i-1})$ is a value of a j-joint from the latest robot joint configuration and $q_{sj}(t_i)$ is a value of j-joint from the inverse kinematics solution.

$$d = \sum_{j=0}^{6} \left| q_{sj}(t_i) - q_{lj}(t_{i-1}) \right| \tag{5}$$

These are all the steps needed to complete the assignment. After completion, students should understand the principles of smooth trajectory generation. For simplicity, collision free trajectory is not required, however the student's solution could be applied as a custom initial trajectory of STOMP algorithm [24], which is implemented in MoveIt! STOMP provides the functionality of trajectory collision checking and if a collision is going to occur, the algorithm tries to generate random trajectories around the initial trajectory, where the resulting trajectory should be closest to the initial trajectory. The selection of the closest trajectory is done by a cost function from the cartesian constrained STOMP, which has been developed in our department and published in [25]. The connection between student's own algorithm and STOMP may give even deeper knowledge of ROS and MoveIt!

4 Course Evaluation

In this section we present the results from the evaluation questionnaire provided to the students after the completion of the course. The evaluation process was focused to analyse and understand feedback from students based on the three main goals:

- To find out and evaluate the benefit of the courses in understanding the theoretical knowledge from lectures (Q1–Q3).
- To find out and evaluate the practical benefit of the subject, and its impact on the motivation to learn the robotics theory and ROS (Q4–Q5).
- To understand how important, the students perceive working with ROS and whether they would prefer ROS or Matlab for practical demonstration of the theoretical knowledge (Q6–Q10).

From the existing course of 2022, 48 members out of a total of 66 participated in the evaluation process. The questionnaire was conducted voluntarily and anonymously. During the evaluation we have asked several questions where the respondents chose the most relevant answer in range from 1(worst) to 10(best) or YES/NO, to describe their feeling from the course and provide feedback. The questions are listed in the Table 3.

The results are shown below (Figs. 4, 5, and 6).

Table 3 The questionnaire provided to graduates of the course

N	Question
Q1	Evaluate the level of your knowledge of robotic manipulator control theory before completing the CIR course and lectures. (1-low, 10-high)
Q2	Evaluate the level of your knowledge of the theory of control of robotic manipulators after attending lectures on the CIR. (1-low, 10-high)
Q3	Evaluate the level of your knowledge of the theory of control of robotic manipulators after attending lectures and courses on the CIR. (1-low, 10-high)
Q4	Evaluate how much the exercises from the subjects you completed within the faculty have been focused on practice. (1-low focus, 10-high focus)
Q5	Evaluate how practice-oriented the CIR exercises are. (1-low, 10-high)
Q6	Do you have any practical work experience with programming outside of the academic sector or as part of your bachelor thesis?
Q7	How do you rate the use of ROS in the exercises from the CIR (regardless of whether it was difficult to learn to work with the ROS platform)? (1-very negative, 10-very positive)
Q8	Regardless of whether you do or do not want to work in robotics, write how do you think the knowledge of ROS concepts can help in robotic career. (1-not helpful, 10-very helpful)
Q9	Which of the listed environments (Matlab, ROS) do you perceive as more suitable for teaching the control of robotic manipulators and gaining experience in robotics. (Judge based on your previous experience)? (1-Matlab preference, 10-ROS preference)
Q10	Did you know ROS before completing the CIR course? (yes/no)

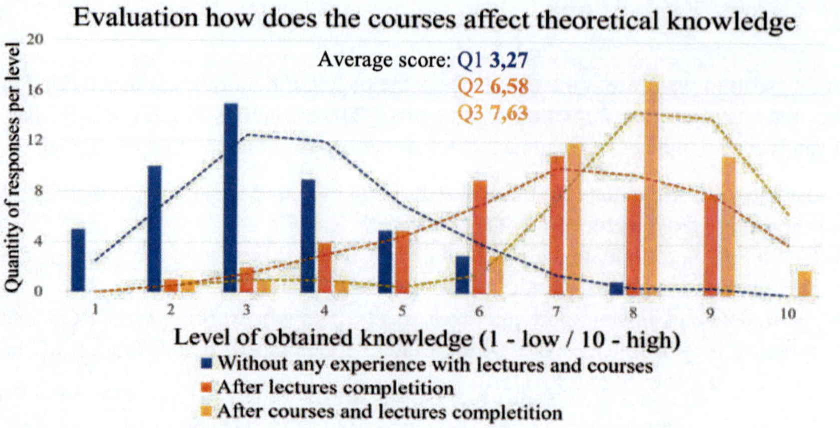

Fig. 4 Responses for the questions Q1–Q3

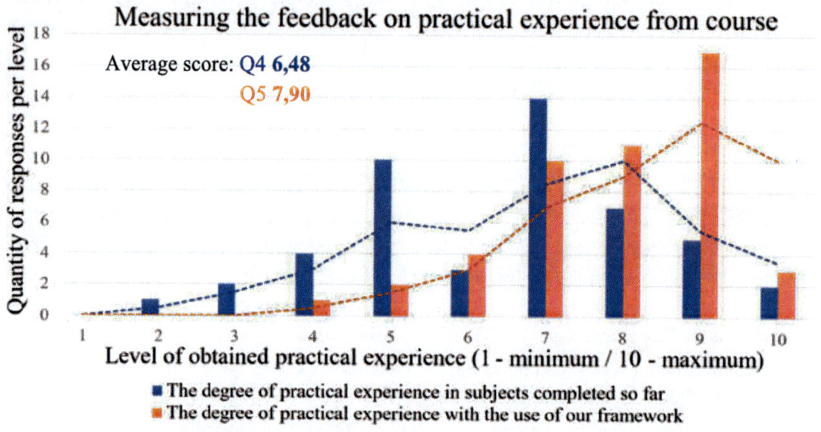

Fig. 5 Responses for the questions Q4–Q5

Based on the results, there is clear improvement in understanding the robotics theory after attending the lectures with the average improvement of 3.3 points, and even more understanding after completing the exercise framework, with further improvement by one point. Also, it could be seen that the graduates also perceive the subject as very practical with the average improvement of 1.42 points compared to other courses at STU. Regarding the sixth question, the questionnaire revealed that 61% of graduates have at least minimal practical knowledge of robotics from which 31% of students program as a part of their bachelor thesis and 30% program as a part of their job. The students evaluated the usage of ROS in the course to be positive with the average score of 8 out of 10 points. Furthermore, they have also expressed

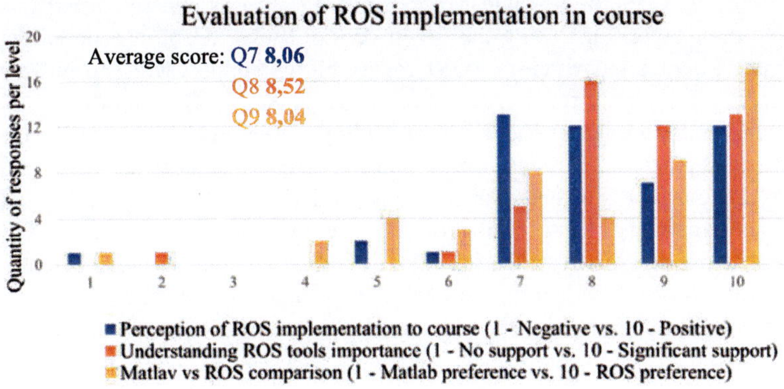

Fig. 6 Responses for the questions Q6–Q9

a positive attitude towards the usefulness of ROS and its concepts in robotics with the average score of 8.52 out of 10 points. Students also prefer ROS framework to Matlab where it is important to note that the respondents used the Matlab tool for most of the time during their studies and still chose ROS as a better framework, even though 60% of graduates stated, that they have not known about ROS prior to taking the CIR course in question 10. From this information we could assume that our framework and the tools chosen, provided sufficient and overall positive setup for teaching the robotics and understanding the robotics theory.

5 Conclusion

Our proposed educational framework using ROS promotes interactive learning of robotics, which directly connects the theory of industrial robotics with the development of real-world robotic application. Majority of ROS courses are focused on understanding the ROS architecture and usage of packages as Rviz, MoveIt, Gazebo and others. In contrast, there are many lectures of industrial robotics theory and learning that are usually done by pen and paper assignments, which may provide effective study of theory, but the connection with real world applications may be lacking.

The aim of our proposed framework is not only to teach the essential principles and ROS but encourage students to write their own algorithms based on the well-known theory to better understand the fundamental robotic concepts, while keeping in touch with real-world practice.

The educational framework was applied in the subjects a part of the CIR course at STU. Participants of the course confirm that they obtain clear improvement in programming skills and understanding of robotics theory and programming skills, backed by the presented evaluation questionnaire.

In the future work, authors would like to prepare exercises, where participants could apply their algorithms using their own control loops. Also, it is important to note that the course is written in ROS1 for which the support will end in May 2025, therefore it is highly recommended to switch to ROS2. On the other hand, vision systems and artificial intelligence have rapidly increased influence and usage in the industry at present. Therefore, the second course could be needed, which will include 3D cameras, collision detection methods, collision free motion planning and other advanced methods of industrial robotics. Moreover, we will prepare similar independent courses for another robotics fields as mobile robotics, drones (currently in the preparation), machine learning or intelligence human–robot interfaces which could be potentially applied and tested in the teaching process of appropriate subjects at STU.

Acknowledgements This paper was supported by the state grant for the support of Excelent Teams of Young Researchers (ETMVP) Complex Collaborative HRI Workplace.

References

1. Študijné plány bakalárskeho a inžinierskeho štúdia pre akademický rok 2022–2023. https://www.fei.stuba.sk/buxus/docs/studium_od_2022/SP_2022_2023_v5.pdf. Last accessed 20 Jan 2023
2. RRM/CIR course syllabus. https://is.stuba.sk/katalog/syllabus.pl?predmet=313716;jazyk=1;lang=en. Last accessed 20 Jan 2023
3. Robotic Manipulation MIT course syllabus. https://manipulation.csail.mit.edu/Fall2021/. Last accessed 20 Jan 2023
4. Robotic Systems Caltech syllabus. https://www.cms.caltech.edu/academics/courses/mecsee-134. Last accessed 20 Jan 2023
5. ETH Zurich robotics courses. https://rsl.ethz.ch/education-students/lectures.html. Last accessed 20 Jan 2023
6. Stanford Introduction to Robotics syllabus. https://cs.stanford.edu/groups/manips/teaching/cs223a/#assignments. Last accessed 20 Jan 2023
7. Introduction to Robotics KTH syllabus. https://www.kth.se/student/kurser/kurs/DD2410?l=en. Last accessed 20 Jan 2023
8. Robot Dynamics & Control TU Delft syllabus. https://studiegids.tudelft.nl/a101_displayCourse.do?course_id=61235. Last accessed 20 Jan 2023
9. Control of Industrial Robots syllabus Milan Polytechnic. https://www11.ceda.polimi.it/schedaincarico/schedaincarico/controller/scheda_pubblica/SchedaPublic.do?&evn_default=evento&c_classe=788169&polij_device_category=DESKTOP&__pj0=0&__pj1=f2d9c1c3221efed323572f6a1b1e4103. Last accessed 20 Jan 2023
10. Fundamentals of Robotics syllabus TU Vienna. https://tiss.tuwien.ac.at/course/courseDetails.xhtml?dswid=5738&dsrid=408&courseNr=376078&semester=2022W. Last accessed 20 Jan 2023
11. Robotics syllabus CVUT. https://cw.fel.cvut.cz/wiki/courses/b3b33rob/cviceni. Last accessed 20 Jan 2023
12. Industrial Informatics and Robotics syllabus TUKE. https://maisportal.tuke.sk/portal/studijneProgramy.mais. Last accessed 20 Jan 2023
13. Robotic Systems Toolbox. https://www.mathworks.com/products/robotics.html. Last accessed 20 Jan 2023

14. Esposito, J.: The state of robotics education: proposed goals for positively transforming robotics education at postsecondary institutions. IEEE Robot. Autom. Mag. **24**(3), 157–164 (2017)
15. Craig, J.: Introduction to Robotics, 3rd edn. Pearson Education, Inc. (2004)
16. ABI Research. https://www.abiresearch.com/press/rise-ros-nearly-55-total-commercial-robots-shipped-2024-will-have-least-one-robot-operating-system-package-installed/. Last accessed 20 Jan 2023
17. Robotic System Toolbox documentation. https://www.mathworks.com/help/robotics/. Last accessed 20 Jan 2023
18. ROS documentation. https://wiki.ros.org/. Last accessed 20 Jan 2023
19. Fogg, B.J.: A behaviour model for persuasive design. In: Proceedings of the 4th International Conference on Persuasive Technology (Persuasive '09), pp. 1–7. Association for Computing Machinery, New York (2009)
20. Cooney, M., Yang, C., Siva, A.P., Arunesh, S., David, J.: Teaching robotics with robot operating system (ROS): a behavior model perspective. In: Teaching Robotics with ROS, pp. 59–68. CEUR Workshop Proceedings, Tampere (2018)
21. Eigen homepage. https://eigen.tuxfamily.org/index.php?title=Main_Page. Last accessed 20 Jan 2023
22. Jazar, R.N.: Path planning. In: Theory of Applied Robotics: Kinematics, Dynamics, and Control, 2nd edn., pp. 729–789. Springer (2016)
23. Diankov, R.: Automated Construction of Robotic Manipulation Programs. Carnegie Mellon University (2010)
24. Kalakrishnan, M., et al.: STOMP: stochastic trajectory optimization for motion planning. In: 2011 IEEE International Conference on Robotics and Automation, pp. 4569–4574. IEEE (2011)
25. Dobiš, M., et al.: Cartesian constrained stochastic trajectory optimization for motion planning. Appl. Sci. **11**(24), (2021)

Simulator-Based Distance Learning of Mobile Robotics

M. Lučan, M. Dekan, M. Trebuľa, and F. Duchoň

Abstract The pandemic period enforced lecturers and students to face new challenges and search for alternatives to the standard presence educational process. This paper presents a simulator-based approach to teaching mobile robotics topics, such as localization, mapping, and navigation. The effectiveness of this approach is measured by various evaluations.

Keywords Kobuki · Robotics education · Navigation · Mapping · Odometry · Navigation · Laser rangefinder

1 Introduction

The field of robotics requires theoretical knowledge and practical hands-on experience with real robotic devices. Curricular projects are tasked to facilitate the conjunction of theoretical background and practical skills and to create quality university graduates. Mobile robotics offers numerous topics to be taught and resolved during lectures. A majority of projects require students' physical presence in laboratories to proceed with assignments. A recent worldwide pandemic situation was restricting standard educational forms and forcing lecturers to introduce alternative approaches to education. Students as well as educators faced a challenging time period in terms of the increased time for class preparations, decreased academic performance, and other aspects [1]. To compensate for the absence of hands-on work with real devices in laboratories, simulators may serve as a sufficient alternative. Working with simulated robots provides access from anywhere, thus this approach is highly suitable for education during a pandemic. Later, as the pandemic retreats, lecturers might keep this tool available for students as a testing environment to prove their code solutions before deployment on real robots.

M. Lučan (✉) · M. Dekan · M. Trebuľa · F. Duchoň
Department of Robotics and Cybernetics, Slovak University of Technology, Bratislava, Slovakia
e-mail: martin.lucan@stuba.sk
URL: https://urk.fei.stuba.sk/

R. Balogh et al. (eds.), *Robotics in Education*, Lecture Notes in Networks and Systems 747, https://doi.org/10.1007/978-3-031-38454-7_26

This paper presents an educational approach based on a Kobuki simulation used in Control of Mobile Robots course. Firstly, the Kobuki simulator is introduced as a digital twin of a physical robot. Subsequently, a paper briefly describes course objectives and assignments. Lastly, the effectiveness of the educational method is represented by the evaluation.

2 Kobuki Robot Simulator

Kobuki is a low-cost robotic platform intended mainly for educational and research purposes. The platform is broadly used in large amounts of research projects as well as educational classes. In some cases, it serves as a base for extension, e.g. Turtlebot 2 [2]. Its differential drive chassis is based on two actuated wheels and two support wheels. Numerous built-in sensors, such as bumper sensors, cliff sensors and wheel drop sensors ensure safe operation. A more detailed description is contained in the online datasheet [3] (see Fig. 1).

In the context of distance teaching, the physical robot platform needs to be substituted by a convenient simulator. For the purpose of the Control of Mobile Robots course (described in Sect. 3), Kobuki robots used during ordinary student attendance are replaced by a customized simulation environment prepared by lecturers. As the students have different programming backgrounds and the lecturers want to focus the lectures on algorithms themselves, they are provided with the simplest possible simulator to work with. Regularly used Robot Operating System was neglected, as it demands extra explanations and hands-on experience. Another benefit of the proposed simulator is the operating system independency, it is fairly easy to install and set up on Windows, Ubuntu, and MacOS. A significant advantage is UDP communication protocol support which enables the direct compatibility of the simulator and real robot. According to these requirements, a number of available simulators

Fig. 1 Kobuki platform with marked onboard equipment

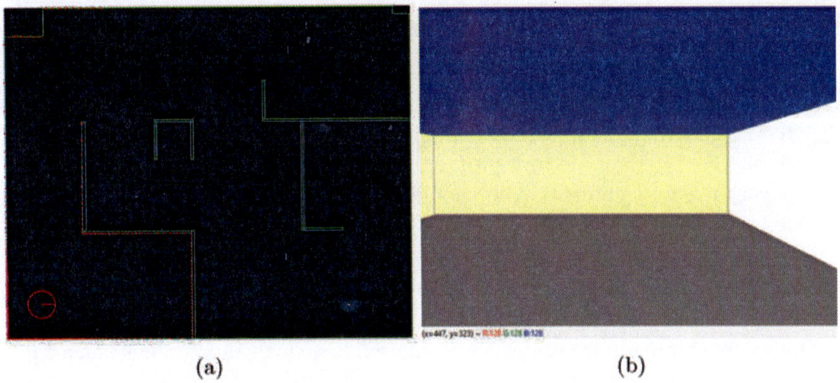

Fig. 2 Simulator Qt GUI window with robot placed in the environment. Red dots illustrate the laser rangefinder spatial data (**a**). The mono camera image stream may be used i.e. for visual odometry (**b**)

were excluded from consideration. Gazebo is natively not supported on Windows, and the installation is not trivial. V-REP [4] provides many useful functions and the installation is fairly straightforward, but the detailed setup of the robot turned out to be complicated and buggy. Webots [5] does not support UDP communication with third-party software.

Natively, the simulator is coded as a C++ language Qt-based framework. It implements Robot class with sensors and features mirrored from the physical Kobuki robot, i.e. the user can access simulated encoders and onboard sensors for further processing. To retain the simulator close to reality, the simulated laser rangefinder borrowed key parameters, such as minimal and maximal range, angular resolution, and noise characteristics from its real counterpart mounted on Kobuki. Moreover, to broaden its capabilities, the simulated robot is equipped with a mono camera (Fig. 2b) whose image stream might be used for e.g. visual odometry, object recognition, or image segmentation. Thanks to the fact that it is a custom simulator programmed by lecturers, it is possible to easily integrate new features if necessary.

The Kobuki simulator has a simple UI (see Fig. 2) as it does not contain any buttons or controls. The simulator immediately starts the necessary communication server after the startup. The simulator uses the same UDP protocol as the real mobile robot with the same data. During the pandemic, students used the simulator on their own local devices. To switch from a simulator to the real robot, one simply exchanges the real robot's IP address with the IP address of the local host in solutions.

3 Control of Mobile Robots Course

The course is a mandatory component for masters students attending Robotics and Cybernetics study program. The only prerequisite to take this course is to be intermediate in C++, Python, or Java programming language. The course is dedicated to first-year master's students who should have gained satisfactory programming knowledge during their bachelor's degree courses. The aim of the course is to teach students the theory and implementation of basic algorithms to localize the robot, build a map, plan the path to a goal, and successfully navigate the robot to a desired destination utilizing multiple algorithms described in Duchon's book [6].

During regular learning, students are required to code the assignments on a computer remotely connected to Raspberry Pi microcomputer located on each robot. Raspberry Pi forwards data from the robot and the laser rangefinder using the UDP protocol. Every student is provided with Qt developed C++ code base allowing them to easily communicate with the robot and its sensors. If preferred, one can use other programming languages, but lecturers do not ensure the support.

Due to pandemic limitations, from the spring term of 2020 the real robot has been replaced by a simulator that substitutes the real robot (Sect. 2). Classes took place in online meetings and students could have consulted their assignment progress during the reserved time. Kobuki robots were prepared in a lab for individual students to test their simulated algorithms (see Fig. 3).

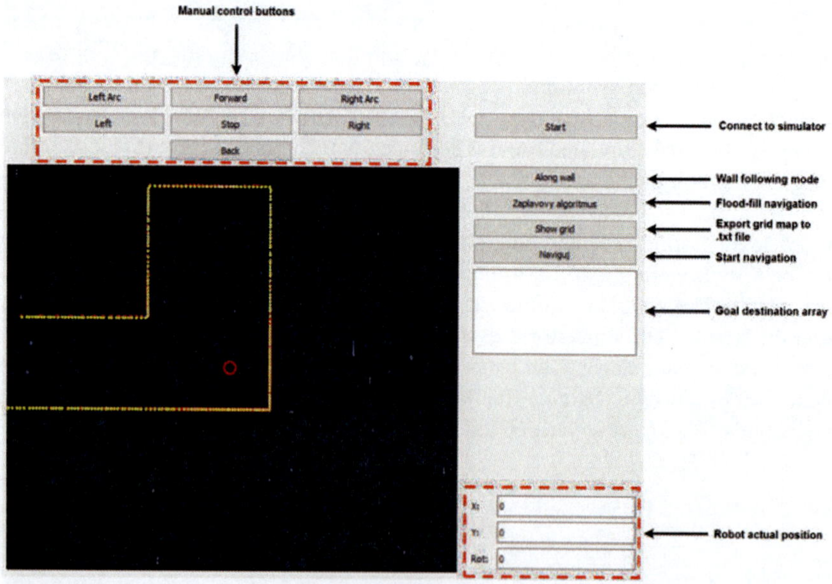

Fig. 3 An example of Qt-based GUI with button annotations created by a student

3.1 Assignment 1: Localization and Positioning

The task of localization is to determine the position w.r.t. startup position. The simplest method for localizing a wheeled robot with a differential chassis is wheel odometry. Odometry is a type of relative localization (it determines the position relative to a certain position), which determines the position increment based on the rotation of the wheels. Since the real and simulated Kobuki robot are equipped with a gyroscope, the user can increase the rotation precision and work in a so-called gyro-odometry mode.

The positioning part of the assignment is to get the mobile robot to the desired coordinates by controlling the speed of individual wheels. To achieve the desired position precisely, it is necessary to design a controller (P, PI or PID). The simplest solution to positioning is to divide robot movement for translation and rotation (i.e. the robot either moves forward or rotates around its vertical axis) and design a controller separately for each of these movements. Alternatively, students might implement more the advanced, conjugated control algorithm, which simultaneously control the translation and rotation components of robot's motion.

3.2 Assignment 2: Reactive Navigation

The navigation of the mobile robot ensures a collision-free passage of the robot through the environment without prior map knowledge. Reactive navigation utilizes current data from laser rangefinder and odometry and desired goal position. A laser rangefinder detects potential obstacles, while odometry serves for localization and positioning purposes Sect. 3.1. The complete path to the destination is unknown, although robot can determine the direction to the navigation goal.

Bug algorithms are a family of simple reactive navigation algorithms of whom only some use a laser rangefinder for their operation. Students in the assignment are tasked to make use of the tangent bug algorithm [7].

3.3 Assignment 3: Environment Mapping

The mapping in context of mobile robotics denotes a process of the process of creating a representation of the environment based on data from the robot's sensors. In other words, mapping is the fusion of robot position information with information on obstacles obtained from a laser rangefinder. The simplest representation of the environment is an occupancy grid, that divides the space into a finite number of elements (cells), where each element contains information about its occupancy (i.e. whether the obstacle is present or not).

The key feature in mapping algorithm, is a transformation of laser rangefinder range and bearing data into the coordinates of the robot, based on homogeneous transformation. If we neglect a movement in Z axis (i.e. we move on a plane), as well as rotations around the X and Y axes (the robot does not tilt in any direction while moving) is possible use a homogeneous transformation for 2 dimensions.

3.4 Assignment 4: Path Planning in a Map

The task of path planning is to generate an optimal global path for the mobile robot based on the existing environment map. For this purpose, the occupancy map from the previous assignment is being used. The proposed assignment suggests using the flood fill algorithm, which is typical for a "robot in maze" problem. The flood fill algorithm assigns the value to each node from the destination to an actual robot's position following certain rules (see Figs. 4 and 5).

4 Educational Method Evaluation

The effectiveness of this distance teaching method has been reviewed by students' evaluation surveys, submitted by the end of the course. The overall class quality and approach of lecturers has been rated by A-E grades. Students are also requested to point out concrete inadequacies and to give improvement suggestions. The evaluation summary of the period from the year 2017/2018 to 2021/2022 (Fig. 6) represents the

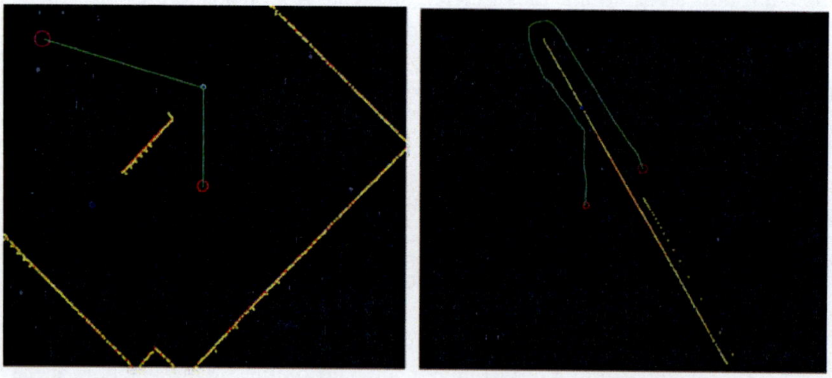

(a) (b)

Fig. 4 Robot navigation by means of a tangent bug algorithm. A robot driving around a short obstacle (**a**) and obstacle longer than a sensor range (**b**). In this case, an algorithm switches to a wall following mode

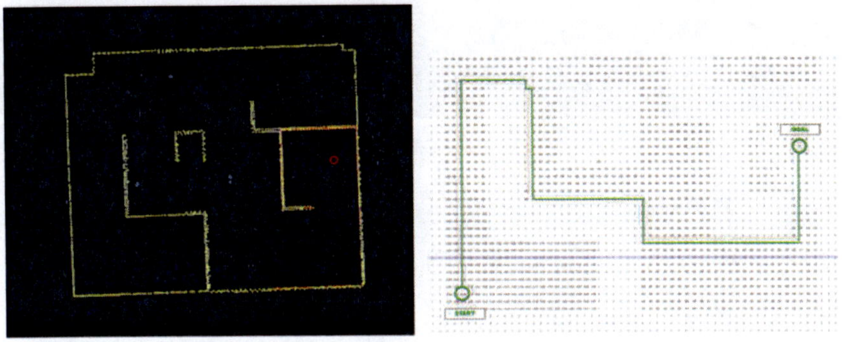

(a) (b)

Fig. 5 Robot builds a map as drives through a maze (**a**). A flood fill algorithm generates a global path based on a previously known map (**b**)

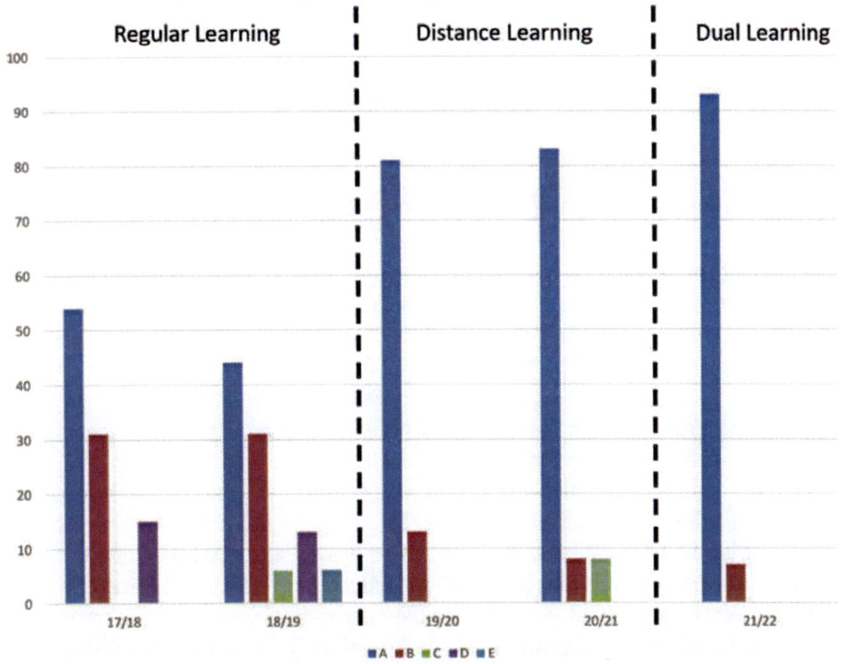

Fig. 6 Graphic visualization of course evaluations submitted by students sorted by years

development of the percentual ratio of evaluation grades. As it reveals, the distance form of education has not decreased the quality of the teaching process. On the contrary, the students' contentment even raised during distance classes. Arising from the evaluation, students appreciated the lecturers' personal approach as well as the opportunity to discuss their progress individually. During the school year 2021/2022,

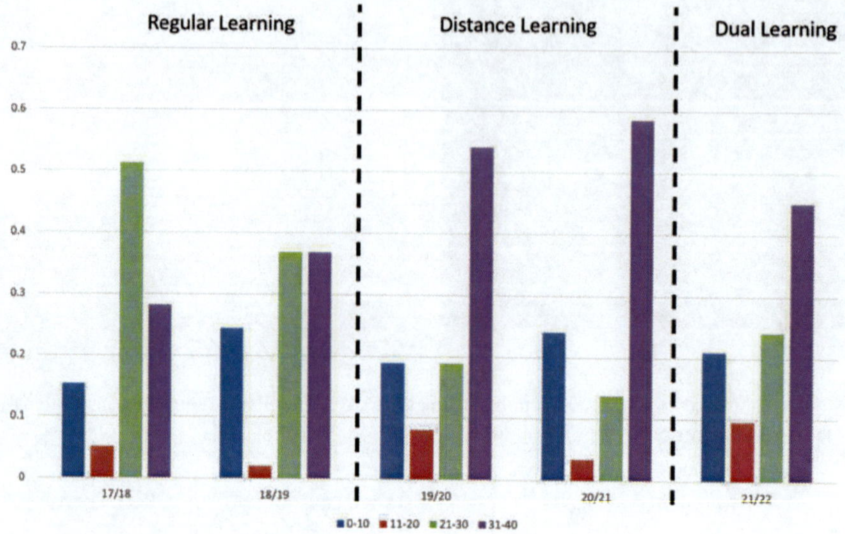

Fig. 7 Graphic visualization of students' evaluations sorted by years and points intervals normalized to 0–1

students had real hardware and a simulator available to test their own algorithms. This dual method proved to be the most effective, as evidenced by further improved evaluation of the subject.

During pre-pandemic years 2017/2018 and 2018/2019, the student evaluations contained complaints regarding the amount of time they had to spend in a laboratory. As the algorithms designed by the students could be verified on real robots only, it was necessary to spend extra time in the laboratory outside of classes. The pandemic years and the simulator introduced in the school year removed this time-consuming issue, but students lacked experience with real hardware.

Another parameter to consider in the teaching process evaluation is the distribution of course attendees' grades, as shown in Fig. 7. During the pandemic years, the course difficulty slightly decreased due to a lack of hardware issues as a result of which the student's score was visibly improved. In dual education, students achieved the best results thanks to the possibility to validate their algorithms from the comfort of their homes and then run them on robots. This offers initiative students a time-unlimited room for prototyping and improving their algorithms, which conduces to improved grades. Overall, the evaluation proves the efficiency of the proposed educational method.

Acknowledgements This research was supported by projects KEGA 028STU-4/2022 and UAVLIFE.

References

1. Vital, L.G.-G., Raul & Rodriguez, J., Paredes-García, W., Zamora-Antuñano, M.A., Elufisan, T., Rodriguez Resendiz, H., Cruz, M.: The impacts of COVID-19 on technological and polytechnic university teachers. Sustainability **14** (2022). https://doi.org/10.3390/su14084593
2. Turtlebot 2. https://clearpathrobotics.com/turtlebot-2-open-source-robot/. Accessed 15 Jan 2023
3. Kobuki datasheet. https://iclebo-kobuki.readthedocs.io/downloads/en/latest/pdf/. Accessed 15 Jan 2023
4. V-REP User Manual. https://coppeliarobotics.com/helpFiles/. Accessed 15 Jan 2023
5. Turtlebot 2. https://cyberbotics.com/. Accessed 15 Jan 2023
6. Duchoň, F.: Lokalizácia a navigácia mobilných robotov do vnútorného prostredia. STU, Bratislava. ISBN 9788022736466 (2012)
7. Mcguire, K., Croon, G., Tuyls, K.: A comparative study of bug algorithms for robot navigation (2018)

The Effectiveness of Educational Robotics Simulations in Enhancing Student Learning

Georg Jäggle, Richard Balogh, and Markus Vincze

Abstract This paper will compare two free versions of robotic simulation programs. All of them aim to offer an entrance into the world of robots in an easy way and with fewer resources. How can it enhance these student learning? This paper will compare two free versions of robotic simulation programs, IDE HedgeHog and Thymio Suite. One system is a web-based simulation in a web browser and does not need an installation. The other is a program on the computer; therefore, a program is to download and install on the computer. Both have different program languages, like text-based or visual-based programming languages, for different difficulty levels. A research design in which 124 students were asked to solve different tasks and responded to an online survey indicates that both simulations are user-friendly and helpful for beginners. They are useful tools to engage students in educational robotics activities or test programme codes in a virtual environment before they are transferred to physical settings.

Keywords Educational robotics simulation · Student learning · User-friendly

---

The authors acknowledge the financial support by the Sparkling Science 2.0 program, an initiative of the Austrian Federal Ministry of Education, Science and Research, under grant agreement no. SPSC01141.

G. Jäggle · M. Vincze
Automation and Control Institute, Vienna University of Technology, Vienna, Austria
e-mail: jaeggle@acin.tuwien.ac.at

M. Vincze
e-mail: vincze@acin.tuwien.ac.at

R. Balogh (✉)
Slovak University of Technology in Bratislava, Bratislava, Slovakia
e-mail: balogh@elf.stuba.sk

R. Balogh et al. (eds.), *Robotics in Education*, Lecture Notes in Networks and Systems 747, https://doi.org/10.1007/978-3-031-38454-7_27

1 Introduction

In recent years, educational robotics has gained significant attention as a teaching tool in Science, Technology, Engineering, and Mathematics (STEM) education. Studies have investigated the impact of educational robotics on students' motivation and learning outcomes. The current research on learning and motivation in educational robotics found that educational robotics can potentially improve students' engagement, motivation, and overall achievement in STEM subjects. They found that educational robotics can increase students' motivation and engagement in STEM subjects and improve their critical thinking and problem-solving skills. It has demonstrated effectiveness, particularly when students opt for advanced STEM studies, even if it goes against their parents' wishes [1]. In conclusion, the studies reviewed suggest that educational robotics has the potential to positively impact students' motivation and learning outcomes in STEM education [2].

Educational robotics has become an increasingly popular field as more individuals look to learn about robotics and gain hands-on experience programming robots. Educational robotics simulations provide a virtual environment for users to program and control robots without the limitations of physical robots. The simulations are accessible to individuals with prior knowledge of robotics and those new to the field. They provide a safe and affordable way for users to learn about robotics and gain practical experience programming robots. An extensive overview of the robotic simulators has been done in the past. There is a consensus that simulators should be favoured against conventional classes or remote labs or, at least, be presented as an optional complementary material [3].

Another important aspect of the simulation is the programming language used. The simulations often used the C/C++, Python and Java. Few provide a kids-friendly programming environment with simple commands or graphical (block) programming environments. A large group of simulators uses Matlab or ROS. Although this is quite standard at universities and industry, this is also not for beginners or kids friendly. Popular simulators like Gazebo, Webots, SimTwo or V-REP offer common characteristics. They had long-term development, support, and many robot variations (wheeled, legged, humanoids and drones). They allow a simulation in different environments and provide high fidelity due to using physics engines. Last but not least, all of them have 3D vision of the robot and the environment, and some offer the ability to use VR and AR techniques for even deeper immersion. Again, these great features can be counteractive for beginners who need to cope with too many variables at the beginning [4, 5].

The current state of educational robotics simulations is one of growth and innovation. Various options are available to suit different needs and skill levels, from visual programming environments, such as IDE Hedgehog,[1] to comprehensive programs, such as Thymio Suite,[2] which offer realistic representations of real-world robotics

---

[1] https://ide.pria.at/.

[2] https://www.thymio.org/.

environments. The simulations are increasingly being adopted in educational institutions to teach robotics and related fields looking to learn about robotics.

The field of educational robotics simulations faces several challenges, including integration into the existing curriculum, cost, lack of standardization, teacher training, and limitations in research. Integrating educational robotics into the curriculum can be challenging and require significant resources [6]. The cost of acquiring and maintaining educational robotics equipment and software can also be a barrier. Teachers may need more technical expertise and training to integrate educational robotics into their curriculum [7]. Furthermore, there is a need for more research to understand the impact of educational robotics simulations.

Section 2 explains the two compared simulations and shows examples of exercises. Further, Sect. 3 includes the objectives, research questions, evaluation design and results. Section 4 gives a conclusion and outlook.

2 Educational Robotics Simulations

This section presents two different educational robotics simulations for which no previous evaluation of their educational impact was done. Both simulations are free and without costs. The difference between these simulations is that the IDE Hedgehog simulation is web-based, and the Thymio suite simulation requires local installation on a personal computer.

2.1 Hedgehog Simulator

Hedgehog IDE is a web-based visual programming environment (see Fig. 1) for beginners. Its purpose is to do experiments with virtual robot programming using an integrated 2D simulator. It allows users to program robots using a drag-and-drop interface, making it accessible to users of all ages and skill levels. Programs can be created in JavaScript, Python, or a visual, Blockly-based language. All the user-written code is saved and executed in the browser. This frame frees the educator from problems with installing, updating and worrying about privacy. The Hedgehog IDE is available either online[3] for testing or can be installed locally. Its open-source code is available from Github.[4]

The IDE Hedgehog makes it possible to modify the included wheeled robot and offers a wide range of pre-prepared tasks and environments to explore. The highly customizable environment allows users to create their own tasks and environments [8, 9].

The robot and environment are also fully customizable and can be created and modified using the same technique, i.e. block programming. For each task, there can

[3] Hedgehog simulator is available at http://ide.pria.at.

[4] https://github.com/PRIArobotics/hedgehog-ide.

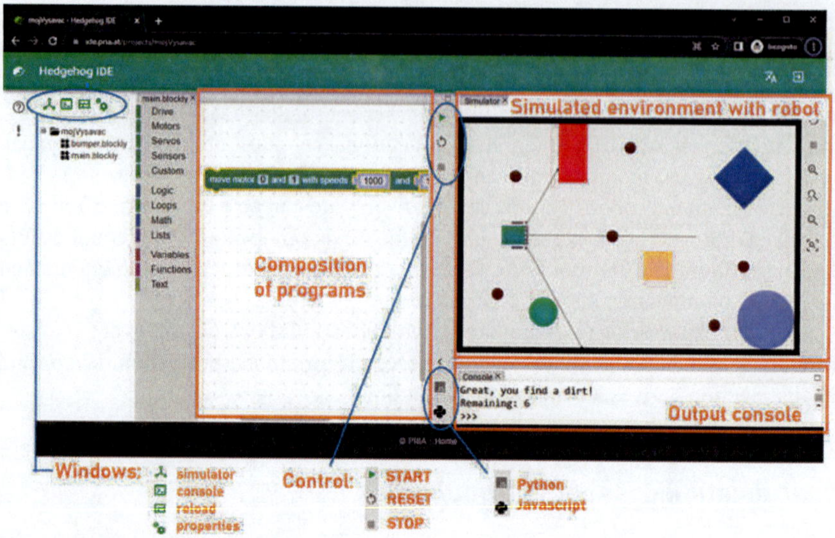

Fig. 1 Appearance of the IDE-Hedgehog simulator with the example of the simple vacuum cleaner problem and description of the control elements

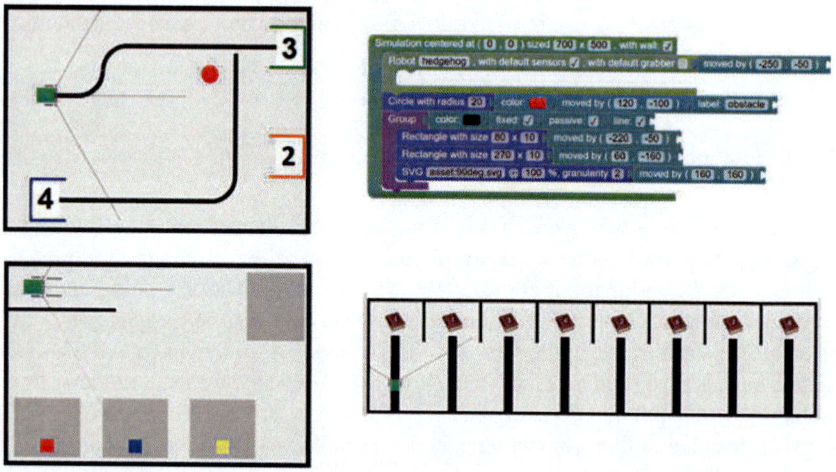

Fig. 2 Up left: simple line-following task, right: excerpt of the configuration for the line-following task. Bottom left: model of the simple restaurant. Right: book sorting task

exist completely different environments, and more robots can be added. The teacher can provide an internal script for interaction with objects and task evaluation, so the student can immediately see the progress. In Fig. 2 two different screenshots of its configuration are shown.

The environment offers the most used simulation features: different pre-prepared tasks, scenes with various objects, three programming languages and a simplified dynamics model. The simulator does not offer 3D scenes or immersive or virtual reality techniques. The IDE Hedgehog provides a user-friendly and accessible educational robotics experience, making it an ideal choice for schools and hobbyists. Its visual programming interface and various tasks and environments make it versatile, while its customizability options allow for a unique and personalized experience.

2.2 *Thymio Suite Simulation*

Thymio Suite is a software simulation tool designed for use in educational settings. It allows users to create and test programs for the Thymio robot in a virtual environment before uploading them to the physical robot. The suite is aimed at educators and students who want to learn about programming and robotics and can be used to teach a variety of concepts related to these fields. The simulation offers five different programming languages.[5] It includes a visual programming interface that allows users to create programs using a drag-and-drop interface. This makes it easy for students of all ages and skill levels to start programming the robot immediately. The suite also includes advanced debugging and code generation features, which are helpful for more experienced programmers. The visual programming interface supports block-based programming like visual programming (VPL) and allows the user to read simultaneously in the Aseba language. The suite is aimed at educators and students who want to learn about programming and robotics and can be used to teach a variety of concepts related to these fields (Fig. 3).

IDE Hedgehog and Thymio Suite are two different educational robotics simulation programs offering users unique experiences. IDE Hedgehog provides a visual and textual programming environment that allows users to program robots using a drag-and-drop interface, making it accessible to users of all ages and skill levels. On the other hand, Thymio Suite is a comprehensive simulation that provides a realistic representation of real-world robotics environments. IDE Hedgehog is more accessible and user-friendly with its visual programming interface, while Thymio Suite is more advanced with its realistic representation of real-world environments and comprehensive resources. Both offer users a manageable level of programming (VPL, blockly, scratch) and an advanced level of programming (Python, Aseba).

3 Evaluation Design and Results

This section presents the evaluation design with the objective and research question. The evaluation design aims to get the pros and cons of the simulations. The design answers the research question: How easily can users learn and operate with the

[5] https://www.thymio.org/products/programming-with-thymio-suite/.

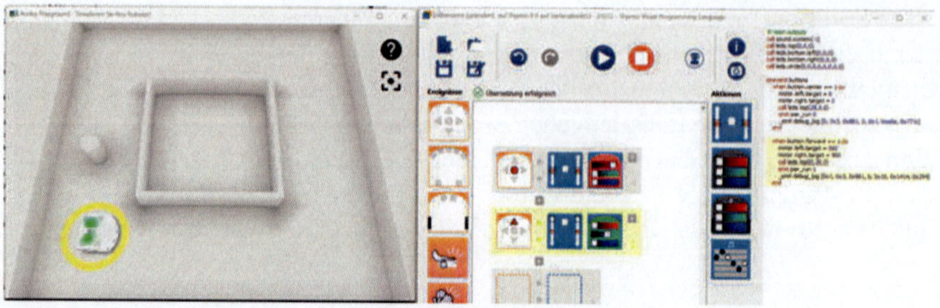

Fig. 3 Thymio suite simulator with the example of VPL-programming

Table 1 The exam questions

Nr.	Questions
Q1	What will be the robot's behaviour with the following program?
Q2	What will be the robot's behaviour with the following program?
Q3	What will be the robot's behaviour with the following program?
Q4	Which of the following conditions will ensure the proper navigation of the robot over the black straight?
Q5	Robotic vacuum cleaner...
Q6	The role of the microprocessor in the robotic device is...

simulation? The evaluation design includes 5-Likert scale questionnaires about how user-friendly it is on one side. Conversely, it was an exam about the content knowledge of programming operations with the simulator, like some topic-relevant questions to solve problems with the robots. We asked four programming-related questions and two additional general ones about robotics (Q1–Q6), as seen in Table 1. At the end of the evaluation, we allowed the participants to fill in their improvements and additional ideas with an open-question. The anonym answers were collected with an online survey after the lecture.

Evaluation of the Thymio suite simulator was based on ten answers from students in the 2nd year of vocational education teacher training at the teacher training University College of Teacher Education Vienna. All participants solved exercises with the simulation and answered the evaluation as an online survey six weeks later. Results show (see Fig. 4) the overall satisfaction with the environment and would also recommend for students.

The correct answers to topic-relevant questions were in the range of 60–90%. The most confusing question was probably about the command to follow a black line. For an example about the topic-relevant question, see Fig. 5 with VPL coding about the behaviour to stop and change the colour of the robot.

Evaluation of the Hedgehog simulator was based on the 114 answers, from students in the 1st year of Automotive mechatronics at the Slovak University of Technology in Bratislava. Results show (see Fig. 6) the overall satisfaction with the environ-

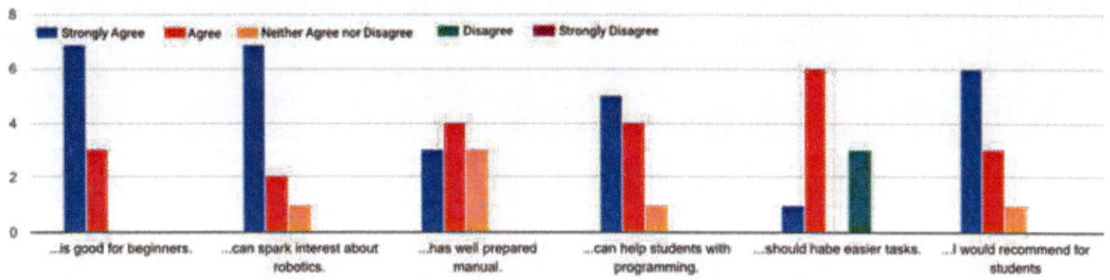

Fig. 4 Evaluation of the Thymio Suite simulation quality

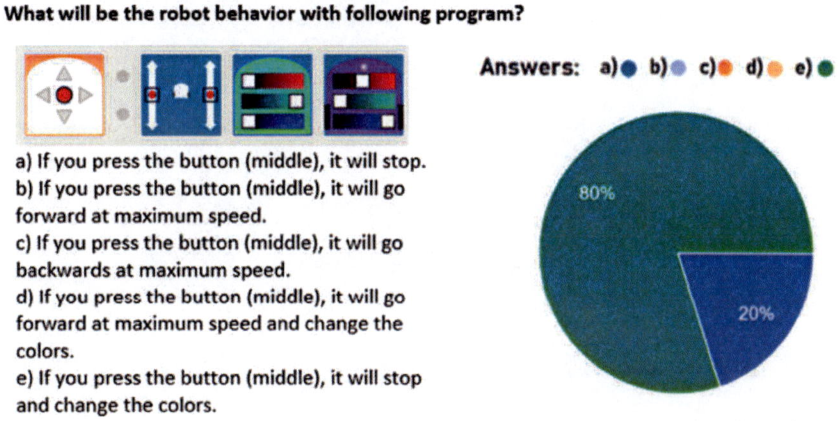

Fig. 5 An example exam question about coding and its evaluation with Thymio Suite

ment, but at a lower level than the Thymio one. We also asked whether they learned something new during the session with the simulated robot. The overall rating was positive, and students learned and understood some concepts in robotics. Of course, as this was only a simulation and not a real robot, they did not feel they could do the same tasks in the real world as they had no experience.

We added some robotics-related questions to the final exam to distinguish the students' feelings from the actual knowledge gained. The correct answers to topic-relevant questions were in the range of 62–92%. The most confusing question was probably not related to the robotics movement but to correctly understanding the repeat loop, as seen in Fig. 7.

The second group of questions was related to the role of sensors and microprocessors in robot construction. We achieved a success rate of 71% for the sensor question (Q5) and 51% only for the microprocessor one (Q6). An additional 33% was close enough to the proper answer, as seen in Fig. 8.

We also asked students for possible improvements and additional ideas for the simulator and its environment. Most often, students liked it as it was and had no complaints or ideas for improvement. They liked the simplicity and ease of programming using the block code. Conversely, they often complained about the simulation's built-

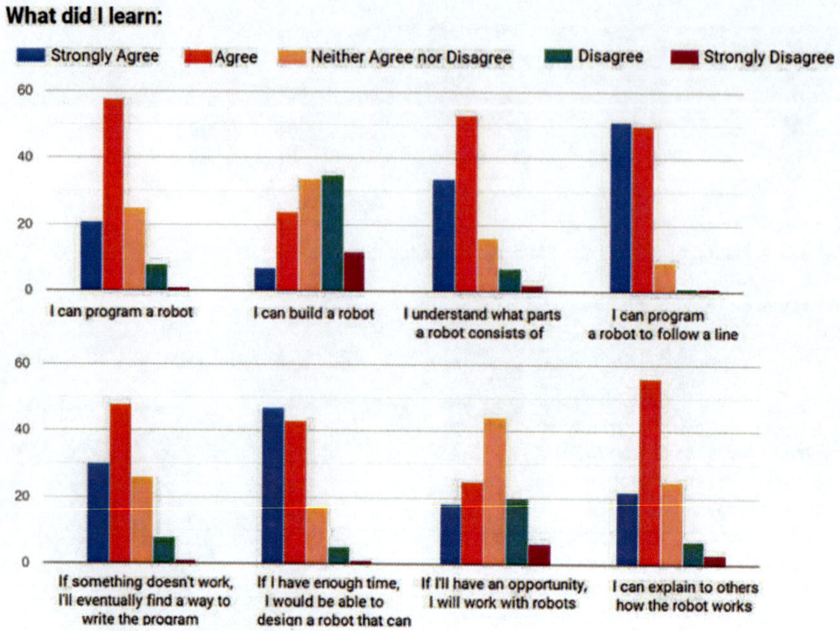

Fig. 6 Is the Hedgehog simulator good for beginners?

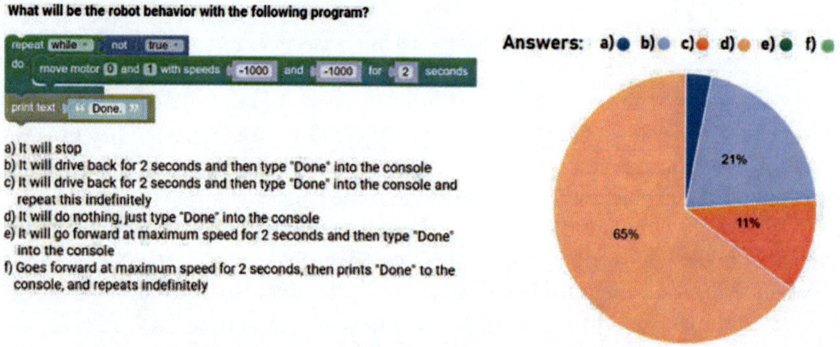

Fig. 7 An example exam question about coding and its evaluation with IDE Hedgehog

in randomness. The robot was sometimes randomly modifying its movements. As a result, no experiment was the same as the previous one (exactly as in real life). Some students were disappointed by it and needed to recognize the importance of sensors. A big group of students also asked for different simulation speeds (as it sometimes takes a long time to try and repeat their programs). As the environment is not yet finished, students also asked for more detailed documentation and help, which will support solving their exercises at home more accessible (in the case the teacher was unavailable).

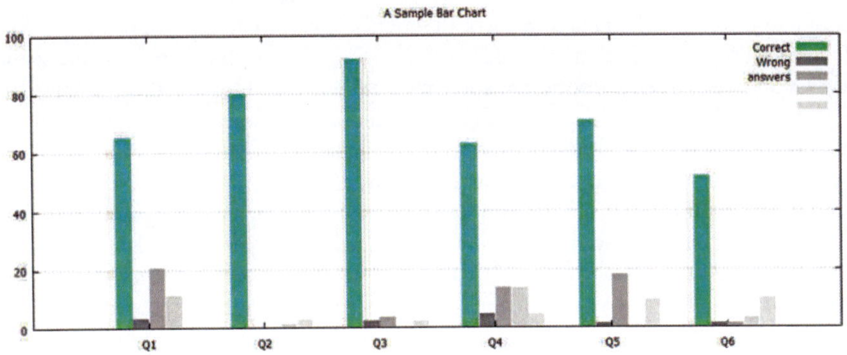

Fig. 8 Results about the exam questions evaluation

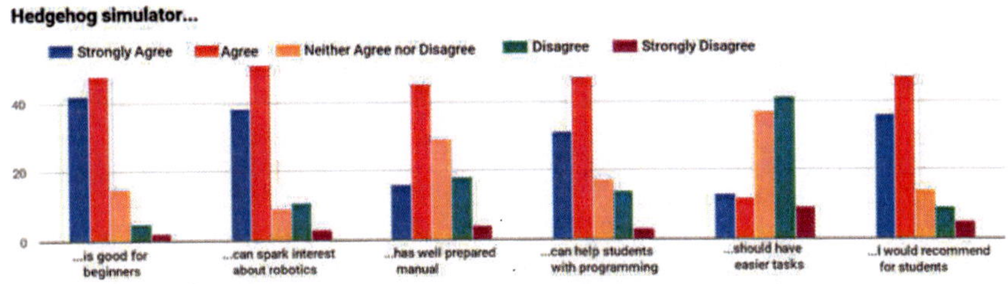

Fig. 9 Evaluation of the IDE-Hedgehoge simulator quality

We also asked whether they think this environment is suitable for young students and beginners. Results in Fig. 9 show that the simulation IDE Hedgehog is a user-friendly program, and can be recommended for students.

A further conclusion is that the virtual setting gave students a higher self-efficacy for performing in the robotics field [9] and students underline that in a study as an easy entry into the world of robotics and programming with VPL from thymio suite simulation [10].

4 Conclusion and Outlook

The study evaluation presents two different online tools for using educational robotics environments. The evaluation indicates that the simulations are an easy entry point for teachers and students because they have low-level programming languages for beginners and advanced-level programming languages for experts and do not need a budget. The comparison shows that both simulators are user-friendly and are an entry into the world of robots. Students' content knowledge about programming was evaluated a long time after the workshop setting. The answers show that most students remembered how robots work. This result leads to the conclusion that the simulation engages students to work with robots and will motivate others too. As the first step, a sustainable strategy could bring all students in touch with robots and, with the second step, offer the option to solve easy exercises and tasks with simulations. Further study could evaluate the limits of the simulations compared with real-world settings. Using robotic simulators for education means dealing exclusively with real robots is only sometimes necessary, which could have high costs. In some cases, such as distance learning, they are the only way students can gain basic programming and robotics experience. Therefore, simulators are a helpful tool that can save resources and help the educational process. This study shows the effect of the simulations, but how is it with effective teaching strategies? The following research could focus on the impact of the teaching strategy of problem-based learning related to the 5-step plan. The problems should be linked to world-life problems like SDGs.[6] Students develop sustainable products with robots in this context. The effect could be compared between real cases with robots and robot simulations.

References

1. Jäggle, G., Merdan, M., Koppensteiner, G., Lepuschitz, W., Posekany, A., Vincze, M.: A study on Pupils' motivation to pursue a STEM career. In: The Impact of the 4th Industrial Revolution on Engineering Education: Proceedings of the 22nd International Conference on Interactive Collaborative Learning, vol. 1 22, pp. 696–706. Springer International Publishing (2020)
2. Coufal, P.: Project-based STEM learning using educational robotics as the development of student problem-solving competence. Math. Multidiscip. Digit. Publ. Inst. **10**(23) (2022)
3. Teixeira, J.V., Hounsell, M.: Educational robotic simulators: a systematic literature review. Nuevas Ideas en Informática Educativa TISE **20**, 340–350 (2015)
4. Camargo, C., Gonçalves, J., Conde, M.Á., Rodríguez-Sedano, F.J., Costa, P., García-Peñalvo, F.J.: Systematic literature review of realistic simulators applied in educational robotics context. Sensors **21**, 4031 (2021). https://doi.org/10.3390/s21124031
5. Tselegkaridis, S., Theodosios, S.: Simulators in educational robotics: a review. Educ. Sci. **11**(1) (2021).https://doi.org/10.3390/educsci11010011
6. Alimisis, D.: Educational robotics: open questions and new challenges. Themes Sci. Technol. Educ. **6**(1) (2013)
7. Khanlari, A.: Teachers' perceptions of the benefits and the challenges of integrating educational robots into primary/elementary curricula. Euro. J. Eng. Educ. (Taylor & Francis) **41**(3) (2016)
8. Koza, C., Wolff, M., Frank, D., Lepuschitz, W., Koppensteiner, G.: Architectural overview and Hedgehog in use. In: Proceedings of the Robotics in Education (RiE 2017) Conference. Advances in Intelligent Systems and Computing, vol. 630 (2018). Springer, Cham. https://doi.org/10.1007/978-3-319-62875-2_21
9. Jäggle, G., Balogh, R., Koza, C., Lepuschitz, W., Vincze, M.: Evaluation of educational robotics activities with online simulations. In: Proceedings of the Austrian Robotics Workshop (ARW'21), Wien (2021)
10. Jäggle, G., Vincze, M.: A conceptual framework for educational robotics activities C4STEM: a virtual educational robotics workshop. In: Handbook of Research on Using Educational Robotics to Facilitate Student Learning. IGI Global (2021). https://doi.org/10.4018/978-1-7998-6717-3.ch011

[6] https://www.sdgwatch.at/en/about-sdgs/.

Machine Learning and AI

Artificial Intelligence with Micro:Bit in the Classroom

Martha-Ivon Cardenas, Lluís Molas, and Eloi Puertas

Abstract This project provides Artificial Intelligence (AI) learning activities based on problem-solving STEAM challenges. Focused on maker philosophy, students are initiated into coding and computer science, exploring its potential and combining robots, microcontrollers and artificial vision cameras. In that sense, students use various strategies that allow the camera to recognize a ball, a face, a colour or a card and also they use Machine Learning (ML) by teaching the robot what to detect. The project not only engages students to build creative solutions but also empowers them to start prototyping their own ideas through exploration.

Keywords Artificial intelligence · Problem-solving · STEAM · Artificial vision camera · Machine learning

1 Introduction

Currently, Artificial Intelligence (AI) has a natural connection with computational thinking (CT) and both have become fundamental skills for our students, immersed in a changing society where STEAM (science, technology, engineering, art, and maths) area plays a fundamental role. After reviewing some related articles that provide existing applications for teaching ML at secondary level [1–3] it was shown that micro:bit [4] kits do not provide ML capabilities on their own. On the contrary, to acquire such capabilities, these kits must be used in conjunction with other microcontrollers like an artificial vision camera and a robot. To cope with this, a

M.-I. Cardenas (✉)
Department of Computer Science, Universitat Politècnica de Catalunya, Barcelona, Spain
e-mail: mcardenas@cs.upc.edu

L. Molas
ESIC Business & Marketing School and RO-BOTICA, Barcelona, Spain
e-mail: lluis.molas@esic.edu

E. Puertas
Department of Matemàtiques i Informàtica, Universitat de Barcelona, Barcelona, Spain
e-mail: epuertas@ub.edu

R. Balogh et al. (eds.), *Robotics in Education*, Lecture Notes in Networks and Systems 747, https://doi.org/10.1007/978-3-031-38454-7_28

microcontroller-based project based on maker philosophy was developed, with the following objectives:

1. To combine CT, programming, creativity, innovation and the achievement of digital skills.
2. To acquire skills in critical thinking and collaboration, building the capacity and confidence to have ideas, share them and make them real.
3. The digital and technological empowerment of boys and girls with different learning capabilities.
4. To share didactic material with teachers and students of the educational community.

2 Methodology

Our project is based on the STEAM methodology, which focuses on structured learning that spans multiple disciplines, but does not emphasise any one discipline in particular, but rather the transfer of content between subjects [5]. To ensure the success of the proposed objectives, the project is developed in a framework where the teacher explains the contextualization of the challenge, offering a space for dialogue and joint reflection in the large group. It is at this point where guidance is given on the blocks that may be necessary and on the structure of the solution proposed by the students [6–8]. Concretely, the teacher has to take the role of guide, in no case directing the students to a single solution, but making them explore different options to achieve the same objective [9]. The phases of the methodology are described as follows:

Stage 0: Knowledge acquisition
Stage 1: Project start-up
Stage 2: Knowledge structuring
Stage 3: Knowledge application
Stage 4: Project evaluation and feedback.

Similarly, the features of an activity that incorporates such methodology are described below:

- It has to provide learning in the areas involved.
- The relationship between the different knowledge involved and their contribution to the resolution of the situation should be explicit.
- It has to involve a diversity of processes and languages.
- It must contribute to the awareness of what has been learned.

Finally, it is convenient to make an evaluation of which of these itineraries has been more optimal.

2.1 *Digital Competences*

The Technology curriculum, and more specifically the Information and Communications Technologies (ICT) curriculum, is focused on teaching new technologies, digital skills and behaviours using all kinds of material. That is why the objectives of this project fit with the structure and competencies of the digital curriculum [10]. Then, the competencies worked on during the development of this project are, from the digital field, mentioned below:

Competence 1. To select, configure and program digital devices according to the tasks to be performed: it is the main competence that is developed and worked when learning to program, encompasses all the skills acquired in the design, implementation and testing of the code using Microsoft MakeCode [11].

Competence 2. To use a basic image, sound and moving image editing applications for digital document production.

Competence 3. Participate in interpersonal communication environments and virtual publications to share information.

Competence 4. Perform group activities using collaborative virtual work environments and tools: throughout these two decades, there has been an increase in investment and popularity in software that allows collaborative work and file storage by sharing files from different accounts and devices.

Competence 5. To act critically and responsibly in the use of ICT curriculum, while considering ethical, legal, security, sustainability and digital identity aspects.

Thus, as referred in [12–14], digital competencies should therefore be ensured and have to allow the transformation of our students from consumers of digital products to creators.

3 Project Description

This project is part of the Programming course addressed to K-10 (15–16 years old) twenty-seven middle school students. It was carried out over an entire first semester, distributed in 3 one-hour sessions per week. It consists of building and coding simple robotic systems which integrate the micro:bit, the Cutebot robot and the hardware extension AI Lens. During the challenge, students were divided into groups of two or three depending on their own rhythms of work. Moreover, cooperation and help among equals were encouraged. The students' worksheets were useful to have a reference of the most relevant aspects of the challenge and can carry out their work autonomously. Furthermore, some extension proposals were given to personalize the different rhythms of work according to their individual needs.

3.1 Materials and Methods

Following the maker philosophy, some interconnectable elements mentioned previously in Sect. 3 were used for the design of the challenges. At the top of the materials, the micro:bit board was the protagonist allowing students to learn how to program physical devices. It is undoubtedly accessible and affordable for the whole academic community. In addition, it can be easily connected to the computer via USB. In full connection with the board, students start programming using the web editor Microsoft MakeCode. Thus, a programme is commonly downloaded in hexadecimal format and uploaded by just dragging it to the editor. Furthermore, it can be translated into javascript or (micro-)python languages. Finally, the AI lens is connected to provide the complement of ML used in the challenges. See Figs. 1 and 2 for an overall picture of the materials used in the challenges.

The micro:bit. This microcontroller board have a number of integrated sensors that allow obtaining data from the environment such as a temperature sensor, accelerometer, light intensity, compass, microphone, or buttons. Data can also be obtained from other external sensors that can be connected to a micro:bit board, such as humidity, atmospheric pressure, CO_2, water level, magnetic field, smoke detector, gas, alcohol and vapour, among others.

Microsoft MakeCode. This is the micro:bit programming environment that has extensions that allow us to work with other components from manufacturers that have created accessories for micro:bit, such as robotics kits. Furthermore, MakeCode provides a window into text-based coding, as the code editor can be switched into JavaScript or Python view to see what lies behind the blocks. Additionally, it has an extension named PlanetX AI lens useful for coding the camera.

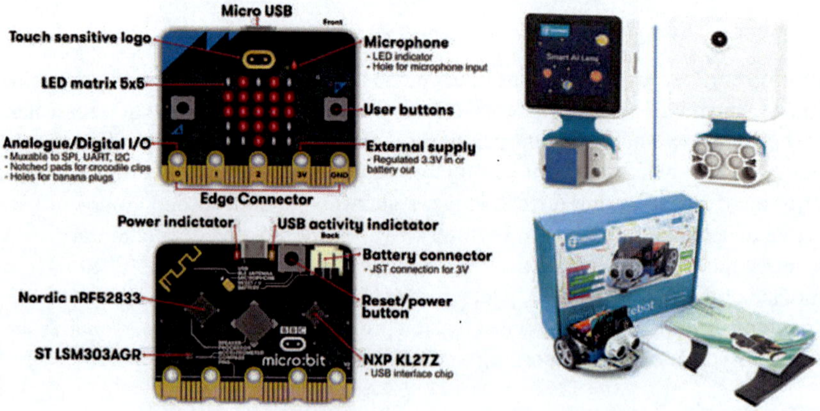

Fig. 1 Material applied in the project: micro:bit V2.2, smart AI lens and Cutebot

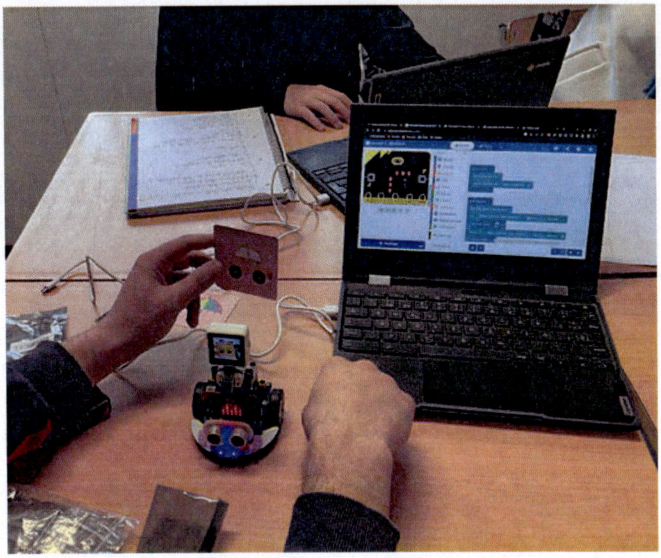

Fig. 2 Assembling and testing the first challenge: recognizing objects

Smart AI lens. Smart AI lens kit is an artificial vision camera compatible with 3.3–5 v micro:bit expansion board which can be programmed graphically. Manufactured by Elecfreacks [15], it contains two red and blue balls as well as a series of visual cards which can be applied in card identification, line-tracking, ball tracking, colour identification and face tracking. The origin point of the AI Lens is on the left top corner (0, 0), and the range of the coordinates X and Y is (0, 224). It needs to be connected to micro:bit via an RJ11 plugin.

Smart Car Cutebot. This robot is a rear-drive smart car driven by dual high-speed motors which include: an ultrasonic sensor, a distance sensor, four RGB LEDs, two line-tracking probes, and an active buzzer as the horn. Easily programmable with the micro:bit board and being compatible with the smart AI lens, it contributes to performing all the challenges of the project.

Python and microPython. The structure and syntax of a microPython program are the same as a regular Python program. Also, a number of the Python standard libraries can still be used but have been "micro-ified" to run on less powerful hardware.

3.2 Project Challenges

In this project we propose several activities sequenced incrementally, described below. The initial ones will help students to become familiar with the micro:bit environment while actively discovering the possibilities it presents. These challenges are

Table 1 A summary of the challenges and their respective filename from our public repository

Challenge	Filename
1	microbit-00_ReconociendoObjetos.hex
2	microbit-02_VidaSaludable.hex
3	microbit-03_Residuos.hex
4	microbit-04_Recogepelotas.hex
5	microbit-05_EnGuardia.hex
6	microbit-07_Carteles.hex
7	microbit-06_Detencion.hex
8	microbit-08_Pista.hex
9	microbit-01_ReconociendoObjetos.hex
10	microbit-09_Cara.hex
11	microbit-10_Heliotropismo.hex
12	microbit-11_Aprender.hex
13	microbit-12_Nube.hex

choosed and partially adapted from the program repository provided in the classroom and available for all the educational community in a public repository [16]. After each practice, the level of difficulty is increased and the focus shift towards textual programming. Overall, challenges were implemented by the students starting from a simple design of the robot structure and code. Then, they incrementally added new functionality to the code, enabling efficient computational solutions. After writing their code, students can pair their micro:bit to the computer via USB and download it to the robot.

Before you start programming, the first thing to do in all cases is to add the Elecfreaks planetAI-X extension. Next, the AI camera needs to be initialised and the mode for capturing an image is also configured in the MakeCode editor. To do this, it is necessary to go to the vision module and select the face or object recognition function as appropriate. Table 1 shows the files created for each challenge which can be download from our public repository and dragged into Microsoft MakeCode editor.

Challenge 1. Recognizing objects. It recognizes each card with the AI camera: cars, umbrellas, cats, and boats by activating the Card recognition function and displaying an icon on the LED matrix that represents the detected shape (see Fig. 3 left).

Challenge 2. A healthy lifestyle. We encourage healthy eating and avoid car journeys in favour of physical exercise so that if we choose to eat fruit, and not drive to school, the icon on the LED matrix will be modified. Then, as shown in Fig. 3 right, this real situation is simulated during the challenge.

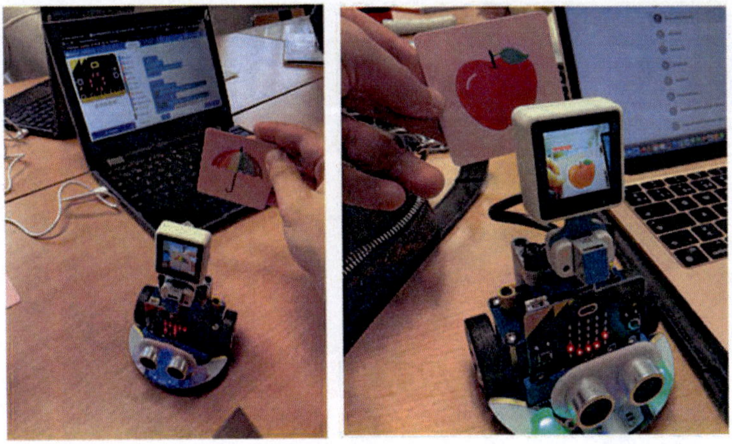

Fig. 3 Challenge 1: recognizing objects (left). Challenge 2: a healthy lifestyle (right)

Challenge 3. Waste separation. The impact of water pollution and how we can identify waste in order to remove it (See Fig. 4).

Challenge 4. The robot collects balls. It is time to collect the balls, locate their position and come to pick them up. The robot moves forward to pick up the ball. The AI camera is programmed on the y-axis (range from 0 to 200) indicating that when it is at an intermediate value it indicates that the ball is just below the robot (See Fig. 6 left and Fig. 5).

Challenge 5. On guard!!!. In fencing it is very important to control the distance of the opponent to be able to make a winning attack: we have worked with the y-axis instead of the motors, we have programmed them to go in the opposite direction (change of direction). We have added some Lego accessories in the form of guides to be able to pick up the balls.

Challenge 6. Vehicle license plates. Imagine that all vehicles with a certain license plate can pass the toll for free. We need to identify 3 different cards. To do this, we create variables that store the correct order of the letters of the license plate we need to detect. To do this, we will use logical operators and then reset the values to 0. As can be seen in Fig. 6 on the right side, many positions of three cards are tested and still it works.

Challenge 7. Intelligent stop. The automatic braking system is an aid to avoid accidents when an indicated signal is detected. We will do this with a stop card and change the colours of the LEDs.

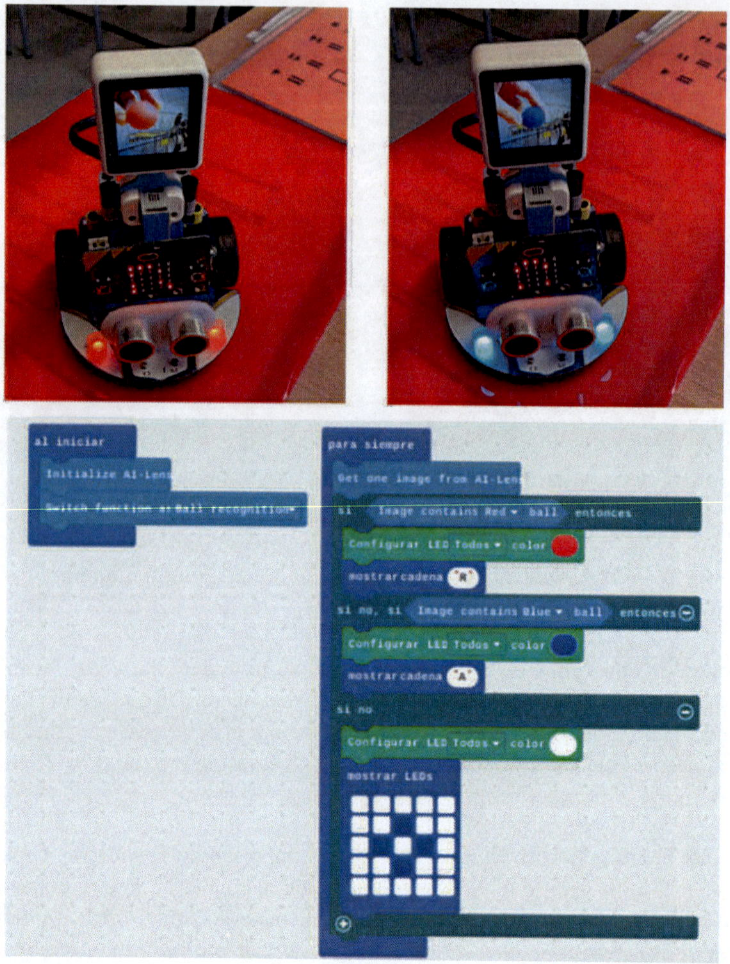

Fig. 4 Challenge 3: final assembling and code

Challenge 8. Attentive to the signs. We can program our vehicle to circulate identifying the visual signs of its surroundings. For this, we will use the LEDs indicating forward or backwards as well as the directional cards. It is important to incorporate 1500 ms pauses so that the robot does not continuously make the previous movement because the AI camera is continuously detecting and thinking that there is still a card (Fig. 7).

Challenge 9. Without losing track. Autonomous vehicles can detect the trajectory to be followed. We cannot lose "sight" of the trajectory. If it goes to the left we will

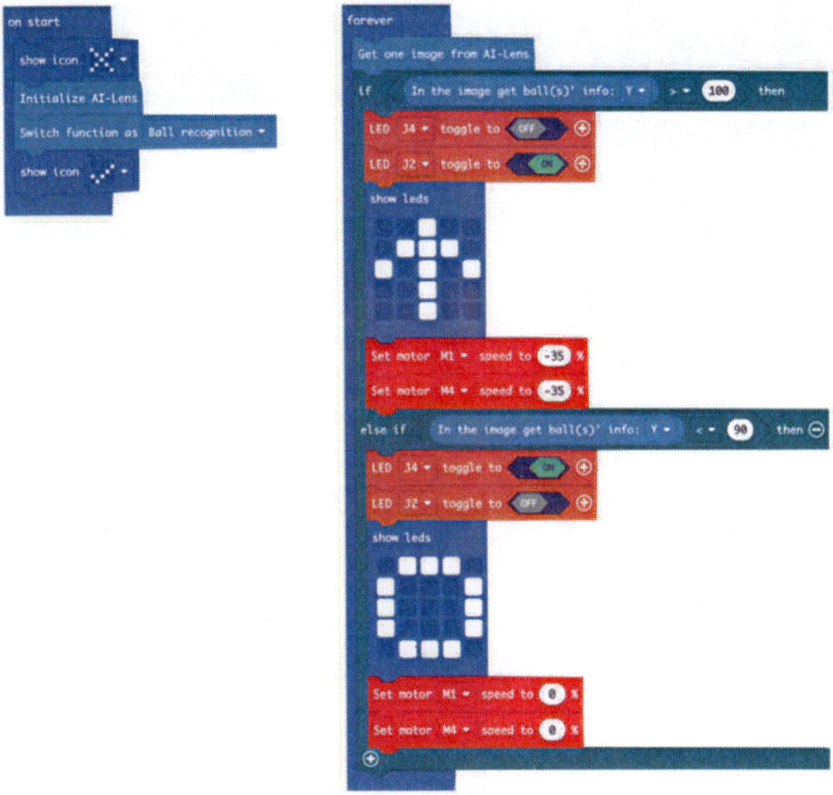

Fig. 5 Challenge 4: robot collects balls (code)

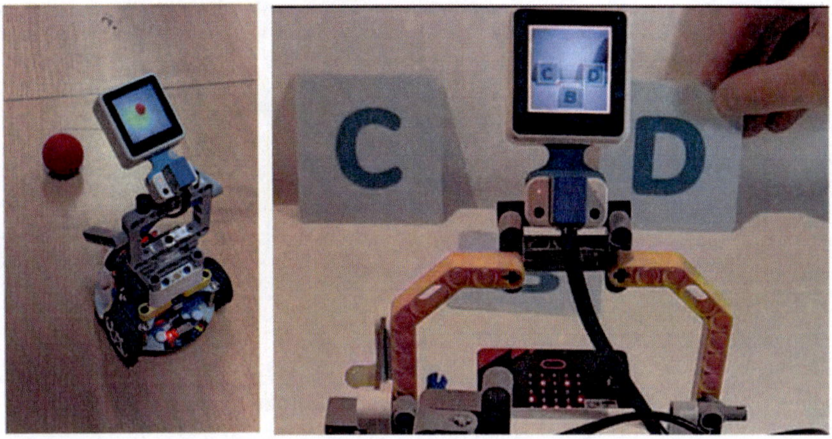

Fig. 6 **Assembling**. Challenge 4: robot collects balls (left). Challenge 6: vehicle license plates (right)

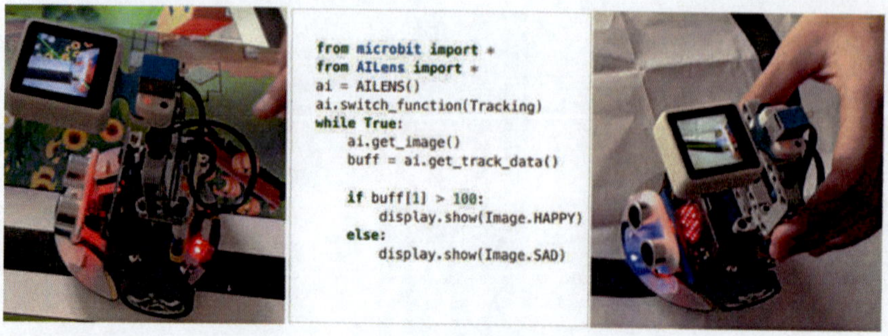

Fig. 7 Challenge 9: a microPython code can be observed at the centre whereas a robot performance is tested in two different panels

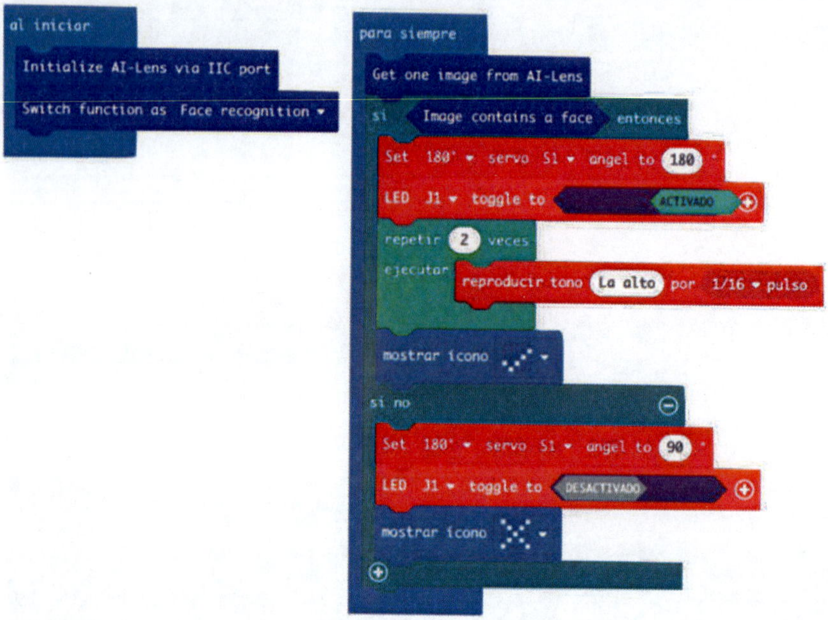

Fig. 8 Challenge 10: facial recognition (code)

increase the speed of the right motor and vice-versa. The AI camera shows a green line to help guide the robot.

Challenge 10. Facial recognition. It is useful in the way it provides comfort and safety when you try to simulate an effective access control. Then, the micro:bit board will activate a buzzer when the smart AI lens detects a human face (Fig. 8).

Challenge 11. Heliotropism. The sunflower has positive heliotropism, which means that it rotates according to the position of the Sun. Then, the robot will rotate according to the position of the ball.

Challenge 12. Respect signals. The AI Lens recognizes the "forward", "turn left", "turn-right" and the "stop" cards. The robot performs the same movements as those depicted in the cards.

Challenge 13. Learning an object. The AI camera will learn to reduce errors by assigning a matching value through our readings. We clean all the identifiers and with button A we activate the camera.

3.3 Self-assessment Rubric

Each of the evaluation criteria has been associated with its relationship to the most relevant competencies previously mentioned in Sect. 2.1.

Figure 9 describes the self-assessment rubric carried out in the project. Then, all the results derived from its application provided pieces of evidence that the majority of the challenges were generally developed and implemented with success. In agreement, Fig. 10 describes the rubrics from the x-axis whereas the y-axis shows the percentage of students' achievement. Consistent with our expectations, it states that the majority of the students achieved a good knowledge in programming blocks (level 2) whereas the category of MakeCode environment had reached the best expert results. Arguably, just a few students didn't achieve programming skills. However, students had some issues with the micro:bit as it was not as flexible as they would like it to be in terms of functionality.

4 Conclusions and Future Work

Programming the micro:bit using AI in the classroom was not an easy task. However, not only was explained to the students that AI combines algorithms to make computers and robots look more human-like, but also that there were different types of AI with a variety of applications. The project was designed to combine a sort of integrated system that allows students to assess their programs through physical performance that they can see and touch. Having explored many lines of actions based on the maker philosophy, that system was the most accessible option, with

Self-evaluation Rubric			
Name and surname:			**Course**
Statements	**Level 1**	**Level 2**	**EXPERT**
Knowledge of micro:bit	I am capable of recognizing some of the components of the micro:bit board.	I am able to recognize most components of the micro:bit board and I know the operation of some of them.	I am able to recognize all components of the micro:bit board and I know how it works.
The Microsoft Make Code programming environment	I had difficulty understanding the operation of the program and I had to ask for help.	I have understood the general operation of the programme, but I do not yet use it quickly.	I have had no difficulty in understanding the operation of the programme and I use it easily.
Programming blocks	I use some programming blocks, but I don't understand their function.	I use most of the programming blocks worked on the module, although I don't understand the function of some of them.	I use all the programming blocks worked on the module and I understand its function.
Challenges	It has taken me hard to understand and carry out the challenges of the module.	I've made most of the module's challenges with ease.	I have carried out all the challenges of the module with ease and have resolved the proposals for enlargement and improvement.
Problem-solving	I build the programs by copying the blocks that we worked on the module, but I have difficulty understanding their operation and resolving the problems.	I build the programs using the blocks worked in the module. I solve the problems by looking at the errors and by debugging the programme.	I am thinking of the answer and I am building the programmes on the understanding of the blocks worked in the module, adding proposals for enlargement and improvements.
General evaluation	It took me a long time to understand the contents of the module and I had to ask for help.	Some of the sessions cost me more than others and I needed some help.	I have been able to follow the sessions without difficulty, and have been autonomous in resolving my doubts.

Fig. 9 Template of the self-assessment rubric used in the project

an online code editor in which to work from visual and textual programming at the same time. Interestingly, strong connections between challenges were found, providing new opportunities for innovation in the project. As future work, we would like to explore storing data on the Cloud through the integration of AI models trained with Teachable Machine into micro:bit, which means a micro:bit connected to a web page in order to share our data or even use another data all across the world.

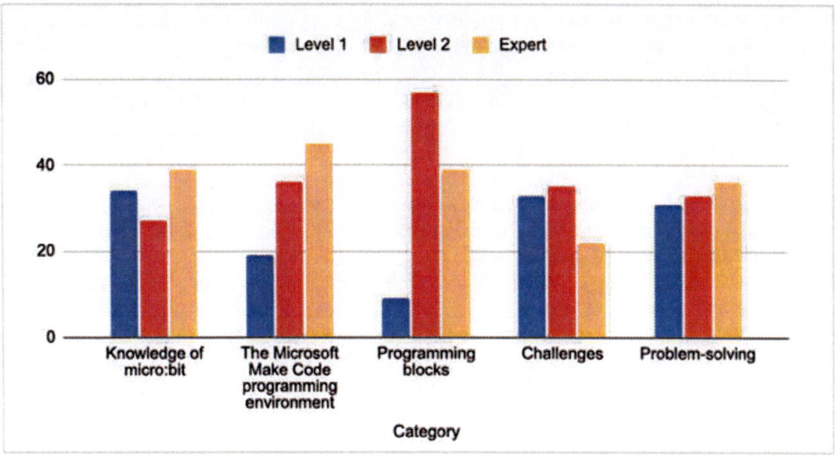

Fig. 10 Average results of the self-assessment rubric: X-axis refers to the category and Y-axis refers to the percentage of achievement

Acknowledgements The project leading to this application has received funding from the MCIUN—Ministerio de Ciencia, Innovaci6n y Universidades through the grants PID2019-104285GB-I00 and PID2019-105093GB-I00

References

1. Karalekas, G., Vologiannidis, S., Kalomiros, J.: Teaching machine learning in K-12 using robotics. Educ. Sci. **13**, 67 (2023). https://doi.org/10.3390/educsci13010067
2. Bellas, F., Guerreiro-Santalla, S., Naya, M., et al.: AI curriculum for European high schools: an embedded intelligence approach. Int. J. Artif. Intell. Educ. (2022). https://doi.org/10.1007/s40593-022-00315-0
3. Llamas L.F., Paz-Lopez A., Prieto A., Orjales F., Bellas F.: Artificial intelligence teaching through embedded systems: a smartphone-based robot approach. In: Robot 2019: Fourth Iberian Robotics Conference. ROBOT 2019. Advances in Intelligent Systems and Computing, vol. 1092. Springer (2020). https://doi.org/10.1007/978-3-030-35990-4_42
4. Micro:bit Educational Foundation Homepage. https://microbit.org/. Accessed 2 Jan. 2023
5. Lozada, R., Escriba, L., Granja, F.: MS-Kinect in the development of educational games for preschoolers. Int. J. Learn. Technol. **13**(4), 277–305 (2018)
6. Kvaššayová, N., et al.: Experience with using BBC micro:bit and perceived professional efficacy of informatics teachers. Electronics **11**(23), 3963 (2022)
7. Lu, S.Y., Wu, C.L., Huang, Y.M.: Evaluation of disabled steam-students' education learning outcomes and creativity under the UN sustainable development goal: project-based learning oriented steam curriculum with micro:bit. Sustainability **14**(2), 679 (2022)
8. Pech, J., Novak, M.: Use Arduino and micro:bit as teaching platform for the education programming and electronics on the stem basis. In: V International Conference on Information Technologies in Engineering Education (Inforino) (2020)

9. Lytle, N., Cateté, V., Boulden, D., Dong, Y., Houchins, J., Milliken, A., Isvik, A., Bounajim, D., Wiebe, E., Barnes, T.: Use, Modify, Create: comparing computational thinking lesson progressions for STEM classes. In: Proceedings of the ACM Conference on Innovation and Technology in Computer Science Education, pp. 395–401, Aberdeen, UK (2019)
10. Lee, I., Martin, F., Denner, J., Coulter, B., Allan, W., Erickson, J., Malyn-Smith, J., Werner, L.: Computational thinking for youth in practice. ACM Inroads **2**, 32–37 (2011)
11. Microsoft MakeCode Homepage. https://makecode.microbit.org/. Accessed 2 Jan. 2023
12. Cederqvist, A.M.: An exploratory study of technological knowledge when pupils are designing a programmed technological solution using BBC micro:bit. Int. J. Technol. Des. Educ. **32**(1), 355–381 (2020)
13. Cederqvist, A.M.: Designing and coding with BBC micro:bit to solve a real-world task—A challenging movement between contexts. Educ. Inf. Technol. **27**(5), 5917–5951 (2021)
14. Shahin, M., et al.: How secondary school girls perceive computational thinking practices through collaborative programming with the micro:bit. J. Syst. Softw. **183**, 111107 (2022)
15. Elecfreaks Homepage. https://www.elecfreaks.com/elecfreaks-smart-ai-lens-kit.html. Accessed 6 Mar. 2023
16. Challenges micro:bit code repository. https://drive.google.com/drive/folders/12aZtJ2Jl2jUf4_nW34MU2Dp5Sh9HA4as

Introducing Reinforcement Learning to K-12 Students with Robots and Augmented Reality

Ziyi Zhang, Kevin Lavigne, William Church, Jivko Sinapov, and Chris Rogers

Abstract As artificial intelligence (AI) plays a more prominent role in our everyday lives, it becomes increasingly important to introduce basic AI concepts to K-12 students. To help do this, we combined physical robots and an augmented reality (AR) software to help students learn some of the fundamental concepts of reinforcement learning (RL). We chose RL because it is conceptually easy to understand but has received the least attention in previous research on teaching AI to K-12 students. We designed a series of activities in which students can design their own robots and train them with RL to finish a variety of tasks. We tested our platform with a pilot study conducted with 14 high school students in a rural city. Students' engagement and learning were assessed through a qualitative analysis of students' behavior and discussions. The result showed that students were able to understand both high-level AI concepts and specific RL terms through our activities. Also, our approach of combining virtual platforms and physical robots engaged students and inspired their curiosity to self-explore more about RL.

Keywords Reinforcement learning · Augmented reality · Educational robot · K-12 education

Z. Zhang (✉) · J. Sinapov · C. Rogers
Tufts University, Medford, MA 02155, USA
e-mail: ziyi.zhang@tufts.edu

J. Sinapov
e-mail: jivko.sinapov@tufts.edu

C. Rogers
e-mail: chris.rogers@tufts.edu

K. Lavigne
Hanover High School, Hanover, NH 03755, USA
e-mail: kevin.lavigne@dresden.us

W. Church
White Mountain Science, Inc., Littleton, NH 03561, USA
e-mail: wchurch@whitemountainscience.org

R. Balogh et al. (eds.), *Robotics in Education*, Lecture Notes in Networks and Systems 747, https://doi.org/10.1007/978-3-031-38454-7_29

1 Introduction

Artificial Intelligence (AI) is progressively transforming the way we live and work. Therefore, it is increasingly important to introduce AI concepts to K-12 students to build familiarity with AI technologies that they will interact with. Reinforcement Learning (RL), a sub-field of AI, has been demonstrated to positively contribute to many fields, including autonomous driving [1], control theory [2], chemistry [3], etc. Recently, ChatGPT, a chatbot fine-tuned with RL and supervised learning, has led to extensive discussions in society [4]. For students, basic concepts of RL are intuitive and attractive to learn since it is similar to our cognition of nature of learning [5]. However, most current platforms and empirical research on introducing AI to K-12 students focused on high-level AI concepts and were mostly based on training supervised learning models [6, 7]. There was limited research on introducing RL to K-12 students [8, 9]. Also, the activities in these RL teaching projects were all fully developed in the simulated world and there is no research on using physical tools like educational robots to introduce RL concepts to K-12 students in the real world.

To address this need, we designed a robot-based RL activity using LEGO SPIKE Prime robot kit. To enrich the activity and provide students with an intuitive way to visualize the RL training process and to interact with their robots, we developed an Augmented Reality (AR) interface to bridge the virtual and physical world. Our activity was designed using constructivist principles, aimed at constructing students' own understanding of RL by having them build and control their own walking robots to go straight. They can also explore human-in-the-loop training through our software and use the RL algorithm to train the robot to complete additional tasks. Our activity covers the following aspects of RL: (1) Key concepts in RL, including state, action, reward, and policy; (2) Exploration and exploitation and how the agent chooses whether to explore or exploit; (3) Q-table and agent's decision making based on it; (4) Episodes and termination rules; and (5) Impacts of human input. By combining virtual interface and physical robot, we aim to provide students with an interactive and engaging learning journey and allow them to develop their own educational experience. Currently, the target group of our research is middle and high school students.

We evaluate our platform in a study with 14 high school students using a three-day curriculum, including a short session of playing with an online RL learning platform we developed in 2020 [10]. Our results showed that students were engaged and excited during the three classes, and they constructed a comprehensive understanding of both general and specific RL concepts.

2 Background and Related Work

2.1 Related Work

K-12 AI education has received increasing attention from institutions and researchers, resulting in a number of approaches and methodologies [11, 12]. The Association for the Advancement of Artificial Intelligence (AAAI) and the Computer Science Teachers Association (CSTA) have collaborated with National Science Foundation (NSF) and Carnegie Mellon University to formulate a guideline for introducing big ideas of AI, like machine learning, to K-12 students [13]. Recent AI education platforms include Google's Teachable Machine [14], Machine Learning for Kids, and MIT's Cognimate [15]. These web-based AI education platforms demonstrate AI-related concepts by implementing web interfaces for students to train and test AI agents to finish different tasks.

LEGO educational robots and others have been studied in many K-12 education contexts including physics [16], mathematics [17] and engineering [18]. These studies have shown that robotics kits can improve the students' engagement and facilitate understanding of STEM concepts. LEGO robots have also been used in studies for teaching AI and robotics to different age group students [19–22]. Particularly, a research in Spain used LEGO robots to teach RL to college level students [23]. These researchers have reported that LEGO robots could make the learning experience more interactive, attractive, and friendly to students who do not have an AI or robotics background, which provided inspiration for our robot-based RL activity design.

At the same time, Augmented Reality (AR) has gradually become a popular tool in K-12 STEM education [24]. Researchers have applied AR technique on teaching robotics [22], physics [25], arts [26], etc. And results from these studies showed that AR could not only increase students' engagement, concentration, and learning outcome, but also reduced the difficulty of learning. However, currently, AR is rarely used in K-12 AI education. Some researchers developed a virtual reality platform that could introduce RL to K-12 students in a digital world [9]. AR also contributed to many HRI research including robotic teleoperation [27], robot debugging [28], etc. Related work has proposed an AR visual tool named "SENSAR" for human-robot collaboration [29], which was also deployed on a LEGO EV3 robot for educational purposes. In our research, we want to leverage the strength of AR and combine it with robots as a visual aid and provide students with an intuitive way to collaborate with their robots.

2.2 Reinforcement Learning Background

Reinforcement learning is a class of problems where an agent has to learn how to act based on scalar reward signals detected over the course of the interaction with the environment. The agent's world is represented as a Markov Decision Process (MDP), a 5-tuple $< \mathcal{S}, \mathcal{A}, \mathcal{T}, \mathcal{R}, \gamma >$, where $\mathcal{S}$ is a discrete set of states, $\mathcal{A}$ is a set of actions, $\mathcal{T} : \mathcal{S} \times \mathcal{A} \rightarrow \Pi(\mathcal{S})$ is a transition function that maps the probability of moving to a new state given action and current state, $\mathcal{R} : \mathcal{S} \times \mathcal{A} \rightarrow \mathbb{R}$ gives the reward of taking an action in a given state, and $\gamma \in [0, 1)$ is the discount factor. We consider episodic tasks in which the agent starts in an initial state s_0 and upon reaching a terminal state s_{term}, a new episode begins.

At each step, the agent observes its current state, and chooses an action according to its policy $\pi : \mathcal{S} \rightarrow \mathcal{A}$. The goal of an RL agent is to learn an optimal policy π^* that maximizes the long-term expected sum of discounted rewards. One way to learn the optimal policy is to learn the optimal action-value function $Q^*(s, a)$, which gives the expected sum of discounted rewards for taking action a in state s, and following policy π^* after:

$$Q^*(s, a) = \mathcal{R}(s, a) + \gamma \sum_{s'} \mathcal{T}(s'|s, a) \times \max_{a'} Q^*(s', a') \tag{1}$$

A commonly used algorithm used to learn the optimal action-value function is Q-learning. In this algorithm, the Q-function is initialized arbitrarily (e.g., all zeros). Upon performing action a in state s, observing reward R and ending up in state s', the Q-function is update using the following rule:

$$Q(s, a) \leftarrow Q(s, a) + \alpha(R + \gamma \max_{a'} Q(s', a') - Q(s, a)) \tag{2}$$

where α, the learning rate, is typically a small value (e.g., 0.05). The agent decides which action to select using an ϵ-greedy policy, which means with small probability ϵ, the agent chooses a random action (i.e., the agent explores); otherwise, it chooses the action with the highest Q-value in its current state (i.e., the agent acts greedily with respect to its current action-value function).

To speed up the RL process, researchers have proposed human-in-the-loop RL methods. For example, in the learning-from-demonstration (LfD) framework, human teachers take over the action selection step, often providing several trajectories of complete solutions before the agent starts learning autonomously. In a related paradigm, the agent can seek "advice" from its human partner, e.g., let humans provide a reward for the action it chooses in a particular state when the Q-values are thought to be unreliable due to lack of experience. One of the goals of our system and proposed activity is to demonstrate to students how human partners can help a robot learn through interacting with it.

3 System Overview

We aim to introduce RL by combining physical robot and virtual interface. To achieve that, we designed a robot-based activity to engage students and inspire their curiosity to learn more about RL. To enrich the activity, we developed an AR mobile application that can communicate with the robot. Students can use the application to control the training of their robots and visualize the learning process through the interface.

3.1 Robot-Based Activity Design

To introduce educational robots to K-12 RL education, we looked for a robot-based RL challenge that is: (1) Intuitive for students to understand; (2) Able to achieve a good training result in less than 10 min; (3) Easy to test some straightforward non-AI solutions, but usually won't have a good result, whereas can be solved quickly and intuitively by RL; (4) Possible for students to customize robot design and explore solutions for their own robots. In order to meet these goals, we used LEGO SPIKE Prime robot kit and designed a robot activity called "Smart Walk". In this activity, students will be asked to build a walking robot with LEGO bricks but without using wheels. An example build is shown in Fig. 1. They will then try to use LEGO SPIKE software to program the robot to go a straight line with block-based coding or MicroPython. Due to the uncertainty of the walking robot's movements, it will be hard for students to explicitly program the robot to go straight. So we will then let them train the robot with a RL algorithm that is installed in advance. The training process is straightforward, students only need to press the left button on the robot to train the robot for an episode, and press the right button to test how well the robot has been trained. The light matrix on the robot shows numbers to indicate how many episodes the robot has been trained, and when students press the right button, it will show that the robot is in test mode. After playing with the RL algorithm, we will ask students to make some modifications to the structure of their robots (e.g., make one leg shorter than the other), then try the RL algorithm and their own code again to compare the results. In this part, we would like to show the adaptability of RL algorithm to the change in environment. At the end of the activity, we will gather students around and let them share their findings, questions, and thoughts about this RL challenge. Through the whole activity, we want students to build up

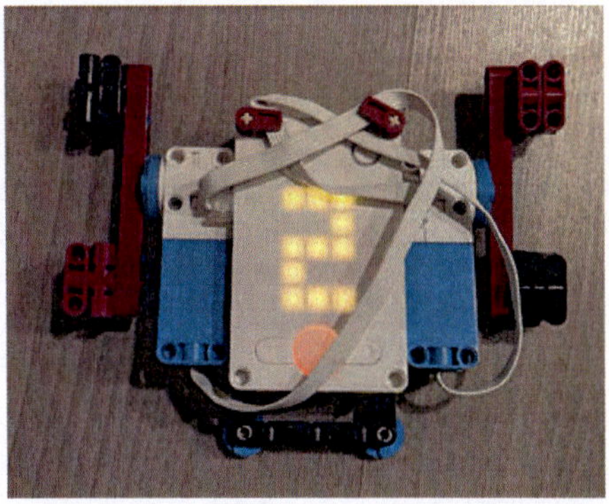

Fig. 1 An example build of a "Smart Walk" robot

a general understanding of the process and some strengths of RL, as well as inspire their curiosity to learn more about how RL works.

We used a ϵ-greedy Q-learning algorithm to train the walking robot to go straight. Based on the gyro sensor data, we defined five states (too left, a little to the left, right in the middle, a little to the right, too right) to describe the robot's orientation. Corresponding to the states, we have five different actions that control the speed of two motors on the robot. At each training step, the robot would choose an action using a ϵ-greedy policy, then run itself for 0.5 s with the new speed. After coming to a stop, the robot would then read data from its gyro sensor to confirm its new state. The algorithm implemented would then assign a reward to the robot based on this data, with possible values of –10, –2, or +10. The robot then employed the received reward to revise the Q-value, allowing for the integration of this learning experience into future decision-making processes.

3.2 AR Application Overview

We understand educational robots have limitations. For example, they usually lack ways to help students visualize abstract information and concepts that are important in understanding AI. Therefore, to enrich the robot-based activity and demystify more specific RL concepts, we developed an AR application to bridge the virtual and physical world using Unity[1] and Vuforia.[2] The application can be easily deployed on Android or iOS devices. It communicates with LEGO robots through Bluetooth Low Energy (BLE). To give students an immersive experience, the user interface (UI) is designed with a science fiction theme.

The interface contains two main parts, the static UI part and the AR part. As shown in Fig. 2, the static UI contains 4 pages that students can navigate through using a menu located at the bottom-left of the screen. The main scene provides students with a straightforward way to train or test their robots. The "Mode Switch" toggle allows students to switch between regular and a "break-down" mode, where each training step is divided into smaller, clearer stages so that students can understand the robot's decision-making and learning process more visually. On the main scene. we also have a BLE connection indicator and a button for students to restart the training. In addition to letting the robot train itself, students can also choose to manually train the robot using the UI components on the "Human Training" page. In this part, students can reward the robot with different values after it makes an action. They can also tweak the ϵ value to change the robot's decision-making strategies during training. With these functionalities, students have the opportunity to customize training strategies for their own robots or attempt to teach the robot to finish more tasks. On the "Training Result" page, we provide an intuitive visual aid to illustrate the current Q-value of each state-action pair, so that students can comprehend the foundation of the robot's

[1] https://unity.com/.

[2] https://developer.vuforia.com/.

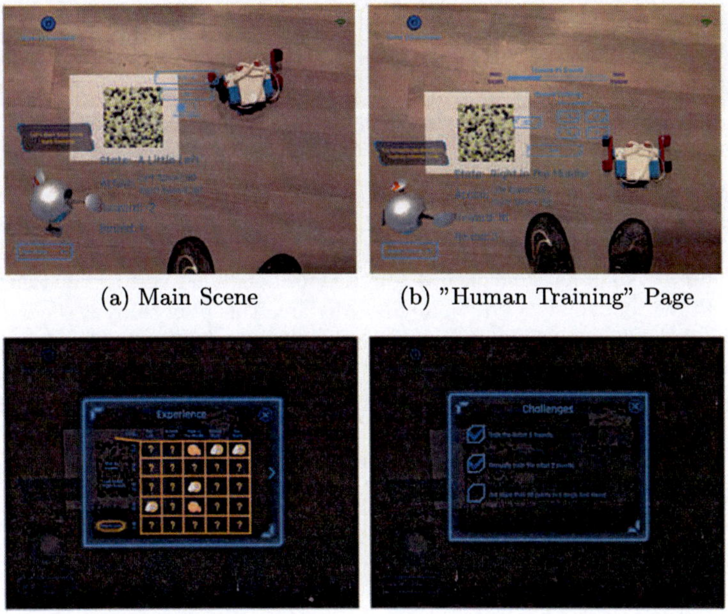

(a) Main Scene (b) "Human Training" Page

(c) "Training Result" Page (d) "Challenges" Page

Fig. 2 Interface of the AR software

decision-making process. Some extra information related to the training, like total training steps, is also included on this page. To prevent students from feeling unsure of where to begin when they start playing the software, we have added three tasks for students to undertake on the "Challenges" page. This page will automatically pop up when the software is launched. These tasks will motivate students to explore all the functionalities of the software. In the AR part of the UI, we focus on showing key RL concepts in real-time, including state, action, reward, and training episodes. We also added a 3D robot model that has a dialogue box on top of it, the virtual robot can "chat" with students using the dialogue box to update task progress, or suggest students try other functions like manual training. Since AR is activated, the background of the application is the camera view, which makes it easier for students to simultaneously keep track of the robot and the RL information. The AR components are superimposed around an image target which will be handed out to students before the activity.

4 Pilot Study and Results

This pilot study was conducted in a rural public high school located in New Hampshire. The class consisted of fourteen students who were seniors enrolled in a year-long engineering design course that focused on developing innovative problem-solving skills. All of the students had one year of chemistry and physics courses previously. Four of the students had a specific semester course in either computer science or a DIY course focused on designing devices using the Arduino and sensors platform. All of the students had a limited introduction to the LEGO Education SPIKE Prime robot and its programming software. This pilot study was conducted over three days in a single week in November 2022. The time spent on this project included ninety minutes on Tuesday and Friday. Students also had a fifty-minute class on Wednesday to play with the web-based RL learning platform we developed in 2020.

4.1 Lesson Structure

The first session was designed around the robot-based activity we described in the previous section. The goal of this session was to introduce students to the AI and RL world, and to help them build a general understanding of RL by comparing it to other regular approaches that students were more familiar with. Specific RL concepts were not highlighted in this session to avoid overwhelming the students and to spire their curiosity to learn more about RL. The students worked in groups of two. To start with, students were asked to build the same walking robot and then spent 20 min trying to program the robot to go straight between two pieces of tape that were parallel to each other, using LEGO SPIKE software. Next, the students spent 15 min trying to get the robot to perform the same task using RL algorithm. After that, we gave students another 20 min to modify the structure of their robots to make them asymmetrical and adapt their own code to the new design. Then they had another 15 min to apply the RL algorithm to train the robot and compare the result to their own code. At the end of the session, we encouraged all the groups to share their robot modifications, perform a test run to show their training results, and discuss their first impressions and questions about the RL training process.

The second class was only 50 min and focused on the web platform we developed before. The web platform contained two treasure-hunting-themed RL challenges as shown in Fig. 3. In each challenge, students needed to train a virtual robot to solve a maze by finding treasure and avoiding traps. The web platform has the following features: (1) An interface for students to visualize how key RL variables (e.g., state, action, reward, Q-value) change during the training process; (2) Opportunities for students to participate in the training process by providing rewards to the robot; (3) An interface for students to tweak the ϵ value to learn the concepts of exploration and exploitation in the RL context. At the beginning of the second class, we gave

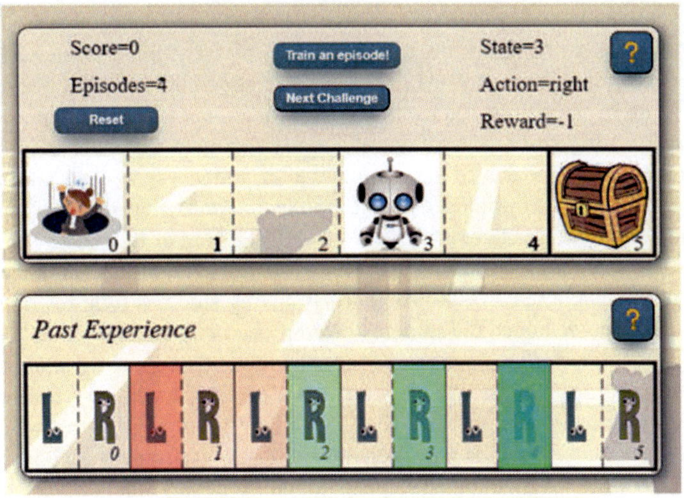

(a) GUI layout of the 1-D treasure hunting challenge

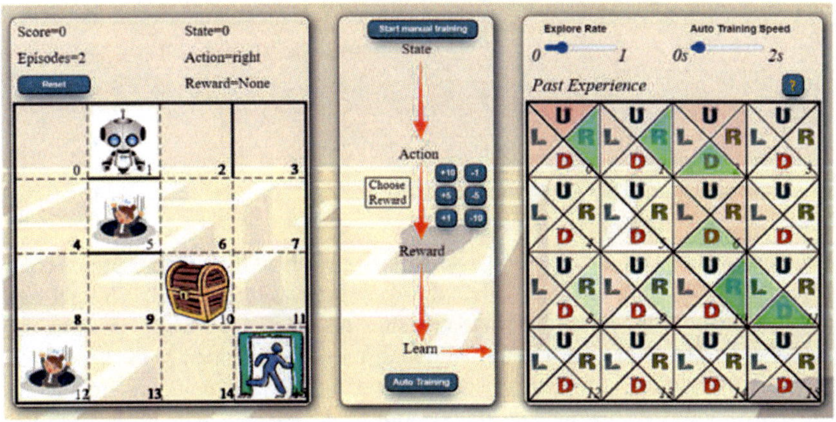

(b) GUI layout of the 2-D treasure hunting challenge

Fig. 3 GUI layout of the online platform

the students a 5 min presentation that included the definition of AI, machine learning, and RL, as well as their relationship. After that, we asked students to explore individually on the web platform. They were encouraged to discuss with each other while solving the RL challenges. They started with the 1D maze-solve challenge (shown in Fig. 3a). The training was automatically accomplished by the embedded RL algorithm. Our goal in this part was to let students get familiar with the interface and establish an understanding of key RL concepts by observing their change during the training process. Next, we let them explore a more complicated 2D maze environment (shown in Fig. 3b). In this challenge, students could choose both to train the robot automatically or to teach the robot manually by providing a numerical reward after the robot made each move. In this part, we hoped students to have a deeper understanding of the key RL concept and could learn about how humans could be part of the learning process and more efficiently facilitate the agent in identifying desirable actions. At the end of the session, we let students share their takeaways.

In the third class, we introduced the AR application to students and asked them to train their walking robots to reach a treasure chest in front of the robot. The goal of this class was to enhance students' learning of RL by letting them implement their knowledge to solve a real-world problem. By giving students little restriction and direct instruction, we hoped the students could proactively explore the RL concepts that they were interested in or confused about, then construct their own understanding of these concepts and disseminate it to other students. The students were divided into the same groups as in session one. After spending 10 min helping every group set up, we gave students 30 min to play with the application and figure out how to finish the task. After that, we encouraged students to redesign their robots and try to adjust their training strategies for the new robots. After another 30 min, we asked students to gather around to share their work, and discussed their takeaways and questions. Since students were curious to know more about the AR technique, at the end of the session, we had a 5 min demo to show the students how to create an AR experience using Unity and Vuforia.

4.2 *Class Observation and Analysis*

In the first session, we had a very structured lesson with all the students progressing at approximately the same pace. When trying to improve their robot's straight line movement between the two tape markers, students elected to change the code by tweaking the speed of two motors. Although they all struggled with the task as we predicted, various approaches were tested by the students including moving the legs asynchronously, slowing down the overall speed, etc. When first trying our RL algorithm, some students were confused when the robot performed worse than the previous episode, and this was mainly caused by the ϵ-greedy policy which made the robot sometimes choose random behavior instead of always choosing the current optimal solution. But after 7–15 min of training, the RL algorithm succeeded on all the robots. After changing the structure of their robots, most groups failed to modify their code to control the new robot going straight. However, after several training episodes, the RL algorithm successfully adapted to the new designs and achieved good training results. Students also spontaneously started over the training and compared the result with previous training sessions. When the final test run happened, the number of training episodes for all groups of robots ranged from 4 to 18. In the debriefing discussion, several students noticed that more training episodes correlated with increased performance of finishing the task, and the robots also had greater resistance to external disturbances (e.g., can turn back and go straight again if someone pushed them in one direction.). A side observation was that many students showed high curiosity to know more about intimate details of the RL technique and programming. Student S[3] asked about how the robot learned, and was further interested to know what type of reward was given and how the robot interpreted the reward. Student N and student R were curious to know how the robot decided what to do next. Overall, at the end of this session, students were able to build a general understanding that RL is a loop of making a choice, receiving feedback, and then adjusting its decision-making. They also showed a high level of engagement and curiosity during the whole session.

In the second session, mostly, students were learning through their own exploration and discussion with others, while the instructors were observing and answering their questions. The majority of the class was able to finish the 1D maze training in 5–7 min. From our observation and their discussion, most students could understand how the key RL variables changed during the training process. As they moved to the exploration of the 2D maze, students started to have more interactions with each other, and they used many RL concepts to communicate. These included "positive reinforcement", "exploration rate", "reward", etc. As the class progressed, students started to apply various strategies to dig deeper into the RL challenge and concepts. Some students focused on training the robot automatically and observing how the robot performed differently in each episode while some other students spent most of the time playing with the manual training part. Students discussed different training strategies with each other. Student L tried to replicate a specific case on student M's computer and they had a discussion about how it happened. Another student explained how to use the Q-table visualization to evaluate current training result to his neighbor. Students also came up with some insightful questions. For example, student S asked us about a case where the map showed the robot was not visiting all places "evenly". He then used this observation to ask more specifically about exploitation versus exploration. Some students implemented interesting experiments and built a deeper understanding of some RL concepts beyond our expectations. A curious student J made some changes to the back end of the platform to test a higher training speed and ϵ value. Another student proposed that we could set a high ϵ value at the beginning and decrease it as the training progresses to accelerate the training process, unknowingly explaining how a decayed ϵ-greedy policy works. The variety of all these explorations and dialogues showed that students were engaged, observant, challenged, and curious to figure out how RL works.

The third session was best characterized as the most creative. Within the first half, most students were able to train their robots to straightly go to the treasure chest using the software. The AR interface clearly excited the students. A few students moved their iPads around and tried to see how the AR components were superimposed in the real world, and we heard a few students comment on how cool it was to their teammates. After the original task was finished, students proactively pursued different directions to test all their ideas and to push the limit of the algorithm. They generated different robot designs and implemented various experiments on them. Student J's

[3] All the names are pseudonym names.

group focused on testing how well the RL algorithm could adapt to different robot designs. Two teams successfully trained their robot manually to go a circle instead of going straight before testing the same training policy on multiple robot designs. Student E proved her assumption that giving a high ϵ value at the beginning and gradually decreasing it could help the robot learn faster. During this process, we noticed that the robot served as a perfect medium for students to interact with each other. When the robots performed some fascinating movements or achieved a goal, it would engage students' attention and generate discussion so that everyone's unique experience was shared around the whole classroom. For all the groups, we found the AR application successfully kept them engaged and students can visualize the "whole picture" directly through the screen. Specifically, student E expressed her preference for the manual training part since she thought it was cool to impact the training with her own input. Also, in this session, students discussed and asked more penetrating RL questions, which demonstrated their deeper understanding of RL. One group discussed the difference between rewarding the action itself and rewarding the result caused by the action. Another group discussed on how to measure the robot's capability of correcting itself from the Q-table. Student K asked that after the robot had chosen an action in a state, how did it interpret the reward it received to an evaluation of the long-term gain from this action. At the end of the session, students were excited to see how an AR experience was made and two students stayed late to ask more in-depth questions about AR and RL.

4.3 Results and Limitations

Overall, students showed a high level of engagement and curiosity during all three sessions. Though very little direct instruction was given, all the students were able to construct an understanding of both general RL ideas and specific RL concepts. In some aspects, like the concept of exploration and exploitation, students could think further than we expected. Besides AI and RL, the students also gained hands-on experience in robotics and AR.

The educational robot played an important role in engaging students and motivating them to explore more about RL. Compared to session two which was an activity based on a pure virtual platform, we observed higher solution diversity in session one and session three. However, some students argued that they "learned more" in web-based session two than in robot-based session one. So we believed that providing students with a straightforward approach to help them understand the training process and visualize the important AI concepts was also necessary. The AR application successfully filled this gap by providing an intuitive interface for students to collaborate with their robots. Students were excited about the AR technique, and it helped demystify the RL process and made it easy for students to monitor their robots and the RL training at the same time.

There were also some limitations in this study. Due to the limited number of students, we did not implement quantitative analysis to measure students' learning outcomes more precisely. And because many of the students had prior experiences with computer science and robotics, we predict a different outcome if this three-session RL/AR curriculum is applied, unmodified to another group of high school students. During session one, we noticed that the way we provided students to train their robots with the RL algorithm could be further improved. In session three, we found that the BLE connection between the robot and the software was unstable at some times. Besides, more AR components and options could be offered to the students to provide them with a more immersive learning experience.

5 Conclusion and Future Work

In this paper, we presented a robot-based activity to introduce educational robots to K-12 RL education. To compensate for the limitations of robots in terms of information visualization and to enrich the robot-based activity, we developed an AR software to provide an intuitive interface for students to follow the training process and to facilitate their understanding of RL through exploring human-in-the-loop training. By combining the virtual and the physical world, we aim to provide students with an engaging and interactive learning journey and help them develop their own educational experience. A pilot study was conducted with 14 high school students over a three-day period. The results indicated that students were able to grasp both general and specific RL concepts through our activity, and they showed a high level of engagement and curiosity during the classes. The AR part excited students and helped them easily keep track of their robots during training. With the opportunity to design their own robots and self-explore different training strategies, students constructed a deeper understanding of some RL concepts than we anticipated.

Our AR implementation is still in its initial stages. Our goal for the future is to improve the system by utilizing AR to directly track robots so we can add more virtual components to enhance the training experience. For example, we can create virtual obstacles and train the robots to avoid them, or show the robot's past trajectories in previous training episodes using AR. Additionally, we aim to evaluate the platform's effectiveness among younger students and measure their learning progress. We also plan to host a workshop for K-12 STEM teachers, providing them with the opportunity to implement the system and activities with more students.

References

1. Kiran, B.R., Sobh, I., Talpaert, V., Mannion, P., Al Sallab, A.A., Yogamani, S., Pérez, P.: Deep reinforcement learning for autonomous driving: a survey. IEEE Trans. Intell. Transp. Syst. **23**(6), 4909–4926 (2022)
2. Zamfirache, I.A., Precup, R.E., Roman, R.C., Petriu, E.M.: Policy iteration reinforcement learning-based control using a grey wolf optimizer algorithm. Inf. Sci. **585**, 162–175 (2022)

3. He, Z., Tran, K.-P., Thomassey, S., Zeng, X., Jie, X., Yi, C.: A deep reinforcement learning based multi-criteria decision support system for optimizing textile chemical process. Comput. Ind. **125**, 103373 (2021)
4. Aljanabi, M., Ghazi, M., Ali, A.H., Abed, S.A.: Chatgpt: open possibilities. Iraqi J. Comput. Sci. Math. **4**(1), 62–64 (2023)
5. Sutton, R.S., Barto, A.G.: Reinforcement Learning: an Introduction. A Bradford Book, Cambridge, MA, USA (2018)
6. Vartiainen, H., Tedre, M., Valtonen, T.: Learning machine learning with very young children: Who is teaching whom? Int. J. Child-Comput. Interact. **100182**, 06 (2020)
7. Sakulkueakulsuk, B., Witoon, S., Ngarmkajornwiwat, P., Pataranutaporn, P., Surareungchai, W., Pataranutaporn, P., Subsoontorn, P.: Kids making AI: integrating machine learning, gamification, and social context in stem education. In: 2018 IEEE International Conference on Teaching, Assessment, and Learning for Engineering (TALE), pp. 1005–1010. IEEE
8. Dietz, G., King Chen, J., Beason, J., Tarrow, M., Hilliard, A., Shapiro, R.B.: Artonomous: introducing middle school students to reinforcement learning through virtual robotics. IDC '22, pp. 430–441. Association for Computing Machinery, New York, NY, USA (2022)
9. Coppens, Y., Bargiacchi, E., Nowé, A.: Reinforcement learning 101 with a virtual reality game. In: Proceedings of the 1st International Workshop on Education in Artificial Intelligence K-12 (2019)
10. Zhang, Z., Willner-Giwerc, S., Sinapov, J., Cross, J., Rogers, C.: An interactive robot platform for introducing reinforcement learning to k-12 students. In: Robotics in Education, pp. 288–301. Springer International Publishing, Cham (2021)
11. Su, J., Zhong, Y., Ng, D.T.K.: A meta-review of literature on educational approaches for teaching ai at the k-12 levels in the Asia-pacific region. Comput. Educ.: Artif. Intell. **3**, 100065 (2022)
12. Vartiainen, H., Tedre, M., Valtonen, T.: Learning machine learning with very young children: who is teaching whom? Int. J. Child-Comput. Interact. **25**, 100182 (2020)
13. Touretzky, D., Gardner-McCune, C., Martin, F., Seehorn, D.: Envisioning AI for k-12: what should every child know about AI? Proc. AAAI Conf. Artif. Intell. **33**(01), 9795–9799 (2019). Jul
14. Carney, M., Webster, B., Alvarado, I., Phillips, K., Howell, N., Griffith, J., Jongejan, J., Pitaru, A., Chen, A.: Teachable machine: approachable web-based tool for exploring machine learning classification. In: Extended Abstracts of the 2020 CHI Conference on Human Factors in Computing Systems, CHI EA '20, pp. 1–8. Association for Computing Machinery, New York, NY, USA (2020)
15. Druga, S.: Growing up with AI. Cognimates: from coding to teaching machines. Ph.D. thesis. Massachusetts Institute of Technology (2018)
16. Petrovič, P.: Spike up prime interest in physics. In: Robotics in Education, pp. 146–160. Springer International Publishing (2021)
17. Mandin, S., De Simone, M., Soury-Lavergne, S.: Robot moves as tangible feedback in a mathematical game at primary school. In: Robotics in Education. Advances in Intelligent Systems and Computing, pp. 245–257. Springer International Publishing, Cham (2016)
18. Laut, J., Kapila, V., Iskander, M.G.: Exposing middle school students to robotics and engineering through LEGO and MATLAB. In: 120th ASEE Annual Conference and Exposition, 2013. 120th ASEE Annual Conference and Exposition; Conference date: 23-06-2013 Through 26-06-2013
19. Whitman, L., Witherspoon, T.: Using LEGOs to interest high school students and improve K-12 stem education. Change **2**, 12 (2003)
20. Williams, R., Park, H.W., Breazeal, C.: A is for artificial intelligence: the impact of artificial intelligence activities on young children's perceptions of robots. In: CHI '19: Proceedings of the 2019 CHI Conference on Human Factors in Computing Systems, pp. 1–11 (2019)
21. van der Vlist, B., van de Westelaken, R., Bartneck, C., Hu, J., Ahn, R., Barakova, E., Delbressine, F., Feijs, L.: Teaching machine learning to design students. In: Pan, Z., Zhang, X., El Rhalibi, A., Woo, W., Li, Y. (Eds.) Technologies for E-Learning and Digital Entertainment, pp. 206–217. Springer, Berlin Heidelberg (2008)

22. Cheli, M., Sinapov, J., Danahy, E.E., Rogers, C.: Towards an augmented reality framework for K-12 robotics education. In: Proceedings of the 1st International Workshop on Virtual, Augmented, and Mixed Reality for HRI (VAM-HRI) (2018)
23. Martínez-Tenor, Á., Cruz-Martín, A., Fernández-Madrigal, J.-A.: Teaching machine learning in robotics interactively: the case of reinforcement learning with LEGO® mindstorms. Interact. Learn. Environ. **27**(3), 293–306 (2019)
24. Sırakaya, M., Alsancak Sırakaya, D.: Augmented reality in stem education: a systematic review. Interact. Learn. Environ. **30**(8), 1556–1569 (2022)
25. Techakosit, S., Nilsook, P.: Using augmented reality for teaching physics (2015)
26. Huang, Y., Li, H., Fong, R.: Using augmented reality in early art education: a case study in Hong Kong kindergarten. Early Child Dev. Care **186**(6), 879–894 (2016)
27. Lee, D., Park, Y.S.:. Implementation of augmented teleoperation system based on robot operating system (ROS). In: 2018 IEEE/RSJ International Conference on Intelligent Robots and Systems (IROS), pp. 5497–5502 (2018)
28. Ikeda, B., Szafir, D.: An AR debugging tool for robotics programmers. In: International Workshop on Virtual, Augmented, and Mixed-Reality for Human-Robot Interaction (VAM-HRI) (2021)
29. Cleaver, A., Faizan, M., Amel, H., Short, E., Sinapov, J.: SENSAR: a visual tool for intelligent robots for collaborative human-robot interaction (2020). arxiv:2011.04515

Measuring Emotional Facial Expressions in Students with FaceReader: What Happens if Your Teacher is Not a Human, Instead, It is a Virtual Robotic Animal?

Alexandra Sierra Rativa, Marie Postma, and Menno van Zaanen

Abstract Based on new methods and algorithms to improve the detection of emotions from facial expressions employing sophisticated software during certain situations, we explore if a virtual robotic animal appearance design with the role of a virtual instructor can affect such emotions in users. A total of 131 students from two secondary public schools in Bogota, Colombia, participated in this study. We used an established facial emotion recognition software called "FaceReader" to analyze data recorded during their class. The results showed that the virtual robot animal's appearance could affect participants' emotional facial expressions when it was a virtual instructor in this study. The study contributes to our understanding of the visual appearance of virtual animals related to emotional facial expressions and their future enforcement concerning advances in artificial intelligence and educational robotics for accurate recognition of the emotions of users.

Keywords Virtual robot animals · Emotional facial expressions · Virtual instructors · Education · FaceReader

A. Sierra Rativa (✉)
Erasmus University Rotterdam, Rotterdam 3062 PA, The Netherlands
e-mail: profealexandrasierra@gmail.com

M. Postma
Tilburg University, Tilburg 5000 LE, The Netherlands
e-mail: mariepostma@uvt.nl

M. van Zaanen
South African Centre for Digital Language Resources, SADILAR, North-West University, Potchefstroom, South Africa
e-mail: menno.vanzaanen@nwu.ac.za

R. Balogh et al. (eds.), *Robotics in Education*, Lecture Notes in Networks and Systems 747, https://doi.org/10.1007/978-3-031-38454-7_30

1 Introduction

An application of virtual robotic animals as virtual instructors has previously been considered a possibility [1]. In the previous research, they explored whether a virtual instructor's appearance can influence students' knowledge recall outcomes. We manipulated three distinct virtual instructor appearances: (1) a Virtual robot animal, (2) a virtual animal, and (3) a human. Moreover, we considered that the topic of video instruction can also affect results, and therefore proposed two topics: (1) a topic related to robotics, and (2) a topic unrelated to robotics (i.e. to Dutch culture). We found that the virtual animal and human appearance scored significantly higher on knowledge recall for both topics compared to the virtual robot animal. Although these results put virtual robots in trouble as great allies as virtual tutors in the cognitive role, we wonder if the same effect happens effectively. Surprisingly, the effects of facial emotional expressions on teaching robotics in education have rarely been closely examined. Continuing the previous research results and using the data about the detection of emotions from facial expressions obtained by Facereader, in this present study, we explore which facial expressions were displayed during the virtual class in the different conditions of this experiment.

1.1 Emotional Facial Expressions

Emotion is a process that can be communicated by facial expressions. Our brain is endowed with the ability to instinctively recognize another person's facial expressions and instantaneously evaluate their emotional state. Several studies in education have assessed children's ability or inability to recognize facial or bodily expressions. For instance, Witkower et al. [2] discovered that children can recognize emotional states from the bodies of other people, with this capacity increasing with age. Likewise, Lierheimer and Stichter [3] considered it important to teach students about facial expressions of emotion to increase their awareness of their externalizing behaviors and expressions that occasionally exhibited overreaction, anger, aggression, and self-adverse behaviors. These negative facial or bodily expressions of the students could lead to difficulties in the harmonic relationship required for collaborative tasks in the classroom. Likewise, for teachers, it is crucial to identify students' emotional states to determine their level of motivation or engagement, or attitudes that might affect their learning. However, most observations from teachers are derived from their own opinions. Still, the latest developments in technology can help us to analyze students' emotional facial expressions from an objective and precise perspective.

Currently, research is underway to improve the methods and algorithms used in the detection and identification of emotions from facial expressions and speech recognition [4]. Facial expressions can be a primary indicator for the classification of emotions and are thus employed in human communication research. Traditionally, a large body of literature has identified facial expressions using facial action coding systems (FACS) to allow their classification into emotions. Original studies of Ekman and Keltner [5] proposed a theoretical framework for describing possible universal facial expressions.

Their proposal was inspired by Darwin's [6] studies of facial expressions in humans and animals. A standard consideration of universal facial expressions is the influence of social settings in different countries, such as Western versus Asiatic culture. For this, it is considered that software or tools can aid in facial recognition and detection and can be adjusted by algorithms that modify the evaluation scale, depending, for example, on age, gender, country, or culture. For instance, six universal emotions established across multiple cultures are happiness, sadness, surprise, disgust, anger, and fear [4]. Some studies have considered a neutral expression among such categories of expression, giving a total of seven [7, 8]. However, other studies consider "neutral expression" to not be an emotion [4, 9]. These new methods and algorithms can quickly recognize and classify emotions, and detect changes in expression, such as micro-expressions.

The relative importance of micro-expressions is widely discussed in the literature on facial expressions. These micro-expressions frequently occur in a small fraction of a second [10, 11]. A recent study has provided evidence to define genuine and involuntary micro-expressions, which are determined by their duration. For instance, a good indicator of an involuntary micro-expression is its occurrence for a duration of between 65 and 500 ms [12] or 1/15th of a second [13]. Such micro-expressions can be modified in three situations: (1) The neutralized expressions can suppress the genuine expression; (2) A simulated expression is not generated by a genuine emotion; and (3) A mask expression can intend to falsify a genuine expression [14]. For the detection of such microexpressions, the software can provide millisecond precision in capturing microexpressions and processing them into percentages of possible emotional states of the users. In this research, we consider "time" as a key value for analyzing facial expressions, in order to negate the impact of masking or simulated expressions in the experiments.

Recent evidence suggests that software can help us to use facial expressions for automated emotion detection (e.g., FaceReader, OpenFace, and others) during a specific period. Previous work by Pardàs et al. [15] has established that the mouths and eyebrows of subjects can be highly informative about facial expressions that are potentially being displayed. Specifically, the overall recognition rate of the eyebrows is 50% and the mouth is 78%. The most identifiable emotions are joy, surprise, and anger, with 98% accuracy levels. However, higher confusion rates are reported for sadness, anger, and fear. Bourel et al. [16] explained that if a neutral expression can be considered a valid class for automatic recognition of emotion, then joy and neutral expressions had similar indicators. For the facial-expression-detecting software FaceReader, previous studies have determined that it accurately identified 88% of the emotional descriptors in the Amsterdam Dynamic Facial Expression Set and Warsaw Set of Emotional Facial Expression Pictures [17]. Research has shown that humans recognize 85% of emotions for both datasets, with this software having greater accuracy performance in emotional automization and recognition in comparison. Moreover, FaceReader can generate a frame rate of between 15 and 30 frames per second, which provides strong data generation for the post-analysis of videos or recordings. This software employs several artificial intelligence applications, such as neural networks and machine learning, for facial expression recognition and categorization, collectively termed Deep Face Model [18, 19]. Another facial analysis toolkit is called OpenFace. It is commonly used as a tool for machine learning, affective computing, and computer vision for computer and mobile applications [20, 21]. However, this software has limited studies validating its ability to recognize happiness, and no validation for recognizing other types of expressions has been conducted so far [22]. Nevertheless, valid and reliable emotion classification and automatization systems have emerged as powerful platforms, with potential applications including educational ones, such as virtual animal studies and virtual worlds.

Virtual animals are usually found in applications as virtual pets or film characters for artificial human companionship or entertainment [23, 24]. Generally, virtual animals can be designed to incorporate characteristics similar to their biological version, but they can also have anthropomorphic characteristics, or be mixed with objects [25, 26]. Likewise, in previous studies, virtual animals and virtual robotic animals have been considered excellent entities through which to explore the uncanny valley theory [27]. Surprisingly, the effects of a virtual robotic animal's design appear on the emotional facial expressions of users have not been closely examined to date. One purpose of this study is to assess which emotional facial expressions are displayed by participants during virtual animal interactions.

1.2 Research Questions

This study explores empirically how software for the detection and processing of emotional facial expressions can give us more information about the appearance of virtual robotic animals in their role as virtual instructors. For this study, we considered which participants display emotional facial expressions during the experiment through the use of FaceReader. The two main questions that we expect to answer are as follows:

(1) Can the design appearance of the virtual robot animal affect the emotional facial expressions of the participants compared with other virtual instructors' appearances?
(2) Which emotional facial expressions are displayed during each condition?

2 Methods

The original experiment of this first study has previously been described by Sierra et al. [1]. The description of the methodology and materials used can be consulted in this previous article. Here, we will address the elements that were added to the videos' dataset that was to be analyzed with a deeper focus on facial expression recognition.

2.1 Participants

Participants were recruited from two schools in Colombia: Institución Educativa Distrital Almirante Padilla and Institución Educativa Distrital Prado Veraniego. The original dataset was gathered from 131 young students who participated in this research. However, 124 videos in total were selected for analysis with Facial Expression Recognition Software. Participants ranged in age from 11 to 17 years. The sample included 52 female (41.9%) and 72 male (58.1%) participants. The participants were distributed between the six experimental conditions as follows: for Condition 11, a total of 21 participants were assigned (M = 14.1, SD = 1.411, Max year = 16 Min years = 11); for Condition 12, a total of 22 participants (M = 13.23, SD = 1.066, Max year = 15 Min years = 11); for Condition 13, a total of 17 participants (M = 13.41, SD = 0.712, Max year = 15 Min years = 12); for Condition 21, a total of 21 participants (M = 14.10, SD = 1.221, Max year = 16 Min years = 12); for Condition 22, a total of 23 participants (M = 13.96, SD = 1.745, Max year = 17 Min years = 11), and for Condition 23, a total of 20 participants (M = 13.2, SD = 0.768, Max year = 15 Min years = 12).

2.2 Ethics Approval

All students either gave their written consent or, in the case of participants younger than 18 years of age, received consent from their legal guardians to be recorded. The directors of the schools I.E.D Almirante Padilla and I.E.D Prado Veraniego gave the teacher and researcher authorization to conduct this study. The Research Ethics and Data Management Committee of the Tilburg School of Humanities and Digital Sciences Ethics approved this experiment with reference REC#2019/89. The open database of videos processed in FaceReader is avalable here [28]: https://dataverse.nl/dataset.xhtml?persistentId=doi:10.34894/O1S3N9.

2.3 Design and Materials: Conditions

This experiment had six different conditions, all combining the three instructor appearances (e.g., virtual robot panda, virtual panda, and virtual human) and the two topics (The Netherlands and Robotics), identified as follows (Fig. 1):

Fig. 1 Six conditions of the experiment

In the study, the experiment encompassed six distinct combinations or experimental conditions, denoted as: (11) The Netherlands with a virtual robot instructor, (12) The Netherlands with a virtual animal (panda) instructor, (13) The Netherlands with a human instructor, (21) Robotics with a virtual robot instructor, (22) Robotics with a virtual animal (panda) instructor, and (23) Robotics with a human instructor. The experimental design followed a 2X3 factorial structure involving three independent variables, each featuring two levels. Employing a between-subject experimental design, each participant was exposed to a specific experimental condition. Notably, the instructional material for the experiment was developed exclusively in the Spanish language, chosen due to its status as the students' native language.

2.4 Procedure

Each participant filled out a pre-test called "domain knowledge" about Dutch culture or Robotics. The teacher and researcher confirmed the permission of the student's parent to participate in this study, and that they were aware that the student's face would be recorded. At the beginning of the experiment, students were asked their age and gender and were given an identification number to preserve their anonymity. Each student watched one of the six (different) video instructions for a duration of 540 seconds. The video instructor only started when students pressed F9, with their faces being synchronically recorded. After the experiment, the researcher manually deleted any recording generated in the experimentation beyond the 540 seconds limit. After the experiment, the student completed a post-test "knowledge recall" questionnaire, with other aspects of the questionnaire covering the perception of the virtual instructor. The questionnaire was administrated online and took approximately 10 minutes to complete. The experiment took less than 30 minutes between the start and completion of each student. The experiment was developed within the schools with the teachers' supervision.

2.5 *Measure*

FaceReader 8.0 (Noldus) was used to code the emotional facial expressions of the participants (see Fig. 2). This software can be used to research effective computing, education, artificial intelligence, machine learning, human-computer interaction, and the design of adaptive interfaces in real-time or off-time videos and images. FaceReader has three primary characteristics that make it suited to this research: (1) This application uses the Viola-Jones algorithm for recognition of the presence of participants' facial expressions [29]; (2) This application uses an algorithmic model based on the Active Appearance Method (AAM) [30] with this model describing 500 vital points in the face such as eyebrows, eyes, nose, and lips; (3) This application uses a method based on Deep Learning that has developed a classification of the patterns of facial expressions through the use of an artificial neural network [31]. Moreover, this application recognizes cultural differences in facial expressions, especially the East Asian and Children models. Likewise, other aspects that must be input into FaceReader are age and gender, which are used in the model of classification.

A total of 124 videos were processed by FaceReader (Noldus). Each video has a duration of 540 s. We deleted seven videos of the original dataset for the following reasons: (1) The software did not recognize the student's face, because the student placed their hands on their face; and (2) Their posture did not allow facial recognition to generate a sufficiently error-free facial recognition percentage. The deleted videos were identified as 3, 8, 12, 34, 85, 89, and 96 from the original dataset. We recorded the faces of the students with a synchronization system to precisely coincide with the start of the video instruction, and we manually ensured that the recording finished after 540 seconds.

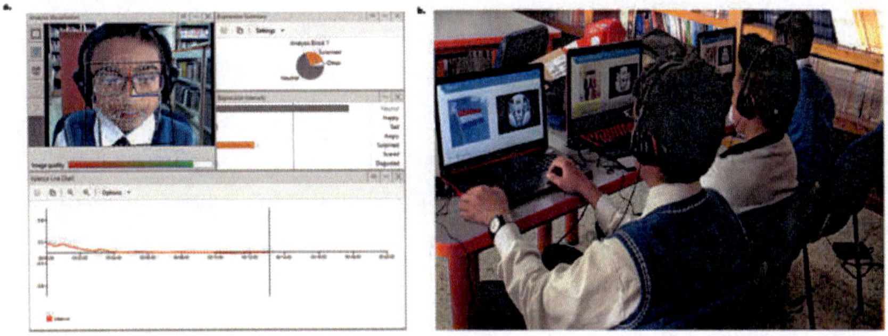

Fig. 2 Video of the student participants analyzed by FaceReader

3 Results

Can the design appearance of the virtual robot animal affect the emotional facial expressions of the participants compared with other virtual instructors' appearances? The first research question investigates whether there is an effect of the visual appearance of the virtual instructor and topic upon the emotional facial expressions displayed by students during virtual robot-animal interactions. As illustrated in Fig. 3, a Kruskal-Wallis H test showed that there was a statistically significant difference in sad emotion between the different appearances of the virtual instructor, $\chi^2(2) = 7.588$, $p = 0.023$, with a mean rank sad score of 75.08 for the virtual human, 61.38 for the virtual robot panda, and 53.20 for the virtual panda. Moreover, there was a statistically significant difference in arousal between the different appearances of the virtual instructor, $\chi^2(2) = 7.981$, $p = 0.018$, with a mean rank sad score of 76.35 for the virtual human, 58.14 for the virtual robot panda, and 55.18 for the virtual panda. There was no statistically significant difference in happy emotion between the different appearances of the virtual instructor $\chi^2(2) = 0.360$, $p = 0.835$, angry emotion $\chi^2(2) = 0.553$, $p = 0.758$, surprised emotion $\chi^2(2) = 2.651$, $p = 0.266$, scared emotion $\chi^2(2) = 0.515$, $p = 0.773$, disgust emotion $\chi^2(2) = 0.906$, $p = 0.636$, and valence $\chi^2(2) = 4.511$, $p = 0.105$. There was no statistically significant difference in happy emotion between the two topics of the virtual instruction $\chi^2(2) = 2.310$, $p = 0.129$, sad emotion $\chi^2(2) = 0.198$, $p = 0.656$, angry emotion $\chi^2(2) = 0.403$, $p = 0.525$, surprised emotion $\chi^2(2) = 1.550$, $p = 0.213$, scared emotion $\chi^2(2) = 0.281$, $p = 0.596$, disgusted emotion $\chi^2(2) = 0.384$, $p = 0.535$, valence $\chi^2(2) = 0.123$ $p = 0.726$, and arousal $\chi^2(2) = 0.518$ $p = 0.472$.

Which emotional facial expressions are displayed during each condition? The second question explores which emotions are displayed in the conditions of this study. Figure 4 shows six conditions when the student had a virtual robot animal as a virtual instructor and learned about the culture of the Netherlands and introduction to robotics, covering a period from 0 s to 540 s (9 min) (see Fig. 4). We can see that, generally, sad, surprised and happy are the emotions with the greatest fluctuations evident in these six conditions. The sad emotion is distinct from the other emotions, and it presents a significant mean, except in Condition 11 (The Netherlands with robot panda instructor) and Condition 22 (Robotics with panda instructor). Remarkably, for Condition 11 (The Netherlands with robot panda instructor), when sad emotion declines their mean, alternatively surprise emotion increases their value during all video instruction. To sum up, for all conditions, the sad emotion bore the most relevant score compared with other emotions. As illustrated in Fig. 5, we also analyze valence and arousal in the six conditions. The valence describes the level of pleasantness that a virtual instructor and topic can evoke from students [32]. We can observe that valence in all six conditions is negative. This means that the conditions are unpleasant for participants. Likewise, arousal (or intensity) describes the level of "autonomic activation" that is evoked for the six conditions. We found that in all conditions, participants reacted very calmly to video instruction.

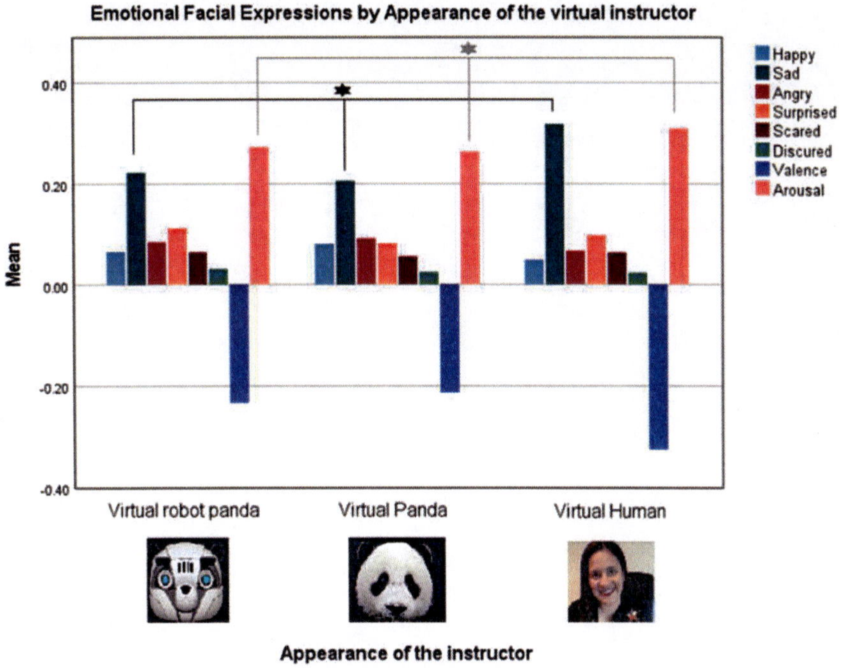

Fig. 3 A bar chart of the mean emotions concerning the virtual instructor's appearance

4 Discussion

Virtual animals as virtual instructors and emotional facial expressions. Currently, e-learning and teaching through digital media have led teachers in the direction of becoming virtual instructors and presenting their classes digitally. Our primary aim in this study was to discover if, instead of using the present method of teachers' faces presenting a class, the same effect can be obtained using avatars with shapes very different from humans. We decided to use robot animals like the virtual robot panda and its visual design to determine its effect on students.

The results indicated that the appearance of the virtual instructor affects the display of the emotion of sadness by students during a class. The students displayed greater levels of sadness when their virtual instructor was a human. In contrast, they displayed less sadness when the instructor was a virtual panda or a virtual robot panda. In Condition 13 (the Netherlands and virtual human) and Condition 23 (robotics and virtual human), as illustrated in the graph, the sad emotion can be observed to have greater relevance compared to the other conditions. We explore if the class topic can influence emotional facial expressions, and we do not find that this has a significant effect on students' emotions. Surprisingly, a study conducted before this experiment found that students had better recall knowledge of the class with the human teacher than with virtual robotic animals [1]. Although there may be better knowledge recall with

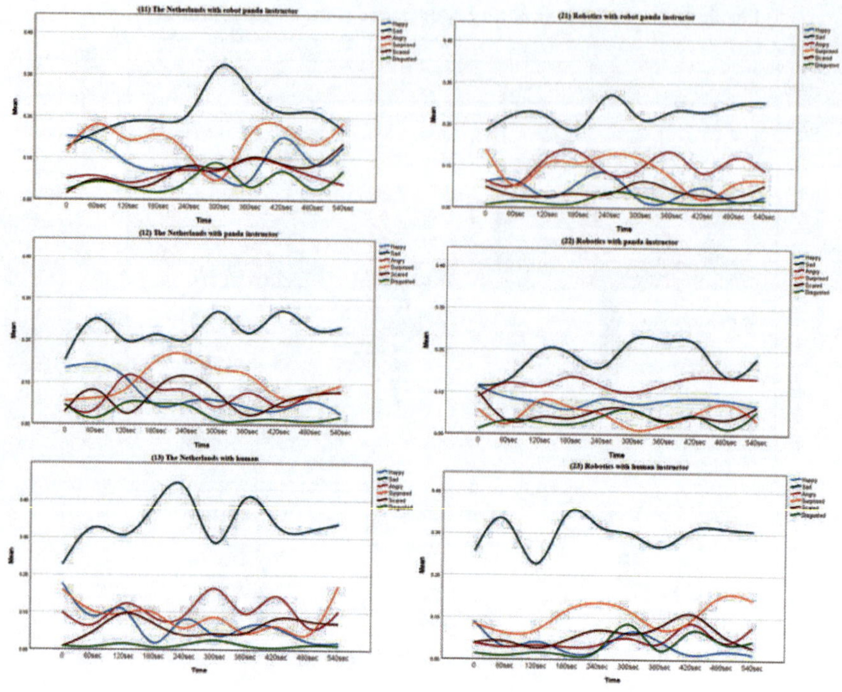

Fig. 4 Graphs of the emotional facial expressions according to virtual instructor appearance and topic

traditional teachers, another type of instructor, such as a virtual robot animal, might influence certain emotions in students that can contribute to reducing their sadness, boredom, and disinterest when undertaking traditional online classes. Moreover, we found that the level of valence in the six conditions was negative or unpleasant, and the level of arousal was low or calm during 540 seconds of the class. A possible explanation for the unpleasant and calm results of the emotional expressions may be that the class was designed linearly, with no interaction present and the students assuming a passive role in its visualization.

5 Conclusions

The primary aim of this research was to determine whether or not the visual appearance of virtual robot animals affects participants' emotional facial expressions. We found that visual appearance significantly affected their role as virtual instructors, where sad, surprised, and happy emotions were most evident. The young learners

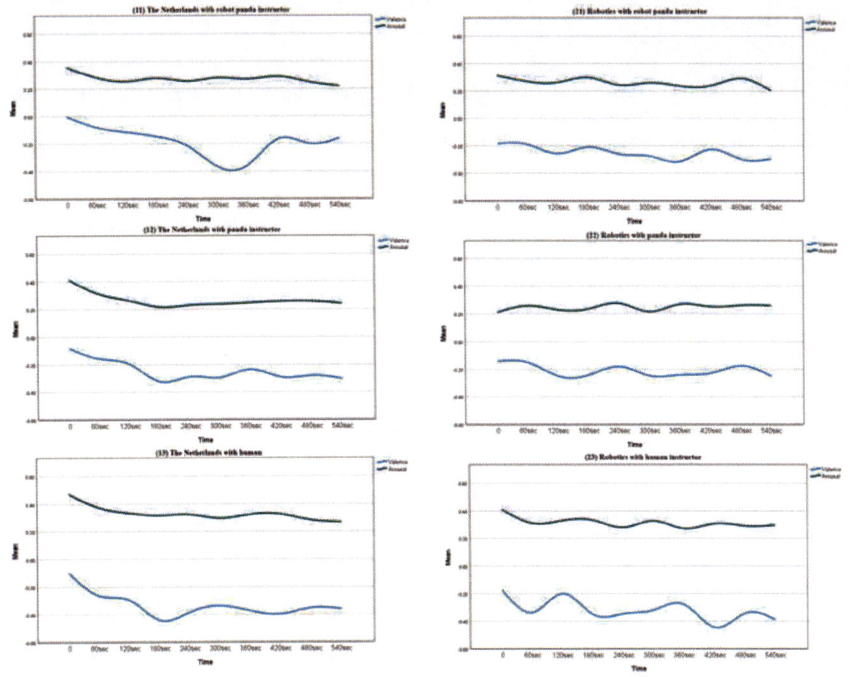

Fig. 5 Graphs of valence and arousal of the emotional facial expressions according to virtual instructor appearance and topic

displayed less sadness when their virtual instructor was a virtual robot animal or a virtual animal compared with human traits during class. For generalization of the results, we recommend applying this research with another zoomorphic robot appearance.

Acknowledgements We would like to thank director Argemiro Pinzón Arias of the school Prado Veraniego and director Wilson Suarez Parrado of the school Almirante Padilla for their help in allowing us to perform this research in their high schools in Colombia. Moreover, we would like to thank to teachers Cindy Carolina Vasquez, Fernando Martinez, Wily Orejuela Ramirez who allows to do this study in their classroom. The authors very much appreciate the support to this research by: (1) MinCiencias, Colciencias (Colombia), (2) the research group called "Destino" validated by Colciencias (Colombia), and (3) the Dutch NWO/NRO Senior Comenius Fellowship 'Schola Ludus' (PI: Marie Postma).

References

1. Sierra Rativa, A., Vasquez, C.C., Martinez, F., Orejuela Ramirez, W., Postma, M., van Zaanen, M.: The effectiveness of a robot animal as a virtual instructor. In: Lepuschitz, W., Merdan, M., Koppensteiner, G., Balogh, R., Obdržálek, D. (eds.) Robotics in Education RiE 2020. Adv. Intel. Syst. Comput. **1316**. Springer, Cham (2021). https://doi.org/10.1007/978-3-030-67411-3_30
2. Witkower, Z., Tracy, J.L., Pun, A., Baron, A.S.: Can children recognize bodily expressions of emotion? J. Nonverbal Behav. 1–14 (2021). https://doi.org/10.1007/s10919-021-00368-0
3. Lierheimer, K., Stichter, J.: Teaching facial expressions of emotion. Beyond Behav. **21**(1), 20–28 (2012)
4. Gantayat, S.S., Lenka, S.: Study of algorithms and methods on emotion detection from facial expressions: a review from past research. In: Communication Software and Networks, pp. 231–244. https://doi.org/10.1007/978-981-15. Author, F.: Contribution title. In: 9th International Proceedings on Proceedings, pp. 1–2 (2021)
5. Ekman, P.: Universal facial expressions of emotions. California Mental Health Res. Digest **8**(4), 151158 (1970)
6. Darwin, C.: The Expression of the Emotions in Man and Animals. University of Chicago Press (2015)
7. Zhan, C., Li, W., Ogunbona, P.O., Safaei, F.: Facial expression recognition for multiplayer online games (2006)
8. Carvalhais, T., Magalhães, L.: Recognition and use of emotions in games. In: 2018 International Conference on Graphics and Interaction (ICGI), November, pp. 1–8. IEEE (2018). https://doi.org/10.1109/ITCGI.2018.8602898
9. Hoffmann, H., Kessler, H., Eppel, T., Rukavina, S., Traue, H.C.: (2010)
10. Ekman, P.: Lie catching and microexpressions. Philos. Deception **1**(2), 5 (2009)
11. Bettadapura, V.: Face expression recognition, and analysis: the state of the art. arXiv preprint. arXiv:1203.6722 (2012)
12. Yan, W.J., Wu, Q., Liang, J., Chen, Y.H., Fu, X.: How fast are the leaked facial expressions: the duration of micro-expressions. J. Nonverbal Behav. **37**(4), 217–230 (2013). https://doi.org/10.1007/s10919013-0159-8
13. Svetieva, E., Frank, M.G.: Empathy, emotion dysregulation, and enhanced microexpression recognition ability. Motiv. Emot. **40**(2), 309–320 (2016). https://doi.org/10.1007/s11031-015-9528-4
14. Godavarthy, S.: Microexpression spotting in video using optical strain. Graduate Theses and Dissertations. University of South Florida (2010)
15. Pardàs, M., Bonafonte, A.: Facial animation parameters extraction and expression recognition using Hidden Markov models. Signal Process.: Image Commun. **17**(9), 675–688 (2002). https://doi.org/10.1016/S0923-5965(02)00078-4
16. Bourel, F., Chibelushi, C.C., Low, A.A.: Recognition of facial expressions in the presence of occlusion. In: BMVC, pp. 1–10 (2001)
17. Lewinski, P., den Uyl, T.M., Butler, C.: Automated facial coding: validation of basic emotions and FACS AUs in FaceReader. J. Neurosci. Psychol. Econ. **7**(4), 227 (2014). https://doi.org/10.1037/npe0000028
18. Noldus, L.P.J.J.: Physiological computing reshapes user-system interaction research and its practical application (2018)
19. Suhr, Y.T.: FaceReader, a promising instrument for measuring facial emotion expression? A comparison to facial electromyography and self-reports, Master's thesis (2017)
20. Baltrušaitis, T., Robinson, P., Morency, L.P. Openface: an open source facial behavior analysis toolkit. In: 2016 IEEE Winter Conference on Applications of Computer Vision (WACV), March, pp. 1–10. IEEE (2016). https://doi.org/10.1109/WACV.2016.7477553
21. Baltrusaitis, T., Zadeh, A., Chong Lim, Y., Morency, L.-P.: Openface 2.0: Facial behavior analysis toolkit. In: 2018 13th IEEE International Conference on Automatic Face & Gesture Recognition, pp. 59–66. IEEE (2018). https://doi.org/10.1109/FG.2018.00019

22. Yudiarso, A., Liando, W., Zhao, J., Ni, R., Zhao, Z. Validation of facial action unit for happy emotion detection. In: Proceedings of the 3rd International Conference on Psychology in Health, Educational, Social, and Organizational Settings (ICP-HESOS 2018) (2020). https://doi.org/10.5220/0008589403600363
23. Thirumaran, K., Chawla, S., Dillon, R., Sabharwal, J.K.: Virtual pets want to travel: engaging visitors, creating excitement. Tour. Manag. Perspect. **39**, 100859 (2021). https://doi.org/10.1016/j.tmp.2021.100859
24. Liang, W., Yu, X., Alghofaili, R., Lang, Y., Yu, L.F.: Scene-aware behavior synthesis for virtual pets in mixed reality. In: Proceedings of the 2021 CHI Conference on Human Factors in Computing Systems, pp. 1–12 (2021). https://doi.org/10.1145/3411764.3445532
25. Kokai, J.A.: The nemofication of nature: animals, artificiality, and affect at disney world. In: Performance and the Disney Theme Park Experience, pp. 87–106. Palgrave Macmillan, Cham (2019). https://doi.org/10.1007/978-3-030-29322-2_5
26. Schneider, E., Wang, Y., Yang, S.: Exploring the uncanny valley with Japanese video game characters. In: DiGRA Conference, September (2007)
27. Schwind, V., Wolf, K., Henze, N.: Avoiding the uncanny valley in virtual character design. Interactions **25**(5), 45–49 (2018). https://doi.org/10.1145/3236673
28. Sierra Rativa, A; Postma, M., Zaanen van, M.: Study I: FaceReader data (Emotional Facial Expressions) after watching a virtual instructor, DataverseNL **1** (2022). https://doi.org/10.34894/O1S3N9
29. Viola, P., Jones, M.: Rapid object detection using a boosted cascade of simple features. In: Proceedings of the 2001 IEEE Computer Society Conference on Computer Vision and Pattern Recognition. CVPR 2001, vol. 1, pp. I–I. IEEE (2001). https://doi.org/10.1109/CVPR.2001.990517
30. Cootes, T.F., Taylor, C.J.: Statistical Models of Appearance for Computer Vision. Imaging Science and Biomedical Engineering. University of Manchester, Manchester (2004)
31. Gudi A., Tasli, H. E., Den Uyl, T. M., Maroulis, A.: Deep learning based facs action unit occurrence and intensity estimation. In: 2015 11th IEEE International Conference and Workshops on Automatic Face and Gesture Recognition (FG), vol. 6, pp. 1–5. IEEE (2015). https://doi.org/10.1109/FG.2015.7284873
32. Bestelmeyer, P.E., Kotz, S.A., Belin, P.: Effects of emotional valence and arousal on the voice perception network. Soc. Cogn. Affect. Neurosci. **12**(8), 1351–1358 (2017). https://doi.org/10.1093/scan/nsx059

Gamification and Competitions

Learning Through Competitions—The FIRA Youth Mission Impossible Competition

Jacky Baltes, Reinhard Gerndt, Saeed Saeedvand, Soroush Sadeghnejad, Petr Čížek, and Jan Faigl

Abstract This paper discusses challenges and opportunities when using competitions in robotics education. The authors describe the Federation of International RoboSports Association (FIRA) competition and, in particular, the FIRA Youth—Mission Impossible, an event targeted at overcoming problems of students consciously or thoughtlessly plagiarizing, by forcing students to solve previously unknown tasks. The paper shows an example of the positive influence robot competitions can have on cutting-edge research, even when targeted at younger roboticists. The FIRA Youth—Mission Impossible 2022 competition, where students had to measure the weight of bottles, and hence the wrench applied on the robot using only

The work of J. Faigl and P. Čížek has been supported by the Czech Science Foundation (GAČR) under research project No. 21-33041J and by the OP VVV funded project CZ.02.1.01/0.0/0.0/16_019/0000765 "Research Center for Informatics."

J. Baltes
National Taiwan Normal University, Taipei, Taiwan
e-mail: jacky.baltes@ntnu.edu.tw

University of Manitoba, Winnipeg, Canada

R. Gerndt (✉) · S. Saeedvand
Ostfalia University of Applied Sciences, Wolfenbüttel, Germany
e-mail: r.gerndt@ostfalia.de

S. Saeedvand
e-mail: saeedvand@ntnu.edu.tw

S. Sadeghnejad
Department of Biomedical Engineering, Amirkabir University of Technology (Tehran Polytechnic), Tehran, Iran
e-mail: s.sadeghnejad@aut.ac.ir

P. Čížek · J. Faigl
Department of Computer Science, Faculty of Electrical Engineering, Czech Technical University in Prague, Prague, Czechia
e-mail: cizekpe6@fel.cvut.cz

J. Faigl
e-mail: faiglj@fel.cvut.cz

R. Balogh et al. (eds.), *Robotics in Education*, Lecture Notes in Networks and Systems 747, https://doi.org/10.1007/978-3-031-38454-7_31

proprioception, inspired an initial approach and the creation of a practical test-bed for much more complex wrench estimation on hexapod robots.

Keywords Educational robot · Competition design · FIRA · Mission impossible · Wrench estimation

1 Introduction

Today, many problems threaten quality of life, such as climate change, air pollution, food shortage and poverty. Technology plays an essential role in overcoming these problems. Furthermore, since daily live is becoming increasingly dependent on technology, it is essential to train young people in their use, limits, and dangers. This paper focuses on intelligent robotics as one of the most relevant core technologies for the later part of the 21st century and beyond. Robot competitions can easily motivate teachers, students, hobbyists, researchers, and professors since cutting-edge technology can be developed using fun and challenging application domains.

Robot competitions, possibly based on a few existing real-world applications for may also act as an benchmark to evaluate current and guide future research. The importance of good benchmarks for intelligent robotics and AI research has already been discussed, and their important features are described in [7–9, 13]. Cleaning robots, autonomous driving, and delivery robots are the robotics applications that are probably closest to large-scale deployment. Other worthwhile applications like humanoid robots acting as autonomous firemen may still be many years away from reality.

The main contributions of this paper are as follows: In Sect. 2 we highlight the danger of commercial entities influence robot competitions by providing specially designed hardware and software and thus "unfairly" tilting the playing field. Section 3 introduces the Federation of International Robosports Association (FIRA), the design of the FIRA Youth—Mission Impossible competition and describe the experiences of the Federation of International Robosports Association (FIRA) community with their leagues for those under 14 years old (U14) and under 19 years old (U19) participants [3]. Section 4 emphasises a specific implementation of the FIRA Mission Impossible competition that was developed to particularly ameliorate the danger of commercial influence by forcing students to develop solutions from scratch in a short time using similar hardware. An impact of the robot competition on serious research is exemplified in Sect. 5, where research on external wrench estimation for the hexapod walking robot is sketched. Section 6 concludes the paper and provides ideas for further development of robot competitions.

2 Influence of Companies on Robotic Competitions

The FIRA Youth competition includes leagues for robot sports, disaster recovery, and battle robots. One issue that the organizers noticed is that the rules of these leagues, as in many other robotics competitions too, are relatively static. Each year, the chairs and technical committees, responsible for the rules of their leagues, make only small adjustments. This leads to the problem of companies producing robot kits, specifically targeted at winning robot competitions.

Typically designers of educational robotics games carefully consider what hardware and software is available to the students targeted and what skills the students must develop. For example, building a robot for one of the most widely used competitions, a line follower or line tracker (see Fig. 1 left), can be done with a single or two light sensors. However, line-following with only one or two light sensors requires non-trivial algorithms and programming. Challenges increase for a path with intersections, holes, and color changes. It requires sensor reading, calibration, filtering, and high-level reasoning. On the other hand, several robot education kits from Chinese and Korean manufacturers include a light sensor array with six or more sensors (see Fig. 1 right), which simplifies the control problem significantly. Moreover, the kits also include high performance line-following software. Students can create a competitive entry by building the robot in the kit and by turning it on. The actual competition requires calibration and high level programming (e.g., `follow_line`, `turn_right_at_intersection()`, `follow_line()`).

Companies providing professionally developed and manufactured hardware and software for the exact competition environment tilts the playing field much more than students finding solutions online (i.e., the thoughtless plagiarism problem). Finding plans and software from various sources and of varying quality, often poorly documented, still requires the students to integrate and adapt their solutions.

One solution to the problem is limiting the hardware allowed in the competition by narrow specifications. For example, the FIRST Lego competition and the World Robot Olympics require students to build their solutions using Lego parts only [3, 6]. Another solution is to have different classes in the competition, e.g. one for

Fig. 1 A typical line follower/tracker competition (on the left) requires non-trivial programming using only one or two light sensors but can be trivially solved with six sensor bars and their associated software (on the right)

the ubiquitous Lego educational robot kits and one for custom-developed solutions or introduce classes based on the cost of the robot.

In view of openness and fostering creativity of students, these approaches are unsatisfactory: (i) discriminatory limiting building blocks to a fixed list of parts limits creativity and unconventional solutions. Comparing costs is close to impossible for international competitions as component prices and qualities vary significantly over the world. Moreover, some specific components may not be available in some regions.

Many students are exposed to robotics during extracurricular activities, outside the official curriculum of their education. Still, many schools attract clients with their successes in international robot competitions. Therefore, these schools often have teachers and/or parents developing high-quality solutions, often including custom hardware, for specific events and then teach students to install and tune their existing software rather than develop their own solutions. For example, one of the authors had 12-year-old students show their VHDL code for an FPGA-based hardware platform, which the students claimed to have developed independently. However, upon request, the students could not compile their code independently.

Collaboration between students, their teachers, and parents is a pedagogically valuable element in learning if underpinned with a respective concept. However, in order to acquire robotics knowledge and competencies, students should be be involved in relevant aspects of designing, implementing and debugging their robot and should have the time to explore the solution space of a given problem.

3 The Federation of International RoboSports Association

The Federation of International Robosports Association (FIRA) was founded by Prof. Jong-Hwan Kim from KAIST, Korea, in 1996. It is the oldest soccer robot competition in the world. Adding new leagues such as drones races and autonomous driving and new age categories include FIRA Youth for under 14 years old (U14) and under 19 years old (U19) under the new presidency of Prof. Jacky Baltes it has seen large growth. Now the annual FIRA Robot WorldCup competition regularly brings together about 1,200 participants of different ages. The aim of FIRA is to (a) provide benchmark problems for robotics research, (b) motivate and educate young researchers, and (c) make robotics research accessible to the general public.

The most prestigious of the FIRA competitions is the HuroCup competition for intelligent humanoid robots. Its focus is on providing challenges and benchmarks for humanoid robotics research, particularly active balancing and push recovery, complex motion planning, and human-robot interaction. The humanoid robots in the event must be fully autonomous. Moreover, a single humanoid robot must compete in a decathlon of archery, basketball, triple jump, marathon, obstacle run, Spartan race, sprint, united soccer, weightlifting, and mini-drc, see Fig. 2. A single humanoid robot is capable to perform a vast variety of tasks, which are relevant for humans, unlike, possibly more efficient special-purpose solution for a single task. For example, a wheeled robot can deliver mail in an office more cheaply and reliably. Similarly, a

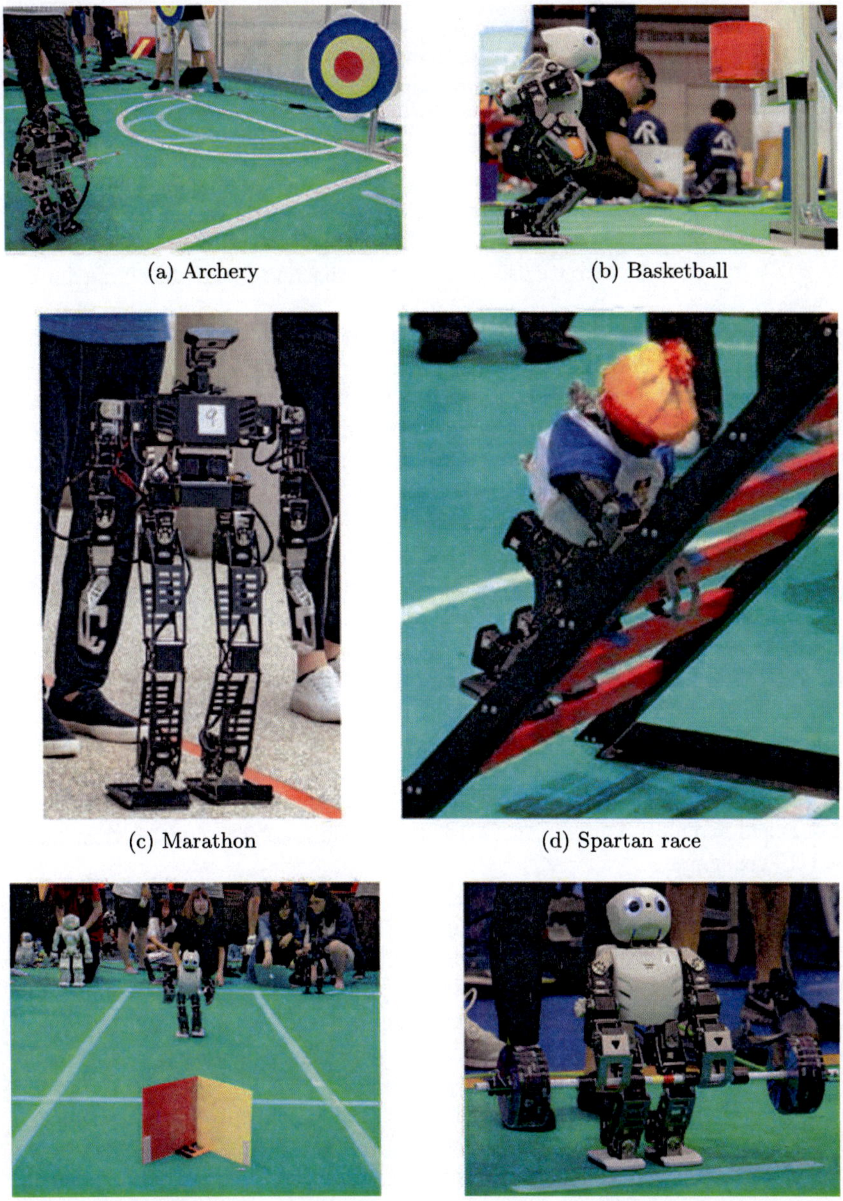

(a) Archery (b) Basketball

(c) Marathon (d) Spartan race

(e) Sprint (f) Weightlifting

Fig. 2 Events of the FIRA HuroCup archery, basketball, marathon, Spartan race, sprint, and weightlifting robot with suction cups is better able to clean windows. The teams are not allowed to modify their robot physically between events, as the creators of the competition believe that intelligence is the ability to adapt and adjust to many different tasks and environments, rather than solving a specific problem optimally. In fact, humans generally do not find the optimal solution but find satisfying solutions, that is, solutions that are good enough in practice. For example, when grocery shopping, people follow the store layout instead of solving the traveling salesman problem in their heads.

FIRA also includes HuroCup Jr., a subset of slightly simplified HuroCup event for U19 participants, where we used a similar approach to design the competition.

Rules of FIRA Youth—Mission Impossible The FIRA Youth—Mission Impossible is another competition targeted to younger participants that especially attempts to alleviate the problems of the commercial or parents-made hardware and software solutions discussed above. Here Students must implement a solution to a previously unknown task within 3 h. FIRA Youth—Mission Impossible uses tasks in many different environments to avoid overspecialization. Sometimes, tasks are more focused on the hardware and mechanics of the device, such as a boat driven by rubber bands, whereas intelligent software is a priority at other times.

For the competition students bring their robot hardware and software. The maximum number of actuators (four continuous revolution motors with or without position feedback, six servo motors) and sensors (six IR sensors, four ultrasound, and four touch sensors) are described in the rules [5]. In addition, students are allowed to use commonly available materials (e.g., cardboard, Lego pieces, wood, and metal) and tools such as pliers, drills, and screwdrivers. For certain events, the rules may include additional restrictions. For example, students may only be allowed to use a maximum of one actuator (any type), and/or may only use a maximum of three light sensors and two ultrasound sensors. U14 students must solve simplified versions of the U19 challenges.

FIRA Youth—Mission Impossible United The challenges in the FIRA Youth—Mission Impossible United are such that multiple robots from different teams must synchronize and collaborate to achieve the goal; see Fig. 3. For the United competition students and their robots are partnered randomly with students from another country. A respective student team typically consists of four students from two different nationalities and two robots.

When introducing the FIRA Youth—Mission ImpossibleUnited event, many skeptics were concerned that the students' mother languages are different and that some of them cannot speak English, as a common language, well. Hence, all participants must provide a one-page introduction in English before being admitted to the FIRA Youth—Mission Impossible. In practice, we realized that students often manage even without English since the topic of discourse is robotics, which is expressed in mathematics or source code and students nowadays also use web-based translation.

Requesting collaboration during the competition fosters networking among the participants. Many friendships were forged in the late hours while trying to prepare the robot for the next day's competition. At the undergraduate and graduate student levels, many students were able to connect with their future supervisors. Our experience is that these friendships and exchanges are common for older students, but the connections are limited to members of the national teams in the U14 and U19 age groups.

Reception of FIRA Youth—Mission Impossible The FIRA Youth—Mission Impossiblecompetition has proven to be very popular with the FIRA Youth participants and is the largest and most prestigious FIRA Youth competition. One of the surprising results of introducing the FIRA Mission Impossible was their reception by the teachers and educators. Occasionally there have been complaints, e.g. when teams had to tape over their extra light sensors in their sensor bars, thus making their standard software useless. However, we learned at many wrap-up meetings with teachers that many teachers really enjoyed preparing their teams. The teachers we interviewed explained that even though it is relatively easy to build a competitive entry for events where commercial solutions are available, creating a winning entry often requires optimizing the calibration of sensors and/or fine-tuning PID control loops to improve a run by a few microseconds. However, they much preferred teaching the scientific method, physics, mechanics, electronics, and real-time programming to sensor calibration and tuning control loops. As a positive aspect, teachers pointed out the possibility of introducing specific topics to their teaching, such as how the center of mass affects the balance, how to build a robot from scrap materials, and how to use a light sensor as a scanner since these may be chosen as the next FIRA Mission Impossible. Indeed, the FIRA Mission Impossible task in 2016 was to design a robot that could balance on a small horizontal rod. In 2017, the participants had to build a small boat that could raise a treasure, and in 2018, the challenge was to distinguish different images on a wall.

4 FIRA Youth—Mission Impossible 2021—Bottle Weights

In 2021, the FIRA Robot WorldCup was organized as a virtual event, due to the COVID pandemic. The event was held and refereed simultaneously in several labs worldwide and coordinated on the FIRA Discord server.[1] Of course, most of the interaction and networking of the participants was limited to their local hub, but he virtual competition still motivated and excited students, especially since the teams still saw the other participants running their robot. In the future, we would try to organize several online sessions, which would allow students to mingle before and after the event.

Teams had to design and implement a robot that can estimate the weight of four bottles with 100, 200, 500, and 1000 mL of water added. The bottles were attached to the robot via a rope and pulley system; see Fig. 3.

[1] https://discord.gg/QDpjK7Gfxe.

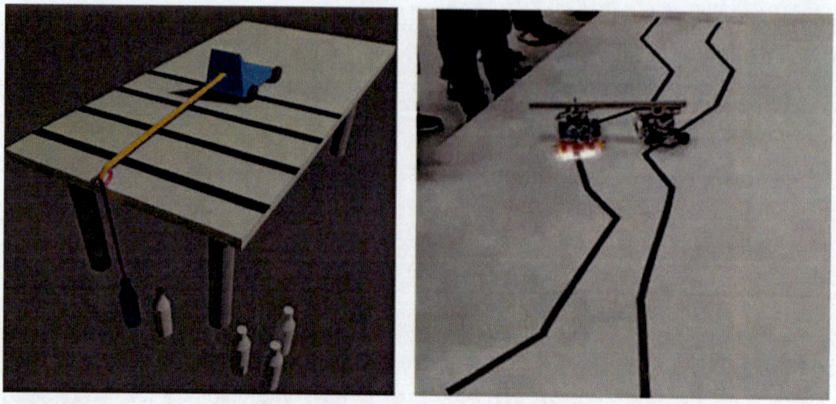

Fig. 3 The FIRA Youth—Mission Impossible 2021 competition required teams to estimate the weight of several bottles (left); the FIRA Youth—Mission Impossible United competition asked students from different countries to carry a beam (right)

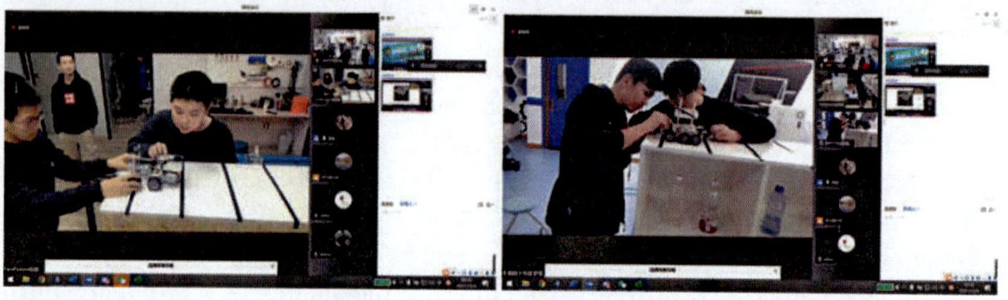

Fig. 4 Participants of the FIRA Youth—Mission Impossible 2021

After announcing the rules of the competition, the teams discussed different methods for estimating the weight of the attached bottle. After a short while, they realized that the force acting on the robot could be estimated by controlling the power/torque settings of the motors (i.e., controlling the pulse width modulation (PWM) duty cycle of the DC motor) and then measuring the time that it takes the robot to drive across the lines marked on the table. It is important that the robot drives in a straight line to make the measurement reliable.

Some teams used the Lego analog motors, which control the velocity directly so that the robot moves at a selected velocity in spite of an external force applied to the robot. Some U19 teams worked around that problem by driving the robot forward across the lines and then by letting the bottle drag the robot backward. A snapshot from the competition is depicted in Fig. 4.

5 Wrench Estimation for Hexapod Robots

This section discusses an example of how robot competitions targeted at young students can provide important and meaningful inspiration for further research. A team of researchers at the Czech Technical University in Prague (CTU) is working on multi-legged walking robots and is experienced with intelligent robotics and developing impressive robots [1]. CTU-teams scored multiple times in the prestigious DARPA Subterranean (SubT) competition [2] or MBZIRC competition [12]. The main hexapod walking robots of the CTU team is the SCARAB II (left) and larger Lily robot (right) depicted in Fig. 5.

The dynamics of multi-legged robots are difficult to model since they experience complex environmental interactions. Apart from the reaction forces of the feet touching the ground on several contact points, interactions due to uneven terrain, collisions with obstacles [11], or user interaction [19] occur. Most previous approaches model these interactions only implicitly [10, 15, 18] that is the control policy compensates for external disturbances.

However, explicit modeling of the external wrench opens new and interesting opportunities for intelligent robotics; since the robot can "sense" an obstacle that is outside of the view of its camera, it can solve "peg in hole" type of manipulation problems, maintaining tension on a tether or leash, or safe collaborative manipulation [14]. Hence, there is increased interest in explicit wrench estimation. For example, the authors of the survey [17] report on the increased interest in explicit wrench estimation, specifically during the DARPA Robotics Challenge. Wrench estimation is non-trivial since the dynamics are difficult to model due to the strong coupling of the robot base with legs and ground contacts [16] (Fig. 6).

In search of a suitable approach to develop and benchmark external wrench on legged robots, the researchers realized that the bottle weight mission is an easy test setup that can consistently apply a known torque without other significant impacts on the kinematics and dynamics. Thus, they created two test beds, one using a flat plane and one using an irregular surface, see Fig. 7 left and right. The test beds used a motion capture system to measure the exact position of the robot with an accuracy of one millimeter.

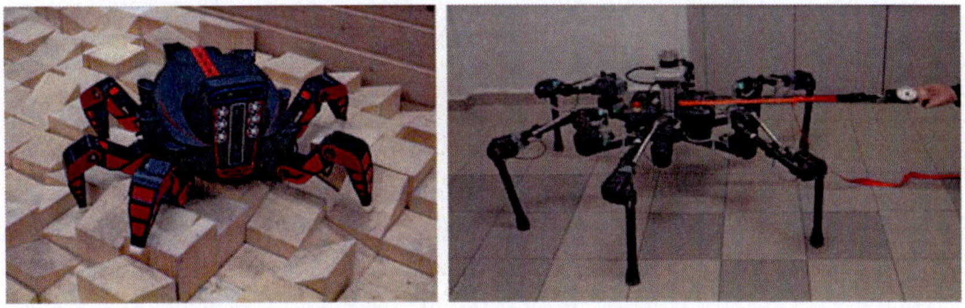

Fig. 5 CTU's SCARAB II (left) and the larger Lily robot (right)

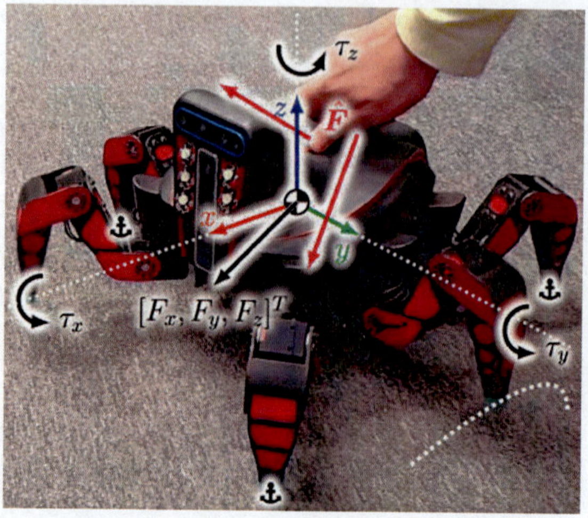

Fig. 6 Illustration of an external wrench acting on the six-legged walking robot SCARAB II in human-robot interaction and ground contact. The operator exerts external forces $\hat{\boldsymbol{F}}$ on the chassis of the robot that is in motion. The robot estimates the resulting wrench $\boldsymbol{\mathcal{F}}_{\mathrm{d}} = [F_x, F_y, F_z, \tau_x, \tau_y, \tau_z]^T$ w.r.t. its center of mass

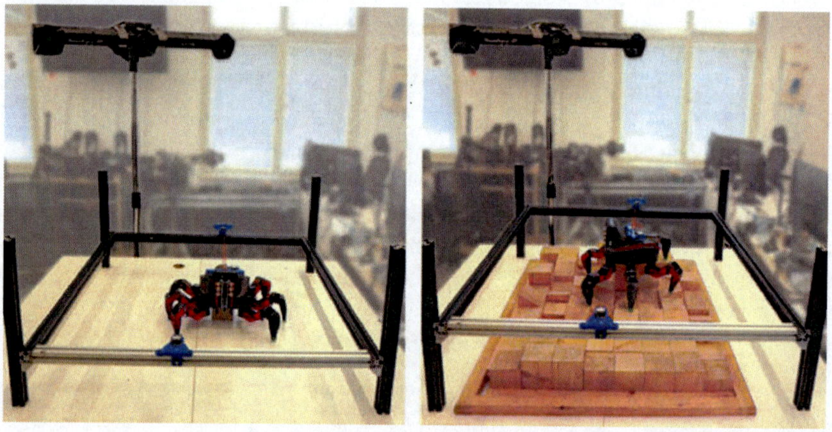

Fig. 7 The test bed for evaluation of the external wrench on a flat (left) and irregular terrain (right). The weight inducing a known external wrench on the robot is suspended on a cord attached to the robot. The position of the robot is tracked using a motion capture system

The exact method of the new wrench estimation is not within the scope of this paper will only be sketched shortly. The new approach for external wrench estimation is based on the formulation of the whole-body dynamic model of the robot used to derive the analytical formulation for the ground reaction forces and external wrench estimation, which are only forces and torques acting at known and unknown contact points of the robot, respectively, as it is shown in Fig. 6, and formulated in the following equation of the robot's rigid-body dynamics model.

$$M \begin{bmatrix} \ddot{\mathcal{X}} \\ \ddot{q} \end{bmatrix} + \begin{bmatrix} \eta_{\mathcal{X}} \\ \eta_q \end{bmatrix} = \begin{bmatrix} 0 \\ \tau \end{bmatrix} + \sum_{i=1}^{N_l} \begin{bmatrix} J_{\mathcal{X}_i}^T \\ J_{q_i}^T \end{bmatrix} \begin{bmatrix} F_{e_i} \\ 0 \end{bmatrix} + \begin{bmatrix} \mathcal{F}_{\mathrm{d}} \\ 0 \end{bmatrix}, \tag{1}$$

where $\mathcal{X} = [x, y, z, \theta_x, \theta_y, \theta_z]^T$ is the robot's position and orientation in the global reference frame, N_j is the number of robot's controllable degrees of freedom, $q \in \mathbb{R}^{N_j}$ is the vector of the generalized joint coordinates, $M \in \mathbb{R}^{N_j+6 \times N_j+6}$ is the system inertia matrix, $\eta_{\mathcal{X}} \in \mathbb{R}^6$ and $\eta_q \in \mathbb{R}^{N_j}$ are the joined Coriolis, centrifugal, and gravity effects of the body and joints, respectively. $\tau \in \mathbb{R}^{N_j}$ is the motors torque, N_l is the number of robot's legs, $J_{\mathcal{X}_i} \in \mathbb{R}^{6\times 6}$ and $J_{q_i} \in \mathbb{R}^{6\times N_j}$ are the contact Jacobian with respect to (w.r.t.) the body and legs, respectively, $F_{e_i} \in \mathbb{R}^3$ represents the individual ground reaction forces, and $\mathcal{F}_{\mathrm{d}} \in \mathbb{R}^6$ is the cumulative external wrench acting on the robot body.

The example shows how robotics competitions can support lead to and support serious research.

6 Conclusions and Future Work

Robot competitions are a powerful tool for teachers to motivate students and evaluate their progress. However, in order to reach these goals organizers must be aware of some potentially problematic issues and design competitions well with a pedagogical concept in mind, as we detailed in this paper. Commercial interests and external ambitions can affect the learning outcomes of competitions, especially if the rules and environment of competitions remain the same with little changes and do not evolve. If competitions are designed such that students can prepare robots and software beforehand, changes during the competition ensure an appropriate involvement of the students. Teachers of participating teams confirmed ad-hoc changes to competitions as being positive, even though students need to be prepared for on-the-spot thinking and quick problem solving. Ad-hoc peer teams, like in FIRA Youth—Mission Impossible United fosters networking and collaboration.

A participant informed us that since about 20% of the teams can solve the entire challenge, it is really not a mission *impossible*. It may be of great value to teach today's youth that solutions are, despite older people in authority trying to tell them that it cannot be done. We close with a famous quote by George Bernhard Shaw [4].

> Reasonable people adapt themselves to the world. Unreasonable people attempt to adapt the world to themselves. All progress, therefore, depends on unreasonable people.

The FIRA Youth—Mission Impossible organizers and the rest of the FIRA community hope to inspire the unreasonable youth of today.

References

1. Center for Robotics and Autonomous Systems, Czech Technical University in Prague Homepage. http://robotics.fel.cvut.cz/cras/. Accessed 02 Jan. 2023
2. DARPA Subterranean Challenge (SubT) Results. https://subtchallenge.com/results.html. Accessed 02 Jan. 2023
3. Federation of international roboathletes association (fira) homepage. https://www.firaworldcup.org. Accessed 02 Jan. 2023
4. George bernard shaw quotes. https://www.goodreads.com/quotes/1036543-reasonable-people-adapt-themselves-to-the-world-unreasonable-people-attempt. Accessed 02 Jan. 2023
5. Rules of FIRA Youth—Mission Impossible. https://firaworldcup.org/leagues/fira-youth/mission-impossible/. Accessed 02 Jan. 2023
6. World robot olympiad homepage. https://www.worldrobotolympiad.de. Accessed 02 Jan. 2023
7. Anderson, J., Baltes, J., Cheng, C.T.: Robotics competitions as benchmarks for AI research. Knowl. Eng. Rev. **26**(1), 11–17 (2011). https://doi.org/10.1017/S0269888910000354
8. Baltes, J.: A benchmark suite for mobile robots. In: IEEE/RSJ International Conference on Intelligent Robots and Systems (IROS). vol. 2, pp. 1101–1106 (2000). https://doi.org/10.1109/IROS.2000.893166
9. Baltes, J., Saeedvand, S.: Rock climbing benchmark for humanoid robots. In: 2022 International Conference on Advanced Robotics and Intelligent Systems (ARIS), pp. 1–4 (2022). https://doi.org/10.1109/ARIS56205.2022.9910449
10. Bledt, G., Wensing, P.M., Ingersoll, S., Kim, S.: Contact model fusion for event-based locomotion in unstructured terrains. In: IEEE International Conference on Robotics and Automation (ICRA), pp. 4399–4406 (2018). https://doi.org/10.1109/ICRA.2018.8460904
11. Buchanan, R., Bandyopadhyay, T., Bjelonic, M., Wellhausen, L., Hutter, M., Kottege, N.: Walking posture adaptation for legged robot navigation in confined spaces. IEEE Robot. Autom. Lett. **4**(2), 2148–2155 (2019). https://doi.org/10.1109/LRA.2019.2899664
12. Dias, J., Lima, P.U., Seneviratne, L., Khatib, O., Tadokoro, S., Dario, P.: Journal of field robotics special issue on MBZIRC 2017 challenges in autonomous field robotics. J. Field Robot. **36**(1), 3–5 (2019)
13. Gerndt, R., Seifert, D., Baltes, J.H., Sadeghnejad, S., Behnke, S.: Humanoid robots in soccer: Robots versus humans in Robocup 2050. IEEE Robot. Autom. Mag. **22**(3), 147–154 (2015). https://doi.org/10.1109/MRA.2015.2448811
14. Haddadin, S., De Luca, A., Albu-Schaffer, A.: Robot collisions: a survey on detection, isolation, and identification. IEEE Trans. Robot. **33**(6), 1292–1312 (2017). https://doi.org/10.1109/TRO.2017.2723903
15. Kalouche, S., Rollinson, D., Choset, H.: Modularity for maximum mobility and manipulation: control of a reconfigurable legged robot with series-elastic actuators. In: IEEE International Symposium on Safety, Security, and Rescue Robotics (SSRR), pp. 1–8 (2015). https://doi.org/10.1184/R1/6555620.v1
16. Mahapatra, A., Roy, S.S., Pratihar, D.K.: Multi-legged robots—A review, pp. 11–32. Springer Singapore (2020). https://doi.org/10.1007/978-981-15-2953-5
17. Masuya, K., Ayusawa, K.: A review of state estimation of humanoid robot targeting the center of mass, base kinematics, and external wrench. Adv. Robot. **34**(21–22), 1380–1389 (2020). https://doi.org/10.1080/01691864.2020.1835532
18. Morlando, V., Teimoorzadeh, A., Ruggiero, F.: Whole-body control with disturbance rejection through a momentum-based observer for quadruped robots. Mech. Mach. Theory **164**, 104412 (2021). https://doi.org/10.1016/j.mechmachtheory.2021.104412
19. Yang, C., Sue, G.N., Li, Z., Yang, L., Shen, H., Chi, Y., Rai, A., Zeng, J., Sreenath, K.: Collaborative navigation and manipulation of a cable-towed load by multiple quadrupedal robots. IEEE Robot. Autom. Lett. **7**(4), 10041–10048 (2022). https://doi.org/10.1109/LRA.2022.3191170

Using a Robot to Teach Python

Minas Rousouliotis, Marios Vasileiou, Nikolaos Manos, and Ergina Kavallieratou

Abstract Educational robotics is rooted in Constructionism and allows learners to investigate and discover new concepts. It would appear that learning a programming language while programming a robot is more motivating and productive than conventional methods. El Greco platform is an educational platform built to teach Python. Users can control El Greco from any computer connected to the Internet due to the platform's web-based interface. El Greco is a social humanoid robot built to be affordable and appropriate for use in education. Potential users can use Python direct code entry or the Blockly library to control El Greco. The Blockly library embeds an editor in an application to represent coding notions like interlocking blocks. Unique functions that control El Greco were created. The inserted code can be executed on the website or by the Robot. The user can view the result of code execution through a live-streaming window. The El Greco platform has been designed with students in mind but is available to anyone at no cost.

Keywords Human–Computer interaction · Robot control · Educational computing · Distance learning · Computer programming

---

M. Rousouliotis (✉) · M. Vasileiou · N. Manos · E. Kavallieratou
Department of Information and Communication Systems Engineering, University of the Aegean Samos, Karlovasi, Greece
e-mail: mrousouliotis@aegean.gr

M. Vasileiou
e-mail: mvasileiou@aegean.gr

N. Manos
e-mail: manos@aegean.gr

E. Kavallieratou
e-mail: kavallieratou@aegean.gr

R. Balogh et al. (eds.), *Robotics in Education*, Lecture Notes in Networks and Systems 747, https://doi.org/10.1007/978-3-031-38454-7_32

1 Introduction

Real-world experience is viewed as the primary source of learning in Constructionism. Using robots as guides, students in the field of educational robotics investigate and uncover previously unknown concepts. The Robot is then used to put these hypotheses to the test; from the results of these experiments, further refinement or expansion of the accumulated knowledge is possible [1]. Educational robots aim to increase students' cognitive abilities and school performance through collaborative problem-solving, research, and knowledge-building [2]. As an added bonus, young people find new technologies, especially robotics, fascinating, which can only help their education [3].

Our primary objectives have been to develop (i) a platform that is freely available to anyone interested in learning Python programming, (ii) a pleasant and enticing interface, (iii) a framework that enables for remote control of a Robot over the Internet, (iv) a programming error feedback system, and (v) a data collection scheme constructed to gather information to enhance the learning process of the Platform. This scheme is entirely automated and anonymous.

Aegean Robotics Team build a Python learning platform [4]. To improve that Platform, a short survey was conducted. Two questionnaires were used, one before and one after using the Platform. Fifteen children answered both of them, one from the first grade, five from the 2nd grade and nine from the 3rd grade of lyceum. The first aimed to examine if the pupils had prior computer programming skills and were keen on using robots at school. The findings were that 65.5% had no previous programming experience, 79.3% believed they could use robotics to study their lessons, and 75.9% thought robotics could be used for educational purposes. In addition, 69% would like to have access to a training platform that utilizes a Robot.

The second questionnaire targeted the usage of El Greco and the children's experience with the Robot and its programming environment. 44.8% believed that the Platform could be used to study various fields of knowledge (e.g., physics or mathematics). In addition, 69% declared that the Platform's environment was user-friendly and easy to use.

The El Greco platform is the upgraded version of the abovementioned Platform. The enhancements are: (i) Running Python directly using Brython [5]. (ii) Detection of errors in the source code and generation of user-friendly error messages. (iii) The execution of code in a step-by-step fashion. (iv) A booking system for making reservations to use the El Greco robot. (v) An Administrator app for remote management, including bookings. (vi) El Greco adventure mode. (vii) Enhancements to the website's user interface and overall appearance. (viii) Elements fundamental to a website, including signup and verification processes, as well as a Python tutorial and the ability to create profiles. And finally, (ix) the creation of log files that take place automatically and retrieves information about the user's performance and interactions with the webpage and the Robot in a completely anonymous way. This information may be put to use in the process of further developing the platform.

In the following section of this paper, similar platforms are compared to the El Greco platform. Section 3 introduces the humanoid Robot El Greco and its capabilities, while Sect. 4 discusses the website implementation. Section 5 presents the El Greco platform and its main characteristics. Finally, in Sect. 6, we draw some conclusions.

2 Similar Platforms

Like Blockly, Scratch is a block-based programming language that lets users create animations, video games, and digital storytelling [6]. Despite the popularity of the Scratch coding environment, the El Greco platform offers several advantages because it uses a real robot to teach the Python programming language.

Blockly is one of the most often used options in code-learning websites. To implement its intriguing features, BlockPy [7] relies on Blockly and is fully committed to Python. Linear plots, dispersion charts, and histograms can all be generated, and guided feedback on the code's output is also available. An exciting feature is a translation method that converts Python code to Python Blocks [8].

Our effort is related to BlockPy, but one of our objectives was to take advantage of educational robotics strengths. Reeborg's World [9] has influenced the development of the El Greco platform. Reeborg World was designed to make learning to code fun for novices. Although Python is the primary focus, JavaScript is also supported. Using Python or JavaScript, users may command a virtual robot named Reeborg. Reeborg can change into various robotic forms and navigate a virtual world filled with diverse objects that can interact with one another and the Robot. The accurate step-by-step execution of Python code is one of the most impressive features of Reeborg's world.

Inspired by Reeborg's World and other projects [1, 6] of a similar nature, we have developed a novel method for teaching Python. However, the key feature of the El Greco platform is that users can remotely control an actual robot, which makes learning more creative, exciting and fun.

Numerous robotic kits permit the programming of an educational robot using Python and Blockly [10–14]. USB, Bluetooth or a Wireless network are most often used to establish the link between the user interface and the Robot. In contrast, the El Greco platform relies on the Internet and, more particularly, a webpage where users engage with the Robot; hence, the El Greco platform enables anyone intrigued by Python to learn while commanding a robot. El Greco platform differs from the robotic kits mentioned above since it incorporates a social humanoid robot. The cost of an educational robot with El Greco's skills is often quite high, and its ability to boost students' creativity, interest, and enjoyment of learning is remarkable.

3 El Greco

The Robot El Greco (Fig. 1) was developed primarily: (i) To be cost-effective, (ii) to make use of prior knowledge and incorporate new features, (iii) to be ideal for education usage and (iv) to incorporate artificially intelligent systems that can be expanded and applied in a variety of applications [15].

To some extent, El Greco looks like a real person. The Robot is meant to look like a young boy about five or six years of age. It can mimic human expressions of joy, sorrow, disagreement, surprise, and even drowsiness. El Greco may also engage in winking and blinking. Eyebrow and eyelid movement contribute significantly to man's non-verbal communication and emotional expression [16].

El Greco's capabilities include interaction in Greek and 160 other languages, saluting, speech and face recognition, and person localization in the crowd. El Greco can access information about the weather, current events, etc., via the Internet. Python was used exclusively to program El Greco. El Greco movements are executed by 25 servomotors of various specifications, giving it 25 degrees of freedom.

The following are the primary components of El Greco's code [15]:

- For speech recognition, the Google Speech Recognition API [17] through the Python speech recognition library [18] are used.
- High-level functions are responsible for the movements of El Greco. GpioZero [19] establishes the link between the hardware and the high-level functions., which is used for the required low-level functions.
- The OpenCV library [20] is utilized for El Greco's image recognition abilities.

Fig. 1 The Robot named El Greco

- Google Text-to-Speech API [21] via gTTS [22] and Google Translate [23] are responsible for the multi-lingual speaking ability of El Greco.
- El Greco is capable of searching the Internet for information requested by the user. Feedparser [24] and Lxml [25] are utilised for Atom and RSS feeds.

4 Website Implementation

There are four essential components to the platform: The El Greco robot, the website, the server and the Playroom. The primary configuration of the server is XAMPP [26], OBS Studio [27], and Nimble Streamer [28]. XAMPP is one of the widely used web servers. From the software supplied with XAMPP, Apache was selected as the site's main server, and MySql is used as the El Greco platform database. The database includes user profiles, login information, and the files needed for El Greco Adventure level solutions and instructions. The live streaming content utilized by the Platform is captured, encrypted, and streamed via OBS Studio in SLDP [29] format, a socket-based web streaming protocol. This live-streaming content consists of audio and video and is served on the website by Nimble streamer, which acts as a secondary server if necessary. The website plays the stream using HTML5 SLDP Player [30]. PHPMailer [31] is used for the website's automatic messaging service requirements. Additionally, Phpseclib [32] is responsible for establishing Secure Shell (SSH) connections between the website, the server, and El Greco. It is important to note that El Greco and the server communicate via WiFi.

HTML, PHP, CSS, Javascript, Python, and Codemirror [33], a text editor implemented in JavaScript for use in browsers, were used to create the webpage. Additionally, Brython, a Python-to-Javascript compiler, was used to replace Javascript with Python as a scripting language for web browsers, and the Javascript library Blockly from Google [34] was also employed.

The Playroom (Fig. 2) is a smooth circular surface with a diameter of 1.60 m, located at the Aegean University robotics laboratory. It is designed to offer El Greco an unobstructed zone to manoeuvre actively. Access to the Playroom via live streaming is limited to users who have made a reservation. The booking system of the website will be discussed later in this paper.

5 El Greco Platform

5.1 Main Features

The El Greco platform enables users to operate remotely the humanoid robot El Greco employing Blockly or by typing Python code. Site features include logging in, registering, recovering a forgotten password, contacting site admins, and learning

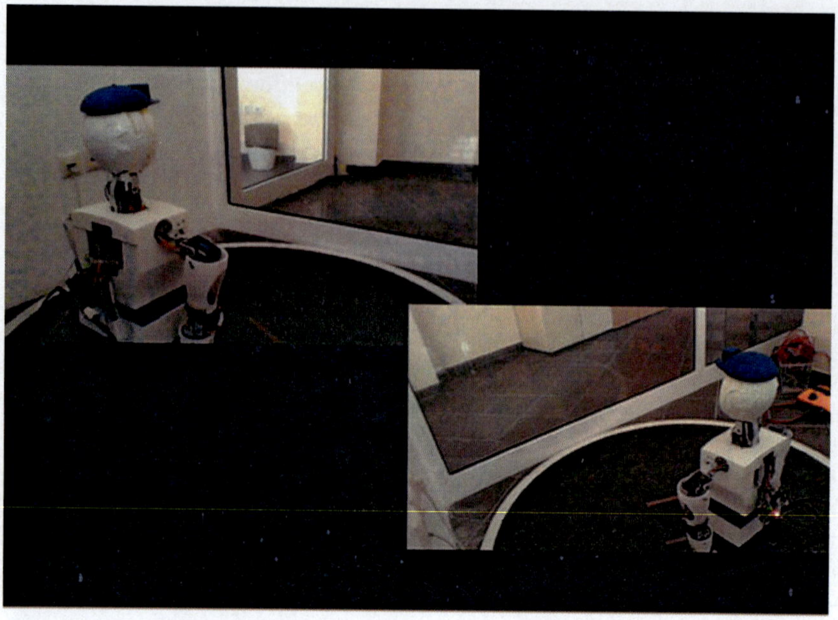

Fig. 2 Image from the playroom

more about El Greco and the Aegean Robotics Team. As part of the registration process, users are asked to create a profile.

Information provided by users during profile creation is used to produce aggregate, anonymous statistics about site visitors to improve it. In addition, an automated log file system was created that captures information about user interactions with the Platform. This data could be used to track instructional markers like user performance and motivation in self-governed online learning environments. Utilizing Web mining techniques, we can improve teaching and learning. The user's performance, preferences and behaviour are the sources of this data [35]. Data acquired from log files is more valuable than online surveys for predicting future user performance in an E-learning setting [36]. El Greco Adventure players' progress is tracked in several ways, including their success rate, the time it takes them to finish a level, whether or not they consult the tutorial, and how many clicks they have made. Combined with a user profile, this data will allow us to better evaluate our website's educational indicators and their potential to enhance education and learning.

Once the users have registered and verified, they can access the website. Once logged in, users can use the reservation service from the site's top menu. To access the Playroom and command the Robot, the user must schedule a session first. Sessions typically last 30 min; however, users can request extra time in a remark section of the booking form.

The El Greco Main Platform and El Greco Adventure options may be found on the Game Types page, which serves as the website's main page and will be covered in greater depth in the subsequent sections of this paper.

5.2 Main Platform

The El Greco main platform is a webpage divided into three sections (Fig. 3). The first is the Blockly area, where programs can be built out of prefabricated Blockly blocks; the second is the code area, a text area that accepts direct Python code entry. To insert code, users can either use the keyboard or convert Blocky blocks to Python code. Users can trigger this translation by clicking the corresponding button. The third and final area is El Greco's Playroom, as seen on a real-time video feed (Fig. 2) access to the Playroom and Robot is unrestricted during user reservations when this window is shown.

Blockly allows the developer to form custom blocks. We crafted the following El Greco functions:

- Demo: El Greco demonstrates its capabilities.
- Salute: El Greco gives the user a salute.
- Dance: El Greco manoeuvres as it dances.
- Wait: El Greco waits for a predetermined amount of time in seconds.
- Walk for: El Greco moves forward for a predetermined amount of time.
- Turn right: El Greco makes a right turn for a predetermined amount of time in seconds.
- Turn left: El Greco makes a left turn for a predetermined amount of time in seconds.

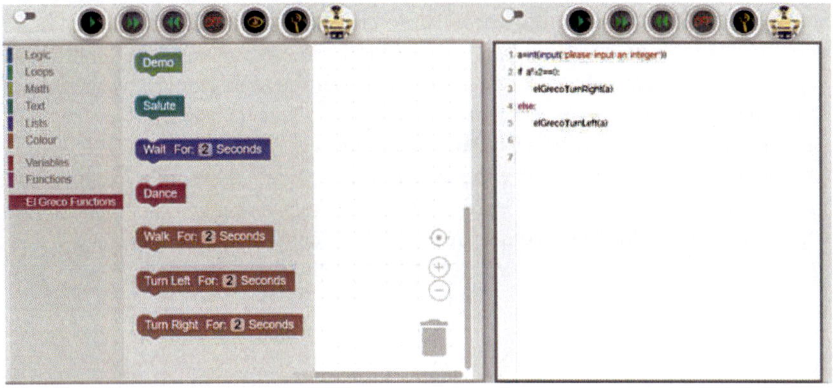

Fig. 3 Blockly and the code area

Code execution varies based on the method used to introduce the code. Blockly utilizes Neil Fraser's Js interpreter [37]. Blocky proactively prevents compiler errors. Unhandled exceptions thrown by the Js Interpreter on runtime faults cause the page to crash. These exceptions are automatically captured by the El Greco Platform, and the relevant data is harvested so that the user may be given helpful feedback to fix the code. A Blockly trap for infinite loops has been developed using the official procedure recommended on the Blockly website [38]. If a single line of code is performed more than 1000 times, the program terminates and displays an appropriate error message.

Brython [5], a Python-to-Javascript compiler, is employed if the code is inserted directly in Python Code. This approach was chosen due to its speedier code execution thanks to eliminating the server round trip. Such an outcome would have been attained if Python had been run on the server instead of the page. Brython handles code errors in debug mode. The collected data is then relayed to the users so they can use it to fix the code. The execution of Python infinite loops is comparable to the behaviour of Blockly.

When the user is in session following code examination for El Greco functions, the code is transferred to El Greco for execution over an SSH connection. The absence of El Greco functions is reported to the user, but no action is taken. The El Greco functions are displayed as print commands in the code output while the user is not in session.It is essential to the learning process to provide the learner with feedback. El Greco platform also allows browsing the code either forwards or backwards while inspecting the code output. We think this feature adds to the educational effectiveness of the El Greco platform. This feature is also available when El Greco executes code.

5.3 El Greco Adventure

El Greco Adventure is a level-based game with eight levels. Each level invites the user to complete a programming task. These tasks evolve the remote control of El Greco. It has the same features as the main platform, adding two more buttons to the page's layout. These buttons are used to check if the code entered by the user is correct and can perform the task necessary to finish a level successfully.

El Greco Adventure can be viewed in a gamification [39] scope, and in this fashion, it challenges the user and enhances the teaching process and the learner's enjoyment. In addition, the objectives of El Greco Adventure levels are meant to necessitate the application of skills learned in other areas, such as mathematics and geometry.

6 Conclusions

El Greco Platform, a platform for learning Python programming, was presented in this paper. It utilizes the humanoid Robot El Greco, which can be remotely programmed via a website. While it is designed with students in mind, anyone can utilize it at no cost. The El Greco Platform Provides feedback to the learner, which is vital in the learning process. Also, it offers a convivial and attractive interface that enhances the learner's enjoyment. In this direction, El Greco adventures challenge the user to enhance the learning process in a gamification aspect.

Furthermore, it provides a new Python framework for Internet-based control of any Python-based Robot. Moreover, a data collection scheme was created to collect data to facilitate the enhancement of the learning process of the Platform. This process is fully automated and anonymous.

Our plans involve making the reservation service independent and using more than one Robot and Playrooms to enable simultaneous multiple-user interactions with the website. In addition, we aim to create playrooms akin to the virtual worlds of Reeborg's universe. Finally, we plan to create more El Greco functions to enrich the user experience.

References

1. Eguchi, A: Bringing robotics in classrooms. In: Robotics in STEM Education: Redesigning the Learning Experience. Springer International Publishing, pp. 3–31 (2017)
2. Blanchard, S., Freiman, V., Lirrete-Pitre, N.: Strategies used by elementary schoolchildren solving robotics-based complex tasks: innovative potential of technology. Procedia—Soc. Behav. Sci. **2**, 2851–2857 (2010). https://doi.org/10.1016/J.SBSPRO.2010.03.427
3. Breazeal, C., Harris, P.L., Desteno, D., et al.: Young children treat robots as informants. Top Cogn. Sci. **8**, 481–491 (2016). https://doi.org/10.1111/TOPS.12192
4. Bakali, I., Fourtounis, N., Theodoulidis, A., et al.: Control a robot via internet using a block programming platform for educational purposes. In: ACM International Conference Proceeding Series. Association for Computing Machinery, pp. 1–2. New York, NY, USA (2018)
5. Developers, B.: Brython (2020). https://brython.info/
6. Fields, D., Vasudevan, V., Kafai, Y.B.: The programmers' collective: connecting collaboration and computation in a high school scratch mashup coding workshop. In: Proceedings of International Conference of the Learning Sciences, ICLS, pp. 855–862 (2014)
7. Bart, A.C., Gusukuma, L., Kafur, D.: BlockPy Editor (2019)
8. Bart, A.C., Tibau, J., Kafura, D., et al.: Design and evaluation of a block-based environment with a data science context. IEEE Trans. Emerg. Top Comput. **8**, 182–192 (2020). https://doi.org/10.1109/TETC.2017.2729585
9. Roberge, A.: Reeborg's world (2020). https://reeborg.ca/index_en.html
10. The LEGO group. MINDSTORMS EV3 Support | Everything You Need | LEGO® Education (2019). https://education.lego.com/en-us/product-resources/mindstorms-ev3/teacher-resources/python-for-ev3
11. Benotti, L., Gómez, M.J., Martínez, C.: UNC++Duino: a kit for learning to program robots in python and C++ starting from blocks. In: Advances in Intelligent Systems and Computing, pp. 181–192. Springer Verlag (2017)
12. Susilo, E., Liu, J., Alvarado Rayo, Y., et al.: STORMLab for STEM education: an affordable modular robotic kit for integrated science, technology, engineering, and math education. IEEE Robot Autom. Mag. **23**, 47–55 (2016). https://doi.org/10.1109/MRA.2016.2546703
13. Khamphroo, M., Kwankeo, N., Kaemarungsi, K., Fukawa, K.: MicroPython-based educational mobile robot for computer coding learning. In: 2017 8th International Conference on Information and Communication Technology for Embedded Systems, IC-ICTES 2017—Proceedings. Institute of Electrical and Electronics Engineers Inc. (2017)

14. Khamphroo, M., Kwankeo, N., Kaemarungsi, K., Fukawa, K.: Integrating MicroPython-based educational mobile robot with wireless network. In: 2017 9th International Conference on Information Technology and Electrical Engineering, ICITEE 2017, pp. 1–6. Institute of Electrical and Electronics Engineers Inc. (2017)
15. Skoupras, P., Upadhyay, J., Fourtounis, N., et al.: El greco: a 3d-printed humanoid that anybody can afford. In: ACM International Conference Proceeding Series. Association for Computing Machinery, pp. 1–2. New York, NY, USA (2018)
16. Doroftei, I., Adascalitei, F., Lefeber, D., et al.: Facial expressions recognition with an emotion expressive robotic head. In: IOP Conference Series: Materials Science and Engineering, p. 012086. Institute of Physics Publishing (2016)
17. Google inc.: Speech-to-Text: Automatic Speech Recognition | Google Cloud (2021). https://cloud.google.com/speech-to-text
18. Python Speech Recognition: SpeechRecognition PyPI. Pypi.org (2019). https://pypi.org/project/SpeechRecognition/
19. Nuttall, B., Jones, D.: GpioZero: a simple interface to everyday GPIO components used with Raspberry Pi.
20. Gary, B.: The OpenCV library. Dr Dobb's J. Softw. Tools **25**, 120–123 (2008)
21. Google Cloud: Text-to-Speech: lifelike speech synthesis (2013). https://cloud.google.com/text-to-speech
22. Nicolas Durette, P.: gTTS PyPI. Pypi.Org (2020). https://pypi.org/project/gTTS/
23. Vollmer, S.: Google translate. In: Figures of Interpretation, pp. 72–76 (2021)
24. Pilgrim, M., McKee, K.: Feedparser: parse atom and RSS feeds in python (2015). https://pypi.org/project/feedparser/
25. Behnel, S.: lxml—Processing XML and HTML with Python (2016)
26. Triandini, E., Suardika, I.G.: Installing, configuring, and developing with Xampp. D Dvorski Dalibor, 1–10 (2007)
27. Jim, Contributors OS: Open Broadcaster Software | OBS. In: Open Broadcast Software (2020). https://obsproject.com/
28. Softvelum LLC Softvelum Nimble Streamer: freeware media server for live and VOD streaming. https://wmspanel.com/nimble
29. Softvelum LLC Softvelum Low Delay Protocol—low latency streaming protocols—Softvelum. https://softvelum.com/sldp/
30. LLC S HTML5 player for SLDP—Softvelum. https://softvelum.com/player/web/
31. PHPMailer Contributors GitHub—PHPMailer/PHPMailer: the classic email sending library for PHP. https://github.com/PHPMailer/PHPMailer
32. Wigginton, J., Monnerat, P., Fischer, A., et al.: phpseclib: pure PHP implementations of SSH, SFTP and RSA
33. Haverbeke, M.: CodeMirror. https://codemirror.net/
34. Google Inc.: Blockly | Google Developers (2017). https://developers.google.com/blockly
35. Hershkovitz, A., Ben-Zadok, G., Mintz, R., Nachmias, R.: Examining online learning processes based on log files analysis: a case study need to cite this paper? Want more papers like this? (2009)
36. Cho, M.H., Yoo, J.S.: Exploring online students' self-regulated learning with self-reported surveys and log files: a data mining approach. Interact Learn. Environ. **25**, 970–982 (2017). https://doi.org/10.1080/10494820.2016.1232278
37. Fraser, N.: JS-Interpreter Documentation. https://neil.fraser.name/software/JS-Interpreter/docs.html
38. Developers G Generating and Running JavaScript | Blockly | Google Developers. https://developers.google.com/blockly/guides/app-integration/running-javascript. Accessed 9 Nov 2022
39. Caponetto, I., Earp, J., Ott, M.: Gamification and education: a literature review. In: Academic Conferences International Limited (2014)

An Overview of Common Educational Robotics Competition Challenges

Eftychios G. Christoforou, Sotiris Avgousti, Panicos Masouras, Andreas S. Panayides, and Nikolaos V. Tsekos

Abstract Educational robotics have become an integral part of STEM education. Robotics competitions are complementary to relevant courses and they have been gaining popularity worldwide. Some common challenges in such events are reviewed using the paradigm of Robotex competition in order to exemplify the concepts and highlight their educational value.

Keywords Educational robotics · Robotics competitions · STEM education

1 Introduction

Robotics is a challenging but also exciting scientific field, which provides a framework for "STEM" (Science, Technology, Engineering, and Mathematics) education [1, 2]. It allows teaching these concepts following an interdisciplinary and integrated approach, which is typically followed when solving modern scientific and engineering problems. The value of robotics as a tool for STEM education has been widely acknowledged in the literature. Many studies have investigated its impact on the academic and social skills of children [3] and expanding further on this topic is

E. G. Christoforou (✉)
Department of Mechanical and Manufacturing Engineering, University of Cyprus, Nicosia, Cyprus
e-mail: e.christoforou@ucy.ac.cy

S. Avgousti · P. Masouras
Department of Nursing, School of Health Sciences, Cyprus University of Technology, Limassol, Cyprus

A. S. Panayides
Videomics FRG, CYENS Center of Excellence, Nicosia, Cyprus

N. V. Tsekos
Department of Computer Science, University of Houston, Houston, TX, USA

P. Masouras
Cyprus Computer Society, Nicosia, Cyprus

R. Balogh et al. (eds.), *Robotics in Education*, Lecture Notes in Networks and Systems 747, https://doi.org/10.1007/978-3-031-38454-7_33

beyond the scope of this work. Relevant courses often embrace robotics competitions as an added stimulus to complement school curricula [2, 4–8].

The purpose of this paper is to review common challenges used in educational robotics competitions, exemplify relevant concepts and highlight the skills gained through participation. The wide range of challenges that exists shows how the creativity and ingenuity of the organizers can contribute towards making the competitions appealing to students, while enhancing their educational value.

The paper is organized as follows. Common challenges in robotics competitions are reviewed in Sect. 2. In Sect. 3, the focus is on the technical skills acquired from robotics competitions. The relevance to practical applications of robotics is also addressed. The last section presents the conclusions.

2 Common Robotics Competition Challenges

Educational robotics competitions include a wide range of challenges [9]. Some competitions are themed (e.g., recycling, soccer, sumo fighting, search & rescue, and firefighting), while others focus on a generic task (e.g., line-following, color-picking, sorting, maze solving). Sometimes a robot competes against a robot opponent (e.g., in sumo) or a team of robots competes against other teams (e.g., robot soccer). Typically, for each challenge teams compete within the same age group in order to ensure fairness. Competitions may engage different robotic platforms, which can be either branded products or the integration of selected off-the-shelf components and microprocessors. In recent years, the scope of robotics has broadened to include ground mobile robots, aerial systems and water (surface or underwater) systems. In response to these technological developments, robotics competitions now embrace new challenges.

Popular challenges are reviewed here and exemplified from two well-established annual competitions in Europe, Robotex International in Talin, Estonia (https://robotex.international/) and Robotex Cyprus in Nicosia, Cyprus (https://robotex.org.cy/en/). The presented photographic material is from the two aforementioned events.

(1) *Educational robotics projects* (Fig. 1a) are themed competitions (e.g., futuristic scenarios, environmental projects, creations addressing social issues and challenges) and they are typically aimed at children of younger ages. They concern basic building projects which may include some functionality, as well as simple motion mechanisms.
(2) *Robotics4All* (Fig. 1a) allows participants to present their innovations and constructions on any theme, while using their imagination and creativity. Their creations are expected to include motion mechanisms and some basic form of automation.
(3) *Folkrace* (Fig. 1b) involves a group of autonomous robots competing in a track field for speed. The closed, curved race track presents difficulties including

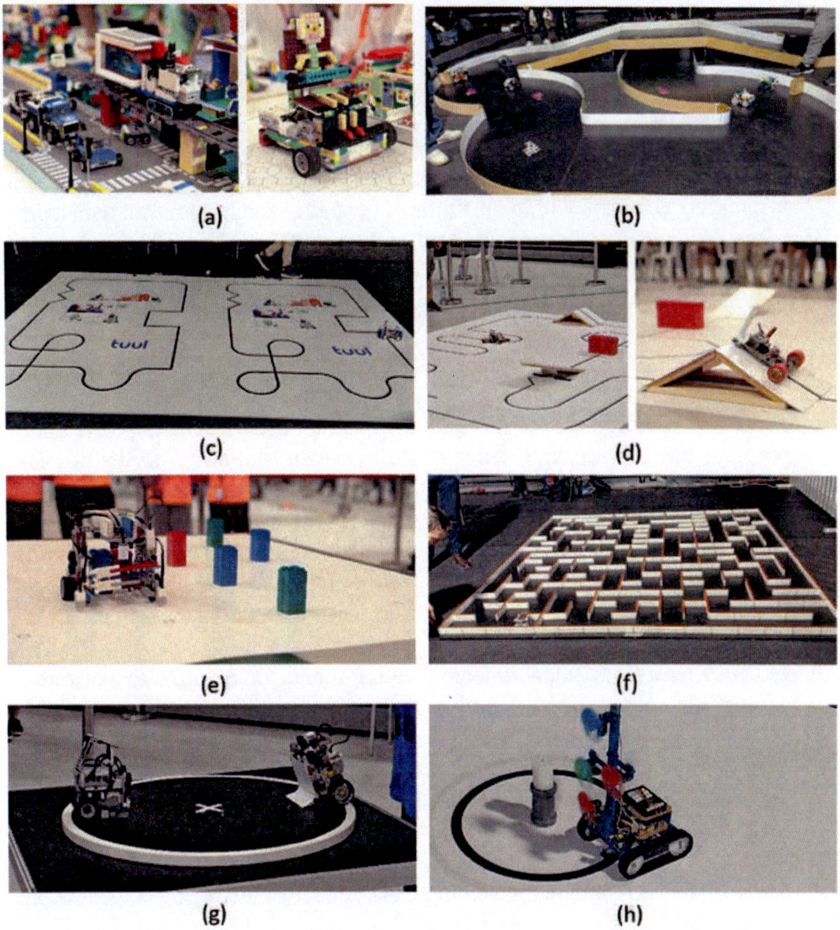

Fig. 1 Common challenges in educational robotics competitions. **a** Educational robotics projects and Robotics4All; **b** Folkrace; **c** Line-following; **d** Enhanced line-following; **e** Colour picking; **f** Maze solving; **g** Sumo fighting; **h** Firefighting

holes, scattered loose materials, and hindering walls along the length of the side walls. The goal is to complete as many laps as possible in the given time.

(4) In *line-following* (Fig. 1c) the objective of the robots is to autonomously drive along a line marked on the floor, as fast as possible. A generic program uploaded on the robot should be able to perform successfully on any track that meets the given specifications and is only revealed at the time of the competition.

(5) A more complex version of the above challenge is the *enhanced line-following* (Fig. 1d). Besides the primary line-following task, the competing robot is required to cope with abnormalities on the track including unexpected line

brakes (dashed line parts), line thickness changes (thinner or wider line), physical obstacles on the track to be avoided, inclined slopes (mountains to pass over) and swings.

(6) *Color picking* (Fig. 1e) requires robots to collect colored objects from within the field and move them to a specific location, considering that each color carries different positive points, while red ones give negative points. The robot is required to remain within the limits of the field, which is marked with a line.

(7) *Maze solving* (Fig. 1f) involves a robot finding its way through a 2-D walled maze (labyrinth) from the start point (e.g., a specified corner) to a target location (e.g., center of labyrinth), without any prior knowledge of the maze, as quickly as possible.

(8) *Sumo fighting* (Fig. 1g) involves two robots attempting to push each other outside of a ring platform. The ring is circular with a line at its perimeter. A distinct characteristic of this challenge is that the participating robot does not have any a-priori knowledge of the opponent's strategy nor its physical construction.

(9) In *Firefighting* (Fig. 1h) the purpose of the challenge is to localize and extinguish some candles, which are placed within a field marked with a black line, without touching them, within a given time. Walls are also placed around the candles blocking some sides. It is common that participating robots use blowers or fans as extinguishers.

(10) In *Drone race* (Fig. 2a), autonomous flying robots are required to navigate in an obstacle course, which includes walls with openings at different heights. The drones have to detect these openings and pass through them (Fig. 2a - top). A more challenging scenario involves autonomous flying robots, which are able to navigate in a warehouse setup to perform inventory control tasks (Fig. 2a - bottom).

(11) *Water rally* (Fig. 2b) involves autonomous boats, which try to complete as many laps as possible in a given time and collect as many points as possible. The race takes place inside a water tank with a closed circular track. Robots are required to go through obstacles found along the way including buoys, walls, tunnels and dividers.

Interestingly, in Robotex competition the Firefighting challenge which was described above, has recently been selected to be a girls-only competition in order to support the notion of attracting girls in STEM related disciplines. A gender imbalance exists in these areas and systematic actions are often taken to tackle this issue [10].

Fig. 2 Common challenges in educational robotics competitions. **a** Drone race; **b** Water rally

3 Technical Skills Learned from Robotics Competitions

Beyond entertainment, a robotics competition is successful as long as it provides a good learning environment with hands-on experience and allows children to develop new skills including: design, integration, critical thinking, creativity and problem-solving, computer programming, teamwork and collaboration, presentation and other general social skills, widely known as twenty-first century skills. The discussion here focuses on technical skills from our perspective as educators and organizers of Robotex Cyprus. Extensive data reflecting the views of participants regarding their engagement and satisfaction during the preparation stage of the above competition, were collected through a questionnaire and analyzed in [5].

Through competitions students learn how a robotic system works, how it is built and controlled. Technology is somehow demystified as they better understand both the capabilities but also the limitations of robots. The connection between software and hardware in this process is clarified. They can also understand the notion of an engineering system. The systematic engineering design process is also among the important topics learned, which includes specific iterative steps, as analyzed in [4].

Learning the fundamentals of automatic control is considered as one of the most valuable outcomes of robotics competitions. Children understand the idea of a closed-loop system and apply the concept of feedback, i.e., the process of constantly monitoring a controlled variable and using this information to appropriately perform control adjustments. For example, in the line-following challenge, a printed line is constantly detected by an appropriate sensor and steering is adjusted for the robot to remain on track. Other important topics in control systems are also encountered including the robustness characteristics of controllers. For example, in the enhanced

Table 1 Technical skills and practical robotics applications related to competition challenges

	Challenge	Technical skills	Practical applications
1	Educational Robotics and Robotics4All	System design, basic mechanisms design, automations, aesthetics	STEM education, entertainment
2	Folkrace	System design, basic control systems	Entertainment, racing
3	Line-Following and Enhanced Line-Following	Mobile robot design, system integration, programming, feedback control systems, sensor technology	Industrial material transfer, service robotics, logistics
4	Color picking	Robot and gripper design, sensing, controller design	Industrial production, sorting, recycling, agricultural robotics
5	Maze solving	Mathematical algorithms, autonomous systems, programming, robot design, robot control, sensing	Search & rescue, exploration
6	Sumo	Robot design, autonomy, programming, motion control, sensing	Entertainment
7	Firefighting	Sensing, autonomy, programming	Firefighting, search & rescue
8	Drone race	UAV design, programming, intelligence, autonomy, safety	Search & rescue, product deliveries, logistics, surveillance
9	Water rally	Autonomy, programming, marine systems	Autonomous surface vehicles, coastal surveillance, environmental monitoring

line-following challenge, the controller is required to perform well in the presence of uncertainty and external disturbances (here introduced by abnormalities encountered along the path). The implementation helps students improve their computer programming skills by designing and testing progressively more effective algorithms. In terms of sensing, the students learn how sensors work, how they are selected and calibrated.

The main technical skills related to the individual challenges which were discussed in Sect. 2, are collected in Table 1. On the same table are included some corresponding practical robotics applications showing the relevance to existing technologies.

4 Conclusions

Robotics is an attractive topic to students and is rich in relevant didactic content. This makes it suitable for supporting the teaching of STEM at different levels, from kindergarten to university. As a complement to relevant courses, students are often engaged in robotics competitions. Robotics competitions provide students with the opportunity for hands-on experience and active learning to maximize the educational benefit and impact. For this purpose, various challenges have been established in competitions but there is always room for creativity to better meet targeted educational needs.

References

1. Afari, E., Khine, M.S.: Robotics as an educational tool: impact of lego mindstorms. Int. J. Inf. Educ. Technol. **7**(6), 437–442 (2017)
2. Eguchi, A.: RoboCupJunior for promoting STEM education, 21st century skills, and technological advancement through robotics competition. Robot. Auton. Syst. **75**, 692–699 (2016)
3. Anwar, S., Bascou, N.A., Menekse, M., Kardgar, A.: A systematic review of studies on educational robotics. J. Pre-College Eng. Educ. Res. **9**(2), 19–42. Purdue University Press (2019)
4. Christoforou, E.G. et al.: Educational Robotics Competitions and Involved Methodological Aspects. In: Merdan, M., Lepuschitz, W., Koppensteiner, G., Balogh, R., Obdržálek, D. (eds) Robotics in Education. RiE 2019. Advances in Intelligent Systems and Computing, vol 1023. Springer, Cham (2020)
5. Christoforou, E.G., Avgousti, S., Masouras, P., Cheng, P., Panayides, A.S.: Robotics Competitions as an Integral Part of STEM Education. In: Lepuschitz, W., Merdan, M., Koppensteiner, G., Balogh, R., Obdržálek, D. (eds) Robotics in Education. RiE 2020. Advances in Intelligent Systems and Computing, vol 1316. Springer, Cham (2021)
6. Murphy, R.R.: 'Competing' for a robotics education. IEEE Robot Autom. Mag. **8**(2), 44–55 (2001)
7. Chew, M.T., Demidenko, S., Messom, C., Sen Gupta, G.: Robotics competitions in engineering education. In: ICARA 2009—Proceedings of the 4th International Conference on Autonomous Robots and Agents, 624–627 (2009)
8. Chatzis, D., Papasalouros, A., Kavallieratou, E.: Planning a robotic competition. Comput. Appl. Eng. Educ. **30**(4), 1248–1263 (2022)
9. Evripidou, S., Georgiou, K., Doitsidis, L., Amanatiadis, A.A., Zinonos, Z., Chatzichristofis, S.A.: Educational robotics: platforms, competitions and expected learning outcomes. IEEE Access **8**, 219534–219562 (2020)
10. Sullivan, A., Bers, M.U.: Vex robotics competitions: gender differences in student attitudes and experiences. J. Inf. Technol. Educ. **18**, 97–112 (2019)

Planning Poker Simulation with the Humanoid Robot NAO in Project Management Courses

Ilona Buchem, Lewe Christiansen, and Susanne Glißmann-Hochstein

Abstract Learning how to predict effort is important for any project team. In agile teams, Planning Poker is one of the most popular estimation game-based techniques used to come to a consensus about how much work should be done in a given iteration. Usually, the product owner or scrum master facilitates planning poker with their team. This paper presents the application of the humanoid robot NAO as a robotic coach and facilitator of planning poker with university students in two project management courses. The paper describes the design of the planning poker simulation, the programming of the robot, as well as the implementation and evaluation of the application "Planning poker with NAO" in two on-campus pilot studies with 29 university students. The evaluation aimed to investigate students' perceptions of the design of the simulation and its effects on students' understanding of agile estimation. The results show that the design of planning poker facilitated by the NAO robot was helpful for students to understand the concept of relative estimation in agile teams.

Keywords Planning poker · Agile estimation · Agile teams · Agile project management · Scrum · Scenario-based design · NAO robot · Educational robots

1 Introduction

Agile methods have been on the rise in software development, project management, and curricula of business management study programs. Agile approaches are characterized by close collaboration in teams, team self-organization, and empowerment of team members in making decisions including the estimation of effort it takes to

I. Buchem (✉) · L. Christiansen · S. Glißmann-Hochstein
Berlin University of Applied Sciences, 13154 Berlin, Germany
e-mail: buchem@bht-berlin.de

L. Christiansen
e-mail: s47389@bht-berlin.de

S. Glißmann-Hochstein
e-mail: susanne.Glissmann-Hochstein@bht-berlin.de

R. Balogh et al. (eds.), *Robotics in Education*, Lecture Notes in Networks and Systems 747, https://doi.org/10.1007/978-3-031-38454-7_34

deliver items and finish tasks. Learning how to predict effort is important for any project team and effort estimation is a critical factor for project success [1, 2]. Since the estimation of relative sizes or complexity of tasks in agile projects is based on subjective assessment and takes a long time, effort estimation remains the main challenge for agile teams [1, 2]. Planning poker is a game-based technique applied by agile teams to leverage effort estimation by empowering all team developers to decide and reach a consensus about how much work should be done in a given iteration or sprint [1]. Reliable estimates can be generated by agile teams playing planning poker given the estimation is done by a team of knowledgeable project experts [2]. Several studies have shown that planning poker in agile teams can provide more reliable estimates compared to the estimation done by single experts [3, 4]. Group discussion in planning poker helps teams identify activities that individual estimators could overlook [4]. Team-based estimation in planning poker tends to provide more accurate and more realistic estimates compared to other expert-based methods [4]. Thus it is not surprising that planning poker was used by 58% of the teams surveyed by the annual 15th State of Agile Report from 2021 [5].

Planning poker as a team-based method of estimation has also entered the curricula of university courses, including engineering, business, computer science, and information management studies. From a didactic point of view, games, and simulations, such as planning poker, can be used as powerful tools for learning [6, 7]. For example, the study by [6] described the application of planning poker with graduates of systems engineering, where it was introduced as an applied estimation methodology in software business development. The study by [7] applied planning poker in the field of software engineering as a follow-up simulation exercise to a lecture on the topic to help students improve their understanding of the relative estimation concept. However, as pointed out by [8], teaching and learning how to perform estimation in agile teams in educational contexts can be challenging since students usually have less experience in working on projects and may encounter starter issues compared to professionals. As pointed out by [9], the lack of available research studies about planning poker in education leaves unanswered questions. This paper aims to fill this gap.

This paper describes a novel approach to applying planning poker as an educational simulation facilitated by a humanoid robot NAO. The simulation "Planning poker with NAO" presented in this paper was tested in two on-campus pilot studies with altogether 29 university students. The evaluation was conducted using an online survey and aimed to investigate students' perceptions about the design of the simulation and its effects on students' understanding of agile estimation. The remainder of the paper is structured as follows. Following this introduction, we present the scenario-based design and programming of the simulation "Planning Poker with NAO". Next, we describe the methodology applied in both pilot studies and the analysis of the evaluation results. The paper ends with conclusions and recommendations for further research and development.

2 Design and Programming

In general, it is recommended to practice planning poker with the help of a facilitator or coach, who guides the process and ensures that the method is used properly and effectively. The scenario presented in this paper applied the humanoid robot NAO as a robotic facilitator. The design of the simulation "Planning Poker with NAO" was informed by scenario-based design (SBD) as a methodology focused on analyzing, prototyping, and problem-solving [10]. Scenario-based designs typically include narrative descriptions of how future users will use a system to be developed [11]. Scenarios are usually designed as stories that consist of a description of a setting or use context, actors involved in the use of the system, and a sequence of actions and events that lead to an outcome [10]. In the case of "Planning poker with NAO", the scenario-based design includes a narrative description of a sequence of facilitation steps performed by the NAO robot and interactions of students with the NAO robot during the game of planning poker. The scenario was designed in three main phases, i.e. (a) analyzing challenges and practices in learning agile estimation at the university, (b) creating an integrated activity, information, and interaction design with detailed descriptions of how the NAO robot will facilitate the planning poker simulation in class, and (c) three stages of iterative prototyping including testing evaluation with students. The scenario was used as a blueprint to guide the programming at a later stage.

The programming was done using the Choregraphe software (Version 2.8.6) and Python. The goal was to implement the designed scenario with its five main parts, i.e. "Introduction", "Explanation", "Planning Poker", "Create protocol" and "Ending". In the "Introduction" guidelines are presented, on how the participants can interact with the robot, for example by speech or by pressing the foot bumpers. In the section "Explanation", the robot explains to students the concept of planning poker, its rules, and the items to be estimated. The next part "Planning Poker" is the game's actual play with an estimation of each item by the team of six students. In the "Create protocol" part, the NAO robot creates a document with the results of the estimation. The estimated numbers are automatically saved and matched with the corresponding items in a PDF file. After every estimation round, the agreed number from the Fibonacci scale is appended to a local text file. At the very end of the simulation, a Python script takes every entry and matches the items to the numbers. Furthermore, the script handles the formatting of the text and where the PDF is saved on the robot. The file can be accessed in Choregraphe in the Connection > Advanced > File Transfer tab. The PDF file can be later accessed and downloaded from Choregraphe. Finally, the "Ending" part includes the reflection question and the review of learnings.

3 Methodology

The simulation "Planning Poker with NAO" was tested in two pilot studies with a sample of 29 students in two different project management courses. The sample comprised 15 bachelor and 14 master students. The 15 bachelor students studied in their third semester of the bachelor program Digital Business (B. Sc.). The 14 master's students studied in their first semester of the master's program Business Systems/ Project Management (M. A.). In both pilot studies, the simulation "Planning Poker with NAO" was integrated into a weekly seminar. Participants in both pilot studies were students, teachers teaching the respective courses (the first and the third author of this paper), and the programmer of the NAO robot (the second author).

The methodology of both pilot studies included a short introduction of the NAO robot to students, followed by the simulation of planning poker facilitated by NAO, and finally an evaluation in the form of an online evaluation. In both pilot studies, one team of six students played planning poker with NAO, while other students observed the game. The limitation to the test with one team only resulted from time constraints during classes. The online evaluation included a range of questions related to the socio-demographics (Table 1), students' perceptions of the design of the simulation, and its effects on students' understanding of agile estimation.

Both groups of students differed in the level of their previous experience in interacting with a humanoid robot. Among the bachelor students from the first pilot study, 73.3% (11 out of 15) already had some previous experience in interacting with a humanoid robot. This may be to a large extent contributed to the fact, that the humanoid robot NAO was introduced in this course earlier during the semester. Compared to this group, master students in the second pilot study had less previous experience in interacting with a humanoid robot, only 14.3% (2 out of 14) reported having some previous experience. Both groups of students were similar when it comes to previous experience in playing planning poker. While only one bachelor student played planning poker before, none of the master students had this experience.

Table 1 Socio-demographics of the study sample, n = 29

	Pilot study 1 (bachelor)	Pilot study 2 (master)
Simple size	15 students	14 students
Gender	73.3% male 26.7% female	57.1% male 42.9% female
Age	60% 20–24 13.3% under 20 13.3% 25–29 13.3% 30–34	50% 25–29 42.9% 20–24 7.1% 30–34

4 Results

The evaluation results from both pilot studies presented in this paper focus on students' perceptions of the design of the simulation "Planning poker with NAO" and its effects on students' understanding of agile estimation. The quantitative results were analyzed using IBM SPSS, version 29.

What did you like about Planning Poker with NAO? Students in the first pilot study praised the engaging and enjoyable interaction with NAO, the fact that they understood how to play planning poker through the simulation, the funny and kind character of the robot, as well as structured answers, helpful feedback, and good gestures of NAO. The students in the second pilot study mentioned they liked "everything" about the interaction with NAO and highlighted some special features such as engaging interaction, good facilitation by the robot, time to discuss in the team, nice gestures and moves, a good structure and a helpful explanation of planning poker.

What did you not like about Planning Poker with NAO? Students in the first pilot study criticized the fact that they had to speak loud to interact with the robot by speech and that the robot can only do things that it is programmed to do. Students also complained that only one team could play planning poker with NAO. Students in the second pilot study also wished for more interaction with NAO. These students also criticised specific features of the interaction with the robot such as the high pitch of NAO's voice which and a lower level of NAO's artificial intelligence than expected.

How interesting was the interaction with NAO during the planning poker? A corresponding statement was rated on a scale from 1 = disagree strongly to 5 = agree strongly. In the first pilot study the average rating for this statement by bachelor students was M = 4.33 (SD = 0.724) and in the second pilot study by master students M = 4.14 (SD = 0.770), both indicating that "Planning poker with NAO" was interesting for both bachelor and master students (high mean values, M) and that students were in agreement (low standard deviation, SD).

How helpful was the simulation with NAO for understanding the planning poker? A corresponding statement was rated on a scale from 1 = disagree strongly to 5 = agree strongly. In the first pilot study the average rating for this statement by bachelor students was M = 4.20 (SD = 775) and in the second pilot study by master students M = 4.14 (SD = 0.663), both indicating that the simulation was helpful for both bachelor and master students to understand the method of planning poker.

How easy was it for you to estimate items during Planning Poker with NAO? A corresponding statement was rated on a scale from 1 = very difficult to 5 = easy. In the first pilot study, the average rating for this statement by bachelor students was M = 3.47 (SD = 0.961), and in the second pilot study by master students M = 3.57 (SD = 0.852), both indicating that the estimation of items was of medium difficulty level for both bachelor and master students and that the students were in agreement on that.

How confident are you in agile estimation after playing Planning Poker with NAO? A corresponding statement was rated on a scale from 1 = totally uncertain to 5 =

totally confident. In the first study, the average rating was M = 3.27 (SD = 0.961), and in the second study M = 3.29 (SD = 0.994), which indicates that a good level of confidence could be reached through this one-time simulation only.

5 Discussion and Conclusions

In this paper, we presented the design, programming, and evaluation of the simulation "Planning Poker with NAO" in two pilot studies with bachelor and master students. The studies aimed to explore to what extent a humanoid robot NAO can effectively facilitate an agile estimation game of planning poker. The results indicate that, in general, the design of the simulation was helpful for students to understand estimation in agile teams and students could gain a good level of confidence in agile estimation through playing one-time only. The criticism from students in both pilot studies towards the capabilities of the NAO robot indicates that students have high expectations towards humanoid robots in relation to speech recognition and respective feedback. The fact that students complained about only one team having the chance to play planning poker with NAO during the pilot studies indicates that students were interested in the experience and wished to interact with NAO. The results related to the difficulty level of agile estimation showed that a medium level of difficulty could be reached, i. e. on average the challenge of estimation in an agile team was assessed to be neither too high nor too low. The results indicating a good level of confidence can be interpreted both as a positive result and as a pointer for improvement. On the one hand, reaching a good level of confidence for the complex and challenging task of agile estimation through one-time-only simulation was possible. On the other hand, the next iterations could further improve the design to increase the level of confidence. As far as further research and development are concerned, we recommend extending the existing capabilities of available robots like NAO with regard to Conversational AI to allow the robot to understand the content of the conversation and generate meaningful feedback for learning. Additionally, simulations like planning poker could be enhanced by emotion recognition allowing the robot to infer and interpret the emotions of members of an agile team and support consensus-building.

References

1. Sudarmaningtyas, P., Mohamed, R.B.: Extended planning Poker: a proposed model. In: 2020 7th International Conference on Information Technology, Computer, and Electrical Engineering (ICITACEE), pp. 179–184 (2020)
2. Alhamed, M., Storer, T.: Playing planning Poker in crowds: human computation of software effort estimates. In: 2021 IEEE/ACM 43rd International Conference on Software Engineering (ICSE), pp. 1–12 (2021)

3. Usman, M., Mendes, E., Neiva, F.W., Britto, R.: Effort estimation in agile software development: a systematic literature review. In: Proceedings of the 10th International Conference on Predictive Models in Software Engineering, pp. 82–91 (2014)
4. Mahnic, V., Hovelja, T.: On using planning poker for estimating user stories. J. Syst. Softw. **85**(9), 2086–2095 (2012)
5. Digital.ai: 15th State of Agile Report (2021). https://info.digital.ai/rs/981-LQX-968/images/SOA15.pdf. Last Accessed 18 Dec 2022
6. Rojas Puentes, M.P., Mora Méndez, M.F., Bohórquez Chacón, L.F., Romero, S.M.: Estimation metrics in software projects. J. Phys.: Conf. Series, 1126 (2018)
7. Chatzipetrou, P., Ouriques, R.A., Gonzalez-Huerta, J.: Approaching the relative estimation concept with planning Poker. In: CSERC '18 (2018)
8. Zhang, Z.: The Benefits and Challenges of Planning Poker in Software Development: Comparison Between Theory and Practice. Auckland University of Technology (2017). https://openrepository.aut.ac.nz/handle/10292/10557. Last Accessed 18 Dec 2022
9. Akilli, G.K.: Games and simulations: a new approach in education. In: Handbook of Research on Effective Electronic Gaming in Education (2009)
10. Rosson, M.B., Carroll, J.M.: Scenario-based design. In: Jacko, J., Sears, A. (eds.) The Human-Computer Interaction Handbook: Fundamentals, Evolving Technologies and Emerging Applications, pp. 1032–1050. Lawrence Erlbaum Associates (2002)
11. Silva, T.R., Winckler, M.: A scenario-based approach for checking consistency in user interface design artifacts. In: Proceedings of the XVI Brazilian Symposium on Human Factors in Computing Systems (2017)

BlackPearl: The Impact of a Marine Robotics Competition on a Class of International Undergraduate Students

Francesco Maurelli, Nayan Man Singh Pradhan, and Riccardo Costanzi

Abstract Robotics competitions are a great tool to address challenging tasks, encouraging young people to be active in STEM and developing a complimentary essential skill set in management, teamwork and complex problem solving. This paper presents the experience of a team of undergraduate students participating in a marine robotics competition, receiving the Best Rookie Award. It also outlines lesson learned about embedding robotics field competitions in an undergraduate program.

Keywords Marine robotics · Robotics competitions · Educational robotics

1 Introduction

Robotics competitions are an excellent platform to provide students with hands-on experience in building and programming robots to perform various tasks. They have become increasingly popular in recent years as a way to promote STEM education and foster innovation in the field of robotics. In addition to encourage students to develop their technical skills, they provide a unique opportunity for them to develop their problem-solving and critical thinking skills as well as their teamwork and communication abilities. Several competitions were organised in the field of marine robotics. Starting from RoboSub,[1] the first underwater robotics competitions in the 1990s, others were organised around the world, each one with different objectives and distinctive characteristics. In Asia, the Singapore AUV Challenge (SAUVC) is a well known and participated competition [1]. The Student Autonomous Underwater Challenge—Europe is the first competition in Europe thought and organised with a specific focus on student teams [2–4]. After a few years organised in France and

[1] https://robosub.org/.

F. Maurelli (✉) · N. M. S. Pradhan
Constructor University, Campus Ring 1, 28759 Bremen, Germany
e-mail: f.maurelli@ieee.org

R. Costanzi
University of Pisa, Largo Lucio Lazzarino, 1, 56122 Pisa, Italy

R. Balogh et al. (eds.), *Robotics in Education*, Lecture Notes in Networks and Systems 747, https://doi.org/10.1007/978-3-031-38454-7_35

in the UK, the NATO CMRE took over the organisation since 2010 [5, 6] in La Spezia, Italy. The competition evolved into euRathlon [7, 8] and, recently, into the RAMI (Robotics for Asset Maintenance and Inspection) competition.[2] An extensive review of marine robotics competitions has been compiled by Ferreira et al. [9]. Often participation to the competition has played a role in scientific publications (e.g. [10, 11]).

This paper aims to present the efforts of a team of undergraduate students from Constructor University[3] in participating in the 2022 edition of the RAMI competition. Section 2 outlines the general framework of the competition in the university setting. Section 3 presents the robot used for the competition and the hardware upgrades led by the team. Section 4 presents the task of the competition. Section 5 analyses the competition field activities, presenting the challenges for the team as well as their approach to those challenges. Finally, Sect. 6 summarises the main lessons from this experience, drawing some general consideration and pointers for future activities.

2 The Framework

Constructor University offers an English-speaking BSc degree in Robotics and Intelligent Systems [12]. Among the electives for the third year, students can choose the specialisation course *Marine Robotics*, during which they have to present a small project in teams of 2–3 people. Additionally, non-curricular activities such as *J-Robotics* were launched, providing students a space where to dive into robotics and intelligent systems, developing own projects and learning from peers. Participating in the competition was therefore aligned with both the curricular and the extra-curricular activities at the university. The Marine Robotics course took place in the Spring semester, running until mid-May. From the end of May—i.e. after the end of the exam session—the competition team fully committed to the competition, which happened in mid-July. The University of Pisa joined the team with two MSc students, who provided valuable input for some of the competition tasks.

3 The Robot

The BlueROV2, visible in Fig. 1 was used in the Marine Robotics course and for the competition. Students had access to the robot through the course and through various J-Robotics activities. The competition and robot provided a platform for the students to encounter real-world problems, discuss practical and priority based solutions, and jointly work on implementing the solutions. The students came up

[2] https://metricsproject.eu/inspection-maintenance/rami-competition/.

[3] At the time of the competition, the university name was *Jacobs University Bremen*. It officially changed its name in January 2023.

Fig. 1 The BlueROV2 vehicle, modified to function as an autonomous underwater vehicle

with ideas to tackle the competition challenges. The time crunch, resource limitation, and technical requirements set by the competition drove the students to modify the hardware of the robot. For example, the pre-existing pre-built RaspberryPi in the BlueROV was replaced with a more powerful one and the low-level software was re-written, the 1-DoF gripper arm was mounted in a unique position and orientation in order to achieve easier control for the competition task by drilling custom holes in the physical housing, a kill-switch was added to the robot for safety purposes, and additional buoyancy foams were attached to the vehicle via custom-made 3D-printed housings, to balance for the increase of weight from the various sensors.

4 The Tasks

The main aim of the competition was to address underwater autonomous Inspection and Maintenance (I&M). The competition was evaluated based on the following three main tasks:

1. **Pipeline area inspection**: Autonomous navigation and inspection of reported damaged pipeline area. This included mapping of area with buoy location, color detection based control, and identification of damaged pipelines.
2. **Intervention on the pipeline structure**: Autonomous identification of manipulation console, real-time valve manipulation, and ring-pole identification and retrieval.
3. **Complete I&M mission at the plant**: Combination of the above two tasks in one run.

Figure 2 shows the competition arena, with the structure in the middle and the buoy area on the bottom left.

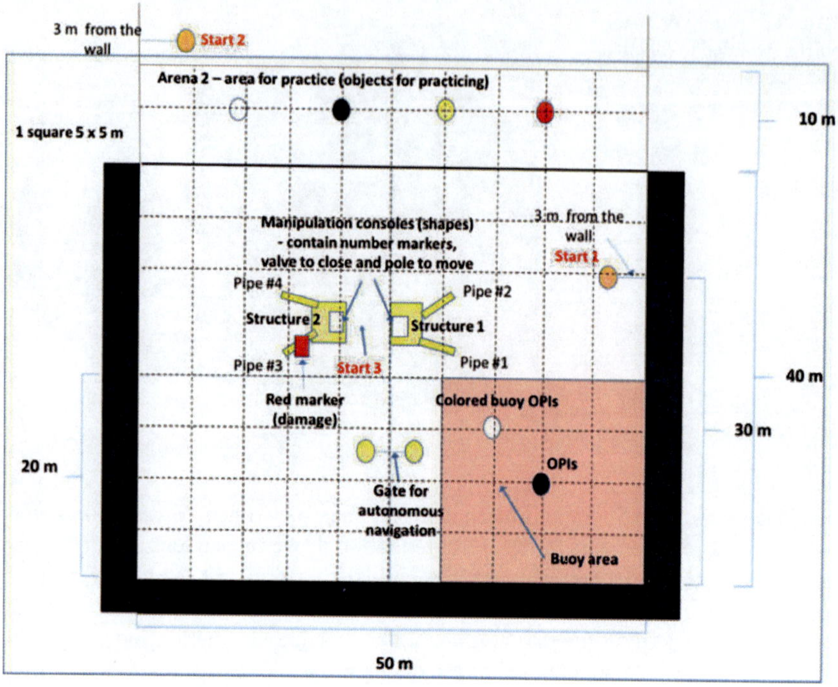

Fig. 2 The competition arena

The software was mostly written in Python, and the Robotics Operating System (ROS) was used as the robotics middleware. The students collaboratively and independently assessed the tasks, proposed solutions, and worked on the implementations. The students were involved in assembling the robot, making hardware modifications based on requirements, and writing software for navigation, perception, and control. The overarching goal of the competition enabled the students to not only focus on their individual task, but to focus on the whole development process and pipeline. This allowed the students to closely collaborate and identify how interlinked everyone's task was and how an individual's task affected the whole system. Initiative was taken by the students to plan meetings, work after school-hours, and dedicate their time and resources.

Technical tasks like reading and making sense of raw data from all robot sensors, writing software for robot control, navigation, manipulation, and using the raw data for object detection, obstacle avoidance, and detection based navigation were tasks that the students enjoyed tackling. Additionally, the field element was particularly important, presening real world-challenges to the team, often under strict time constraints. The international group of students coming from seven different nations managed to co-ordinate and contribute in-person, as well as online, even from different time-zones and world regions in order to work together and successfully participate in the competition.

5 The Competition

The competition tasks challenged the students to apply their text-book knowledge gained through their university degree in a practical and real-world scenario. The Marine Robotics course taught in the Spring semester covered the theoretical overview of underwater sensors and technologies, technical navigation and control algorithms, and challenges in the underwater robotics domain. However, in addition to the theoretical challenges, the students faced several practical and real-world challenges that could only be taught through first-hand experience. The overall integration of everyone's contribution into one robotic system was undoubtedly one of the greatest challenges and main learning outcomes: no one part can exist alone and the integration can be very demanding.

The students reported significant differences between competing in the competition as opposed to working in a simulation environment or with simple data-sets, as they did during the standard course. When working in a non-ideal and highly-dynamic environment, the students reported challenges caused by the real world like broken cables, troubles with a proper GPS fix, issues with wireless connectivity and wifi, controller instability and sensitivity adjustments based on water currents, and last minute bugs. These challenges were not visible at such scale in laboratory conditions, when the students tested the robot in the university tank, compared to the field.

The students were confronted with several unexpected challenges during the competition, such as a tether fracture, which made the control to the robot very difficult, and a DVL (Doppler Velocity Log) failure, an essential sensor for navigation. Even though these were major issues, the students collaboratively managed to find quick and temporary fixes for them with prolonged hardware and software sessions, rewiring a new cable and developing a new time-based navigation system, updating the previous DVL-based way-point navigation system. Being confronted with very hard challenges, the team was able to properly mitigate the consequences of the unexpected hardware failures.

An additional challenge was represented by the access to the competition facilities. Many team members, including the team captain, could not access the competition arena. A constant Zoom-link, plus evening tests in the public harbour helped the team to work together, despite the logistical challenges. Overall, the team succeeded in receiving the *Best Rookie Award* and scored 3rd place overall. Considering that competing teams were mostly formed by postgraduate students, this was an important achievement (Fig. 3).

Fig. 3 The black Pearl team with the best Rookie and 3rd place awards

6 General Considerations

The RAMI competition had a very positive impact on team BlackPearl. The students were exposed to several real-world problems that could only be learned through field activities like competitions. The students were able to take leadership and collaborate in an international team, jointly work on the full development process of a robotic system, and gain first-hand in-field experience in the field of their interest. This experience complemented the formal class activities and gave an opportunity to students to go well over what could be achieved during on-campus initiatives. Third year students had the opportunity to link the competition activities to their thesis and to future career opportunity in this field. Second year students could learn and develop more advanced algorithms during their third year and their thesis. From the structural point of view, this experience has positevely contributed to institutional support to open a toolshop to interested students. Despite many challenges, embedding robotics competition in the educational path is surely a rewarding experience.

References

1. Tay, K.H., Pallayil, V.: The Singapore AUV challenge (SAUVC) 2016. IEEE OES Beacon Newsl. **5**(2), 206 (2016)
2. Djapic, V.: SAUC-E past, present, and future. In: 2nd Workshop on Robot Competitions: benchmarking, Tech. Transfer, and Education of the ERF, vol. 200 (2013)
3. Maurelli, F., Cartwright, J., Johnson, N., Petillot, Y.: Nessie IV autonomous underwater vehicle wins the SAUC-E competition. In: 10th International Conference on Mobile Robots and Competitions (ROBÓTICA2010) (2010)
4. Valeyrie, N., Maurelli, F., Patron, P., Cartwright, J., Davis, B., Petillot, Y.: Nessie v turbo: a new hover and power slide capable torpedo shaped AUV for survey, inspection and intervention. In: AUVSI North America 2010 Conference (2010)

5. Ferri, G., Ferreira, F., Djapic, V.: Boosting the talent of new generations of marine engineers through robotics competitions in realistic environments: the SAUC-E and EURATHLON experience. In: OCEANS 2015—Genova, pp. 1–6 (2015)
6. Ferri, G., Ferreira, F., Djapic, V.: Multi-domain robotics competitions: the CMRE experience from SAUC-E to the European robotics league emergency robots. In: OCEANS 2017—Aberdeen, pp. 1–7 (2017)
7. Ferri, G., Ferreira, F., Djapic, V., Petillot, Y., Palau Franco, M.: The EURATHLON 2015 grand challenge: the first outdoor multi-domain search and rescue robotics competition-a marine perspective (2019)
8. Petillot, Y., Ferreira, F., Ferri, G.: Performance measures to improve evaluation of teams in the EURATHLON 2014 sea robotics competition. IFAC-PapersOnLine **48**(2), 224–230 (2015)
9. Ferreira, F., Ferri, G.: Marine robotics competitions: a survey. Curr. Robot. Rep. **1**, 169–178 (2020)
10. Maurelli, F., Mallios, A., Ribas, D., Ridao, P., Petillot, Y.: Particle filter based AUV localization using imaging sonar. In: IFAC Proceedings Volumes (IFAC-PapersOnline), vol. 42, pp. 52–57 (2009)
11. Maurelli, F., Krupiński, S., Xiang, X., Petillot, Y.: AUV localisation: a review of passive and active techniques. Int. J. Intell. Robot. Appl. 246–269 (2022)
12. Maurelli, F., Dineva, E., Nabor, A., Birk, A.: Robotics and intelligent systems: a new curriculum development and adaptations needed in coronavirus times. In: Merdan, M., Lepuschitz, W., Koppensteiner, G., Balogh, R., Obdržálek, D., (eds.), Robotics in Education. Springer, Cham, pp. 81–93 (2022)

Author Index

R. Balogh et al. (eds.), *Robotics in Education*, Lecture Notes in Networks and Systems 747, https://doi.org/10.1007/978-3-031-38454-7